THE TESLA PAPERS

edited by

David Hatcher Childress

The **New Science Series**:
- THE TIME TRAVEL HANDBOOK
- THE FREE ENERGY DEVICE HANDBOOK
- THE FANTASTIC INVENTIONS OF NIKOLA TESLA
- THE ANTI-GRAVITY HANDBOOK
- ANTI-GRAVITY & THE WORLD GRID
- ANTI-GRAVITY & THE UNIFIED FIELD
- ETHER TECHNOLOGY
- THE ENERGY GRID
- THE BRIDGE TO INFINITY
- THE HARMONIC CONQUEST OF SPACE
- VIMANA AIRCRAFT OF ANCIENT INDIA & ATLANTIS
- UFOS & ANTI-GRAVITY: Piece For a Jig-Saw
- THE COSMIC MATRIX: Piece For a Jig-Saw, Part II

The **Mystic Traveller Series:**
- IN SECRET TIBET by Theodore Illion (1937)
- DARKNESS OVER TIBET by Theodore Illion (1938)
- IN SECRET MONGOLIA by Henning Haslund (1934)
- MEN AND GODS IN MONGOLIA by Henning Haslund (1935)
- MYSTERY CITIES OF THE MAYA by Thomas Gann (1925)
- THE MYSTERY OF EASTER ISLAND by Katherine Routledge (1919)
- SECRET CITIES OF OLD SOUTH AMERICA by Harold Wilkins (1952)

The **Lost Cities Series**:
- LOST CITIES OF ATLANTIS, ANCIENT EUROPE & THE MEDITERRANEAN
- LOST CITIES OF NORTH & CENTRAL AMERICA
- LOST CITIES & ANCIENT MYSTERIES OF SOUTH AMERICA
- LOST CITIES OF ANCIENT LEMURIA & THE PACIFIC
- LOST CITIES & ANCIENT MYSTERIES OF AFRICA & ARABIA
- LOST CITIES OF CHINA, CENTRAL ASIA & INDIA

The **Atlantis Reprint Series:**
- THE HISTORY OF ATLANTIS by Lewis Spence (1926)
- ATLANTIS IN SPAIN by Elena Whishaw (1929)
- RIDDLE OF THE PACIFIC by John MacMillan Brown (1924)
- THE SHADOW OF ATLANTIS by Col. A. Braghine (1940)
- ATLANTIS MOTHER OF EMPIRES by R. Stacy-Judd (1939)

THE TESLA PAPERS

Adventures Unlimited Press

Acknowledgements

Many thanks to the following authors who helped contributed to this book, including Nikola Tesla, Marc Seifer, Gary Johnson, Dr. Patton McGinley, Moray B. King, Steve Elswick, Jeffery Hayes, Harry Osoff, Bruce Perreault, Tom Bearden, and many others.
Many thanks for your work on Tesla Technology.

Dedication

To Scientists and Inventors everywhere,
who believe in themselves
and the incredible workings of nature.

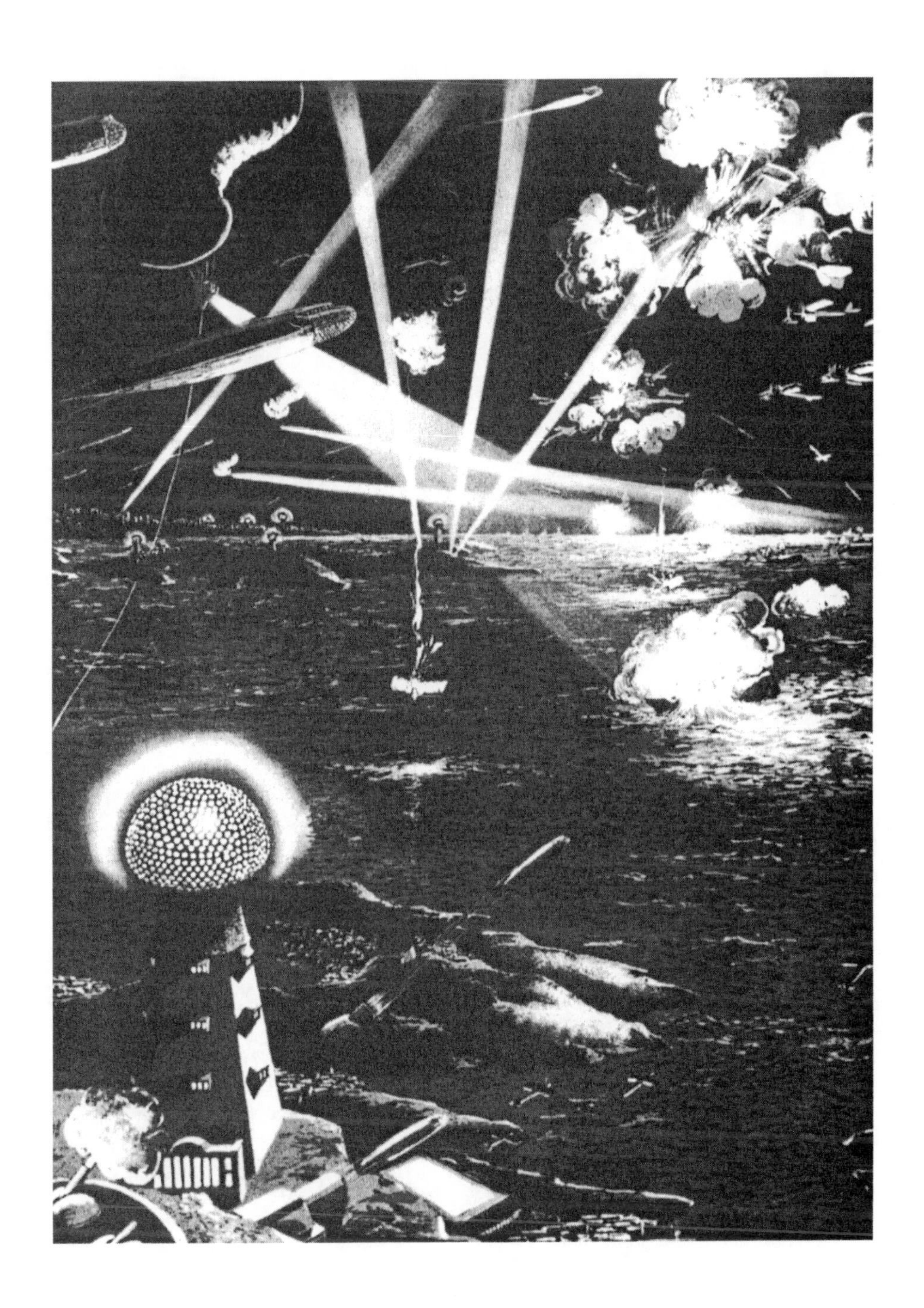

The Tesla Papers

edited by David Hatcher Childress

First Printing

October 2000

ISBN 0-932813-86-0

Printed in the United States of America

Published by
Adventures Unlimited Press
One Adventure Place
Kempton, Illinois 60946 USA
auphq@frontiernet.net

10 9 8 7 6

THE TESLA PAPERS

CONTENTS

Part One
Tesla: Humanitarian 11

Part Two
The Problem of
Increasing Human Energy 19

Part Three
The Wireless Transmission of Power 75

Part Four
Tesla's Electric Car 85

Part Five
The Tesla Papers 99

The History of Lasers & Particle Beam Weapons
by Marc J. Seifer, Ph.D. 115

The Secret History of the Wireless
by Marc. J. Seifer, Ph.D. 131

Tesla's Contributions to Electrotherapy
by Patton H. McGinley, Ph.D. 162

Part Six
Tesla's FBI Files 169

Part Seven
The Marconi—Tesla Trial Transcripts 183

Practical Electrics
Over 200 Illustrations
HOW TO MAKE THE
UNIVERSITY OF CALIFORNIA
1,500,000 VOLT
TESLA EXPERIMENTS
SEE PAGE 309
12

Part One
Tesla: Humanitarian

Nikola Tesla: Humanitarian Genius

Excerpted from vol 6, no. 4, *Power and Resonance,* the "Journal of the International Tesla Society."
Author unknown

Ask any school kid: "who invented radio"? If you get an answer at all it will doubtless be Marconi—an answer with which all the encyclopedias and textbooks agree. Or ask most anyone: "who invented the stuff that makes your toaster, your stereo, the street lights, the factories and offices work?" Without hesitation, Thomas Edison, right? Wrong both times. The correct answer is Nikola Tesla, a person you have probably never heard of. there's more. He appears to have discovered x-rays a year before W. K. Roentgen did in Germany, he built a vacuum tube amplifier several years before Lee de Forest did, he was using fluorescent lights in his laboratory 40 years before the industry "invented" them, and he demonstrated the principles used in microwave ovens and radar decades before they became an integral part of our society. Yet we associate his name with none of them.

For about 20 years around the turn of the century, he was known and respected in academic circles world wide, corresponding with eminent physicists of his day, including Albert Einstein, quoted and conferred with on matters of electrical science, adopted by New York's high society, backed by such financial and industrial giants as J. P. Morgan, John Jacob Astor, and George Westinghouse. He counted as friends eminent artists such as Mark Twain and pianist Ignace Paderewski. His honorary degrees, major prizes, and other citations number in the dozens.

The house where Nikola Tesla was born in Smiljan, Croatia in 1856.

Tesla was born in Smiljan, Croatia in 1856, the son of a clergyman and an inventive mother. He had an extraordinary memory, one that made learning six languages easy for him. He entered the Polytechnic School at Gratz, where for four years he studied mathematics, physics and mechanics, confounding more than one professor by an understanding of electricity, an infant science in those days, that was greater than theirs. His practical career started in 1881 in Budapest, Hungary, where he made his first electrical invention, a telephone repeater (the ordinary loudspeaker) and conceived the idea of a rotating magnetic field, which later made him world famous in its form as the modern induction motor. The polyphase induction motor is what provides power to virtually every industrial application, from conveyer belts to winches to machine tools.

Milutin Tesla, Nikola's father.

Tesla's mental abilities require some mention, since, not only did he have a photographic memory, he was able to use creative visualization with an uncanny and practical intensity. He describes in his autobiography how he was able to visualize a particular apparatus and was then able to actually test run the apparatus, disassemble it and check for proper action and wear! During the manufacturing phase of his inventions, he would work with all blueprints and specifications in his head. The invention invariably assembled together without redesign and worked perfectly. Tesla slept one to 2 hours a day and worked continuously on his inventions and theories without benefit of ordinary relaxation or vacations. He could judge the dimension of an object to a hundredth of an inch and perform difficult computations in his head without benefit of slide rule or mathematical tables. Far from an ivory tower intellectual, he was very much aware of the issues in the world around him, made it a point to render his ideas accessable to the general public by frequent contributions to the popular press, and to his field by numerous lectures and scientific papers.

He decided to come to the USA in 1884. He brought with him the various models of the first induction motors, which, after a brief and unhappy period at the Edison works, were eventually shown to George Westinghouse. It was in the Westinghouse shops that the induction motor was perfected. Numerous patents were taken out on this prime invention, all under Tesla's name.

Tesla worked briefly for Thomas Edison when he first came to the United States, creating many improvements on Edison's dc motors and generators, but left under a cloud of controversy after Edison refused to live up to bonus and royalty commitments. This was the beginning of a rivalry which was to have ugly consequences later when Edison and his backers did everything in their power to stop the devel-

A young Nikola Tesla.

opment and installation of Tesla's far more efficient and practical ac current delivery system and urban power grid. Edison put together a traveling road show which attempted to portray ac current as dangerous, even to the point of electrocuting animals both small (puppies) and large (in one case an elephant) in front of large audiences. As a result of this propaganda crusade, the state of New York adopted ac electrocution as its method of executing convicts. Tesla won the battle by the demonstration of ac current's safety and usefulness when his apparatus illuminated and powered the entire New York World's Fair of 1899.

Tesla's most important work at the end of the nineteenth century was his original system of transmission of energy by wireless antenna. In 1900 Tesla obtained his two fundamental patents on the transmission of true wireless energy covering both methods and apparatus and involving he use of four tuned circuits. In 1943, the Supreme Court of the United States granted full patent rights to Nikola Tesla for the invention of the radio, superseding and nullifying any prior claim by marconi and others in regards to the "fundamental radio patent" It is interesting to note that Tesla, in 1898, described the transmission of not only the human voice, but images as well and later designed and patented devices that evolved into the power supplies that operate our present day TV picture tubes. The first primitive radar installations in 1934 were built following principles, mainly regarding frequency and power level, that were stated by Tesla in 1917.

In 1889 Tesla constructed an experimental station in Colorado Springs where he studied the characteristics of high frequency or radio frequency alternating currents. While there he developed a powerful radio transmitter of unique design and also a number of receivers "for individualizing and isolating the energy transmitted". He conducted experiments designed to establish the laws of radio propagation which are currently being "rediscovered" and verified amid some controversy in high energy quantum physics.

Tesla wrote in *Century Magazine* in 1900: "...that communication without wires to any point of the globe is practicable. My experiments showed that the air at the ordinary pressure became distinctly conducting, and this opened up the wonderful prospect of transmitting large amounts of electrical energy for industrial purposes to great distances without wires...its practical consummation would mean that energy would be available for the uses of man at any point of the globe. I can conceive of no technical advance which would tend to unite the various elements of humanity more effectively than this one, or of one which would more add to and more economize human energy..." This was written in 1900! After finishing preliminary testing, work was begun on a full sized broadcasting station at shoreham,

Long Island. Had it gone into operation, it would have been able to provide usable amounts of electrical power at the receiving circuits. After construction of a generator building (still standing) and a 180 foot broadcasting tower (dynamited in world war I on the dubious pretext of being a potential navigation reference for German U-boats), financial support for the project was suddenly withdrawn by J. P. Morgan when it became apparent that such a worldwide power project couldn't be metered and charged for.

Another one of Tesla's inventions that is familiar to anyone who has ever owned an automobile, was patented in 1898 under the name "electrical ignitor for gas engines". More commonly known as the automobile ignition system, its major component, the ignition coil, remains practically unchanged since its introduction into use at the turn of the century.

Nikola Tesla also designed and built prototypes of a unique fuel burning rotary engine based upon his earlier design for a rotary pump. Recent tests that have been carried out on the Tesla bladeless disk turbine indicate that, if constructed using newly developed high temperature ceramic materials, it will rank as the world's most efficient gas engine, out-performing our present day piston type internal combustion engines in fuel efficiency, longevity, adaptability to different fuels, cost and power to weight ratio.

Tesla's generosity eventually left him without adequate funds to pursue and realize his inventions. His idealism and humanism left him with little stomach for the world of industrial and financial intrigue. His New York laboratory was destroyed by a mysterious fire. References to his work and accomplishments were systematically purged from the scientific literature and textbooks. Driven into a Hermetic exile in a New York hotel during the period between the two wars, 20 years of his potentially rich and productive contribution were taken from us. The only occasions of public appearance were the yearly press interview on his birthday when he would describe amazing and far reaching inventions and technological possibilities. These were distorted and sensationalized in the popular press, particularly when he described advanced weapons systems on the eve of world war II. He died in obscurity in 1943. Only the FBI took note: they searched his papers (in vain) for the design of the "death-ray machine". It is interesting to note that the motivation for our "Star Wars" defense system was based upon fears that the soviets had begun deployment of weapons based upon Tesla high energy principles. Public reports of mysterious "blindings" of U.S. surveillance satellites, anomalous high altitude flashes and fireballs, elf wave radio interference, and other cases lend credence to this interpretation.

Credit must be given where credit is due for the labor saving and humanitarian inventions such as universal ac current that have been incorporated into the very fabric of our daily lives and also the devices who's design have been made available, but have not been utilized by society at large.

A Short History of Nikola Tesla

This is a file to straighten out misconception and disinformation that has occurred over the years, about how supposedly "great" Edison was, and how Nikola Tesla was brushed under the capitalist power rug.

Edison was a thief, employing all kinds of people for their brains, he stole their inventions, their ideas, so much so, that it is unclear today what Edison actually invented, and what was stolen from others.

The Edison Electric Institute was formed to perpetuate the notion that Edison was the inventor of record, and to make sure that school textbooks, etc., only mentioned HIM in connection with these many inventions. Much like Bell Labs does today.

Nikola Tesla was pretty much always a genius, after having made many improvements in the electric trolleys, and trains in his country, he came to America, sought employment, and eventually ended up working for Edison.

Edison had contracted with New York City to build Direct Current (D.C.) power plants every square mile or so, so as to power the lights that he supposedly invented. Street lights, hotel lighting etc. Having trenches dug throughout the city to lay the cables, copper, and as big around as a man's bicep, he told Tesla that if Tesla could save him money by redesigning certain aspects of the installation, that he would give Tesla a percentage of the savings. A verbal agreement. After approximately a year, Tesla went to Edison's office and showed him the savings that had occurred ($100,000 or so, which in those days was quite a piece of change) as a direct result of his (Tesla's) engineering, and Edison pretended ignorance of any agreement. Tesla quit. From that point on, the two men were enemies.

Tesla invented useable Alternating Current (A.C.) that we all use today, in a world where Edison and others already had a huge investment in D.C. power.

Tesla proselytized A.C. power and had some success building A.C. power plants, and providing A.C. power to various entities. One of these was Sing Sing prison, in upstate New York. Tesla provided A.C. power for the "electric chair" there. Edison had big articles printed in the New York newspapers, saying that A.C. power was dangerous "killing" power, and in general, gave a bad name to Tesla.

To contradict this jab, Tesla set out on his own positive marketing campaign, appearing at the 1893 World Exposition in Chicago passing high frequency "dangerous" A.C. power over his body to power light bulbs in front of the public. Shooting huge, long sparks from his "Tesla coil", and touching them, etc. "Proving" that A.C. power was safe for public consumption.

The advantage of A.C. power was that you could send it a long distance through reasonably sized wires with little loss, and if you touched the wires together, "shorted them", you got a lot of sparks, and only the place where they were touching melted until the two wires weren't touching anymore.

D.C. power, on the other hand, needed huge cables to go any distance at all, while using power, the cables heated up. When shorted, the cables melted all the way back to the power house, streets had to be dug up again and new cables laid. If a short occurred in a single light, it usually started a fire, and burned down the hotel or destroyed whatever it was in contact with! This was quite profitable for those in the D.C. power business, and quite good for those into ditch digging, construction, etc.

Tesla invented 2-phase, and 3-phase Alternating Current. He figured motors turned in a circle, so alternately driving separate, 180 degree, sections of the surrounding armature would build up less heat, and use less electricity. He was right.

1929 came, the stock market crashed, bankers, lawyers, everyone who had lost their wealth and hadn't jumped out a window, sought work, many as common laborers if lucky, for a dollar a day. Tesla found himself digging ditches in the company of broke but influential ex-Wall-streeters. During the short lunch period, he would tell his buddies about phased A.C. electricity, and how it was efficient, etc. Along about 1932, he was working at a small generator rebuilding shop in New York, and one of the bankers that he used to dig ditches with, found him, and took him to Mr. Westinghouse, to whom he told his stories. Westinghouse bought 19 patents outright, and gave Tesla a dollar per horsepower for any electric motor produced by Westinghouse using the Tesla 3-phase system.

Tesla finally had the money with which to start building his laboratories, 5 and conducting the experiments with free earth energy. The idea that really made him unpopular.

Something free, that the masters of war and business couldn't control? They couldn't have that! So, the day after Tesla died in 1943, his huge laboratory on Long Island mysteriously burned down, no records saved, and the remnants were bulldozed the day after that to further eradicate any equipment still left. So much for "free energy.".

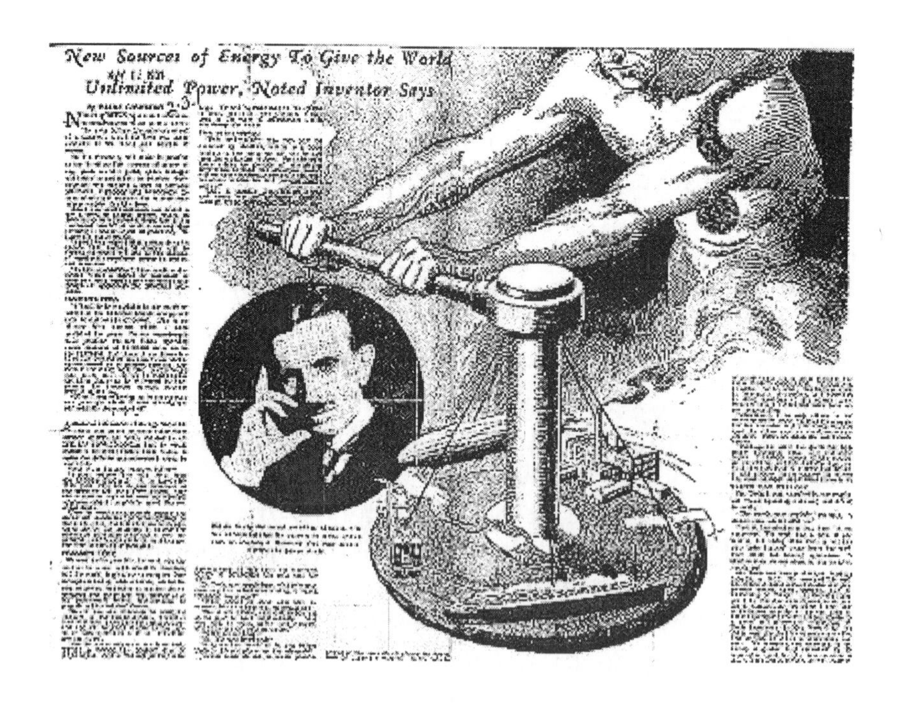

New Sources of Energy To Give the World Unlimited Power, Noted Inventor Says

Mark Twain (Samuel Clemmens) in Tesla's laboratory, 1895.

Part Two
The Problem of Increasing Human Energy

with special reference to the harnessing of the sun's energy.

by Nikola Tesla

OSCILLATION TRANSFORMER
INDUCTION MOTOR
WORLD WIRELESS TELEPHONE TRANSMITTER
TELAUTOMATON
STEAM & GAS TURBINE
TESLA COMPANY, INC.
TELEPHONE
9090 BRYANT
8 West 40th St.
NEW YORK

Electrical oscillator activity ten million Horsepower
Power transmission without wires

THE ONWARD MOVEMENT OF MAN—
THE ENERGY OF THE MOVEMENT—
THE THREE WAYS OF INCREASING HUMAN ENERGY.

Of all the endless variety of phenomena which nature presents to our senses, there is none that fills our minds with greater wonder than that inconceivably complex movement which, in its entirety, we designate as human life. Its mysterious origin is veiled in the forever impenetrable mist of the past, its character is rendered incomprehensible by its infinite intricacy, and its destination is hidden in the unfathomable depths of the future. Whence does it come? What is it? Whither does it tend? are the great questions which the sages of all times have endeavored to answer.

Modern science says: The sun is the past, the earth is the present, the moon is the future. From an incandescent mass we have originated, and into frozen mass we shall turn. Merciless is the law of nature, and rapidly and irresistibly we are drawn to our doom. Lord Kelvin, in his profound meditations, allows us only a short span of life, something like six million years, after which time the sun's bright light will have ceased to shine, and its life-giving heat will have ebbed away, and our own earth will be a lump of ice, hurrying on through the eternal night. But do not let us despair. There will still be left on it a glimmering spark of life, and there will be a chance to kindle anew fire on some distant star. This wonderful possibility seems, indeed, to exist, judging from Professor Dewar's beautiful experiments with liquid air, which show that germs of organic life are not destroyed by cold, no matter how intense; consequently they may be transmitted through the interstellar space. Meanwhile the cheering lights of science and art, ever increasing in intensity, illuminate our path, and the marvels they disclose, and the enjoyments they offer, make us measurably forgetful of the gloomy future.

Though we may never be able to comprehend human life, we know certainly that it is movement, of whatever nature it be. The existence of a movement unavoidably implies a body which is being moved and a force which is moving it. Hence, wherever there is life, there is a mass moved by a force. All mass possesses inertia, all force tends to persist. Owing to this universal property and condition, a body, be it at rest or in motion, tends to remain in the same state, and a force, manifesting itself anywhere and through whatever cause, produces an equivalent opposing force, and as an absolute necessity of this it follows that every movement in nature must be rhythmical. Long ago this simple truth was clearly pointed out by Herbert Spencer, who arrived at it through a somewhat different process of reasoning. It is borne out in everything we perceive—in the movement of a planet, in the surging and ebbing of the tide, in the reverberations of the air, the swinging of a pendulum, the oscillations of an electric current, and in the infinitely varied phenomena of organic life. Does not the whole of human life attest it? Birth, growth, old age, and death of an individual, family, race, or nation, what is it all but a rhythm?

All life-manifestation, then, even in its most intricate form, as exemplified in man, however involved and inscrutable, is only a movement, to which the general laws of movement which govern throughout the physical universe must be applicable.

When we speak of man, we have a conception of humanity as a whole, and before applying scientific method to the investigation of his movement, we must accept this as a physical fact. But can anyone doubt today that all the millions of individuals and all the innumerable types and characters constitute an entity, a unit? Though free to think and act, we are held together, like the stars in the firmament, with ties inseparable. These ties we cannot see, but we can feel them. I cut myself in the finger, and it pains me: this finger is a part of me. I see a friend hurt, and it hurts me, too: my friend and I arc one. And now I see stricken down an enemy, a lump of matter which, of all the lumps of matter in the universe, I care least for, and still it grieves me. Does this not prove that each of us is only a part of a whole?

For ages this idea has been proclaimed in the consummately wise teachings of religion, probably not alone as a means of insuring peace and harmony among men, but as a deeply founded truth. The Buddhist expresses it in one way, the Christian in another, but both say the same: We are all one. Metaphysical proofs are, however, not the only ones which we are able to bring forth in support of this idea. Science, too, recognizes this connectedness of separate individuals, though not quite in the same sense as it admits that the suns, planets, and moons of a constellation are one body, and there can be no doubt that it will be experimentally confirmed in times to come, when our means and methods for investigating psychical and other states and phenomena shall have been brought to great perfection. Still more: this one human being lives on and on. The individual is ephemeral, races and nations come and pass away, but man remains. Therein lies the profound difference between the individual and the whole. Therein, too, is to be found the partial explanation of many of those marvelous phenomena of heredity which are the results of countless centuries of feeble but persistent influence.

Conceive, then, man as a mass urged on by a force. Though this movement is not of a translatory character, implying change of place, yet the general laws of mechanical movement are applicable to it, and the energy associated with this mass can be measured, in accordance with well-known principles, by half the product of the mass with the square of a certain velocity. So, for instance, a cannon ball which is at rest possesses a certain amount of energy in the form of heat, which we measure in a similar way. We imagine the ball to consist of innumerable minute particles, called atoms or molecules, which vibrate or whirl around one another. We determine their masses and velocities, and from them the energy of each of these minute systems, and adding them all together, we get an idea of the total heat-energy contained in the ball, which is only seemingly at rest. In this purely theoretical estimate this energy may then be calculated by multiplying half of the total mass—that is, half of the sum of all the small masses—with the square of a velocity which is determined from the velocities of the separate particles. In like manner we may conceive of human energy being measured by half the human mass multiplied with the square of a velocity which we are not yet able to compute. But our deficiency in this knowledge will not vitiate the truth of the deductions I shall draw which rest on the firm basis that the same laws of mass and force govern throughout nature.

Man, however, is not an ordinary mass, consisting of spinning atoms and molecules, and containing merely heat-energy. He is a mass possessed of certain higher qualities by reason of the creative principle of life with which he is endowed. His mass, as the water in an ocean wave, is being continuously exchanged, new taking the place of the old. Not only this, but he grows, propagates, and dies, thus altering his mass independently, both in bulk and density. What is most wonderful of all, he is capable of increasing or diminishing his velocity of movement by the mysterious power he possesses of appropriating more or less energy from other substance, and turning it into motive energy. But in any given moment we may ignore these slow changes and assume that human energy is measured by half the product of man's mass with the square of a certain hypothetical velocity. However we may compute this velocity, and whatever we may take as the standard of its measure, we must, in harmony with this conception, come to the conclusion that the great problem of science is, and always will be, to increase the energy thus defined. Many years ago, stimulated by the perusal of that deeply interesting work, Draper's "History of the Intellectual Development of Europe," depicting so vividly human movement, I recognized that to solve this eternal problem must ever be the chief task of the man of science. Some results of my own efforts to this end I shall endeavor briefly to describe here.

Let, then, in *Diagram A,* M represent the mass of man. This mass is impelled in one direction by a force f, which is resisted by another partly frictional and partly negative force R, acting in a direction exactly opposite, and retarding the movement of the mass. Such an antagonistic force is present in every movement, and must be taken into consideration. The difference between these two forces is the effective force which imparts a velocity V to the mass M in the direction of the arrow on the line representing the force f. In accordance with the preceding, the human energy will then be given by the product $1/2\ MV^2 = 1/2\ MV \times V$, in which M is the total mass of man in the ordinary interpretation of the term "mass," and V is a certain hypothetical velocity, which, in the present state of science, we are unable exactly to define and determine. To increase the human energy is, therefore, equivalent to increase this product, and there are, as will readily be seen, only three ways possible to attain this result, which are illustrated in the above diagram. The first way, shown in the top figure, is to increase the mass (as indicated by the dotted circle), leaving the two opposing forces the same. The second way is to reduce the retarding force R to a smaller value r, leaving the mass and the impelling force the same, as diagrammatically shown in the middle figure. The third way, which is illustrated in the last figure, is to increase the impelling force f to a higher value F, while the mass and the retarding force R remain unaltered. Evidently fixed limits exist as regards increase of mass and reduction of retarding force, but the impelling force can be increased indefinitely. Each of these three possible solutions presents a different aspect of the main problem of increasing human energy, which is thus divided into three distinct problems, to be successively considered.

THE FIRST PROBLEM:
HOW TO INCREASE THE HUMAN MASS—
THE BURNING OF ATMOSPHERIC NITROGEN.

Viewed generally, there are obviously two ways of increasing the mass of mankind: first, by aiding and maintaining those forces and conditions which tend to increase it; and, second, by opposing and reducing those which tend to diminish it. The mass will be increased by careful attention to health, by substantial food, by moderation, by regularity of habits, by the promotion of marriage, by conscientious attention to the children, and, generally stated, by the observance of all the many precepts and laws of religion and hygiene. But in adding new mass to the old, three cases again present themselves. Either the mass added is of the same velocity as the old, or it is of a smaller or of a higher velocity. To gain an idea of the relative importance of these cases, imagine a train composed of, say, one hundred locomotives running on a track, and suppose that, to increase the energy of the moving mass, four more locomotives are added to the train. If these four move at the same velocity at which the train is going, the total energy will be increased four per cent.; if they are moving at only one half of that velocity, the increase will amount to only one per cent.; if they are moving at twice that velocity, the increase of energy will be sixteen per cent. This simple illustration shows that it is of the greatest importance to add mass of a higher velocity. Stated more to the point, if, for example, the children be of the same degree of enlightenment as the parents,—that is, mass of the "same velocity,"—the energy will simply increase proportionately to the number added. If they

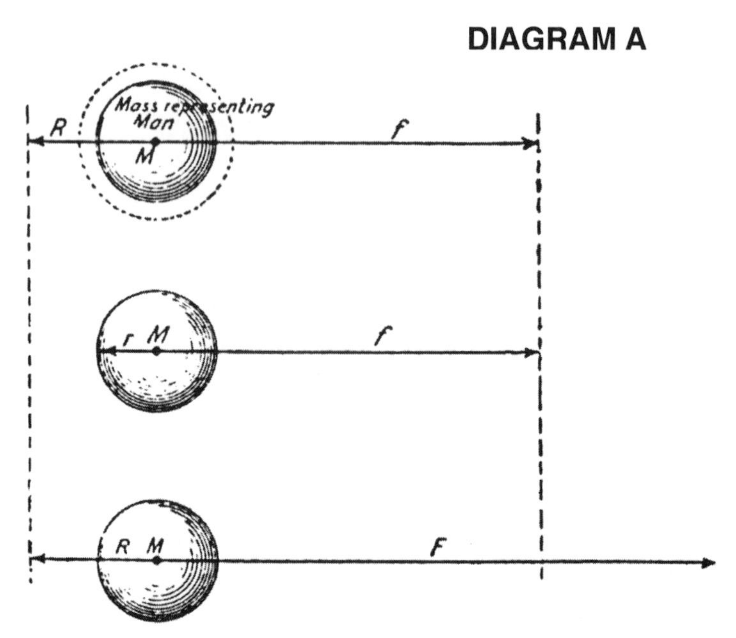

Diagram A. The three ways of increasing human energy.

are less intelligent or advanced, or mass of "smaller velocity," there will be a very slight gain in the energy; but if they are further advanced, or mass of "higher velocity," then the new generation will add considerably to the sum total of human energy. Any addition of mass of "smaller velocity," beyond that indispensable amount required by the law expressed in the proverb, "*Mens sana in corpore sano,*"[1] should be strenuously opposed. For instance, the mere development of muscle, as aimed at in some of our colleges, I consider equivalent to adding mass of "smaller velocity," and I would not commend it, although my views were different when I was a student myself. Moderate exercise, insuring the right balance between mind and body, and the highest efficiency of performance, is of course, a prime requirement. The above example shows that the most important result to be attained is the education, or the increase of the "velocity," of the mass newly added.

Conversely, it scarcely need be stated that everything that is against the teachings of religion and the laws of hygiene is tending to decrease the mass. Whisky, wine, tea, coffee, tobacco, and other such stimulants are responsible for the shortening of the lives of many, and ought to be used with moderation. But I do not think that rigorous measures of suppression of habits followed through many generations arc commendable. It is wiser to preach moderation than abstinence. We have become accustomed to these stimulants, and if such reforms are to be effected, they must be slow and gradual. Those who are devoting their energies to such ends could make themselves far more useful by turning their efforts in other directions, as, for instance, toward providing pure water.

For every person who perishes from the effects of a stimulant, at least a thousand die from the consequences of drinking impure water. This precious fluid, which daily infuses new life into us, is likewise the chief vehicle through which disease and death enter our bodies. The germs of destruction it conveys are enemies all the more terrible as they perform their fatal work unperceived. They seal our doom while we live and enjoy. The majority of people are so ignorant or careless in drinking water, and the consequences of this are so disastrous, that a philanthropist can scarcely use his efforts better than by endeavoring to enlighten those who are thus injuring themselves. By systematic purification and sterilization of the drinking water the human mass would be very considerably increased. It should be made a rigid rule—which might be enforced by law—to boil or to sterilize otherwise the drinking-water in every household and public place. The mere filtering does not afford sufficient security against infection. All ice for internal uses should be artificially prepared from water thoroughly sterilized. The importance of eliminating germs of disease from the city water is generally recognized, but little is being done to improve the existing conditions, as no satisfactory method of sterilizing great quantities of water has as yet been brought forward. By improved electrical appliances we are now enabled to produce ozone cheaply and in large amounts, and this ideal disinfectant seems to offer a happy solution of the important question.

[1] Sound mind in a sound body

Gambling, business rush, and excitement, particularly on the exchanges, are causes of much mass-reduction, all the more so because the individuals concerned represent units of higher value. Incapacity of observing the first symptoms of an illness, and careless neglect of the same, are important factors of mortality. In noting carefully every new sign of approaching danger, and making conscientiously every possible effort to avert it, we are not only following wise laws of hygiene in the interest of our well being and the success of our labors, but we are also complying with a higher moral duty. Everyone should consider his body as a priceless gift from one whom he loves above all, as a marvelous work of art, of indescribable beauty and mastery beyond human conception, and so delicate and frail that a word, a breath, a look, nay, a thought, may injure it. Uncleanliness, which breeds disease and death, is not only self destructive but a highly immoral habit. In keeping our bodies free from infection, healthful, and pure, we are expressing our reverence for the high principle with which they are endowed. He who follows the precepts of hygiene in this spirit is proving himself, so far, truly religious. Laxity of morals is a terrible evil, which poisons both mind and body, and which is responsible for a great reduction of the human mass in some countries. Many of the present customs and tendencies are productive of similar hurtful results. For example, the society life, modern education and pursuits of women, tending to draw away from their household duties and make men out of them, must needs detract from the elevating ideal they represent, diminish the artistic creative power, and cause sterility and a general weakening of the race. A thousand other evils might be mentioned, but all put together, in their bearing upon the Problem under discussion, they would not equal a single one, the want of food, brought on by poverty, destitution, and famine. Millions of individuals die yearly for want of food, thus keeping down the mass. Even in our enlightened communities, and notwithstanding the many charitable efforts, this is still, in all probability, the chief evil. I do not mean her absolute want of food, but want of healthful nutriment.

How to provide good and plentiful food is, therefore, a most important question of the day. On general principles the raising of cattle as a means of providing food is objectionable, because, in the sense interpreted above, it must undoubtedly tend to the addition of mass of a smaller velocity." It is certainly preferable to raise vegetables, and I think, therefore, that vegetarianism is a commendable departure from the established barbarous habit. That we can subsist on plant food and perform our work even to advantage is not a theory, but a demonstrated fact. Many races living almost exclusively on vegetables are of superior physique and strength. There is no doubt that some plant food, such as oatmeal, is more economical than meat, and superior to it in regard to both mechanical and mental performance. Such food, moreover, taxes our digestive organs decidedly less, and, in making us more contented and sociable, produces an amount of good difficult to estimate. In view of these facts every effort should be made to stop the wanton and cruel slaughter of animals, which must be destructive to our morals. To free ourselves from animal instincts and appetites, which keep us down, we should begin at the very root from which they spring: we should effect a radical reform in the character of the food.

There seems to be no philosophical necessity for food. We can conceive of organized beings living without nourishment, and deriving all the energy they need for the performance of their life-functions from the ambient medium. In a crystal we have the clear evidence of the existence of a formative life-principle, and though we cannot understand the life of a crystal, it is nonetheless a living being. There may be, besides crystals, other such individualized, material systems of beings, perhaps of gaseous constitution, or composed of substance still more tenuous. In view of this possibility, —nay, probability, —we cannot apodictically deny the existence of organized beings on a planet merely because the conditions on the same are unsuitable for the existence of life as we conceive it. We cannot even with positive assurance, assert that some of them might not be present here, in this our world, in the very midst of us, for their constitution and life-manifestation may be such that we are unable to perceive them.

The production of artificial food as a means for causing an increase of the human mass naturally suggests itself, but a direct attempt of this kind to provide nourishment does not appear to me rational, at least not for the present. Whether we could thrive on such food is very doubtful. We are the result of ages of continuous adaptation, and we cannot radically change without unforeseen and, in all probability, disastrous consequences. So uncertain an experiment should not be tried. By far the best way, it seems to me, to meet the ravages of the evil would be to find ways of increasing the productivity of the soil. With this object the preservation of forests is of an importance which cannot be overestimated, and in this connection, also, the utilization of water-power for purposes of electrical transmission, dispensing in many ways with the necessity of burning wood, and tending thereby to forest preservation, is to be strongly advocated. But there are limits in the improvement to be effected in this and similar ways.

To increase materially the productivity of the soil, it must be more effectively fertilized by artificial means. The question of food-production resolves itself, then, into the question how best to fertilize the soil. What it is that made the soil is still a mystery. To explain its origin is probably equivalent to explaining the origin of life itself. The rocks, disintegrated by moisture and heat and wind and weather, were in themselves not capable of maintaining life. Some unexplained condition arose, and some new principle came into effect, and the first layer capable of sustaining low organisms, like mosses, was formed. These, by their life and death, added more of the life-sustaining quality to the soil, and higher organisms could then subsist, and so on and on, until at last highly developed Plant and animal life could flourish. But though the theories are, even now, not in agreement as to how fertilization is effected, it is a fact, only too well ascertained, that the soil cannot indefinitely sustain life, and some way must be found to supply it with the substances which have been abstracted from it by the plants. The chief and most valuable among these substances are compounds of nitrogen, and the cheap production of these is therefore, the key for the solution of the all-important food problem. Our atmosphere contains an inexhaustible amount of nitrogen, and could we but oxidize it and produce these compounds, an incalculable benefit for mankind would follow.

Long ago this idea took a powerful hold on the imagination of scientific men, but an efficient means for accomplishing this result could not be devised. The problem was rendered extremely difficult by the extraordinary inertness of the nitrogen, which refuses to combine even with oxygen. But here electricity comes to our aid: the dormant affinities of the element are awakened by an electric current of the proper quality. As a lump of coal which has been in contact with oxygen for centuries without burning will combine with it when once ignited, so nitrogen, excited by electricity, will burn. I did not succeed, however, in producing electrical discharges exciting very effectively the atmospheric nitrogen until a comparatively recent date, although I showed, in May, 1891, in a scientific lecture, a novel form of discharge or electrical flame named "St. Elmo's hotfire," which, besides being capable of generating ozone in abundance, also possessed, as I pointed out on that occasion, distinctly the quality of exciting chemical affinities. This discharge or flame was then only three or four inches long, its chemical action was likewise very feeble, and consequently the process of oxidation of the nitrogen was wasteful. How to intensify this action was the question. Evidently electric currents of a peculiar kind had to be produced in order to render the process of nitrogen combustion more efficient.

The first advance was made in ascertaining that the chemical activity of the discharge was very considerably increased by using currents of extremely high frequency or rate of vibration. This was an important improvement, but practical considerations soon set a definite limit to the progress in this direction. Next, the effects of the electrical pressure of the current impulses, of their wave-form and other characteristic features, were investigated. Then the influence of the atmospheric pressure and temperature and of the presence of water and other bodies was studied, and thus the best conditions for causing the most intense chemical action of the discharge and securing the highest efficiency of the process were gradually ascertained. Naturally, the improvements were not quick in coming; still, little by little, I advanced. The flame grew larger and larger, and its oxidizing action more and more intense. From an insignificant brush-discharge a few inches long it developed into a marvelous electrical phenomenon, a roaring blaze, devouring the nitrogen of the atmosphere and measuring sixty or seventy feet across. Thus slowly, almost imperceptibly, possibility became accomplishment. All is not yet done, by any means, but to what a degree my efforts have been rewarded an idea may be gained from the inspection of *Figure 1,* which, with its title, is self explanatory. The flame-like discharge visible is produced by the intense electrical oscillations which pass through the coil shown, and violently agitate the electrified molecules of the air. By this means a strong affinity is created between the two normally indifferent constitu-

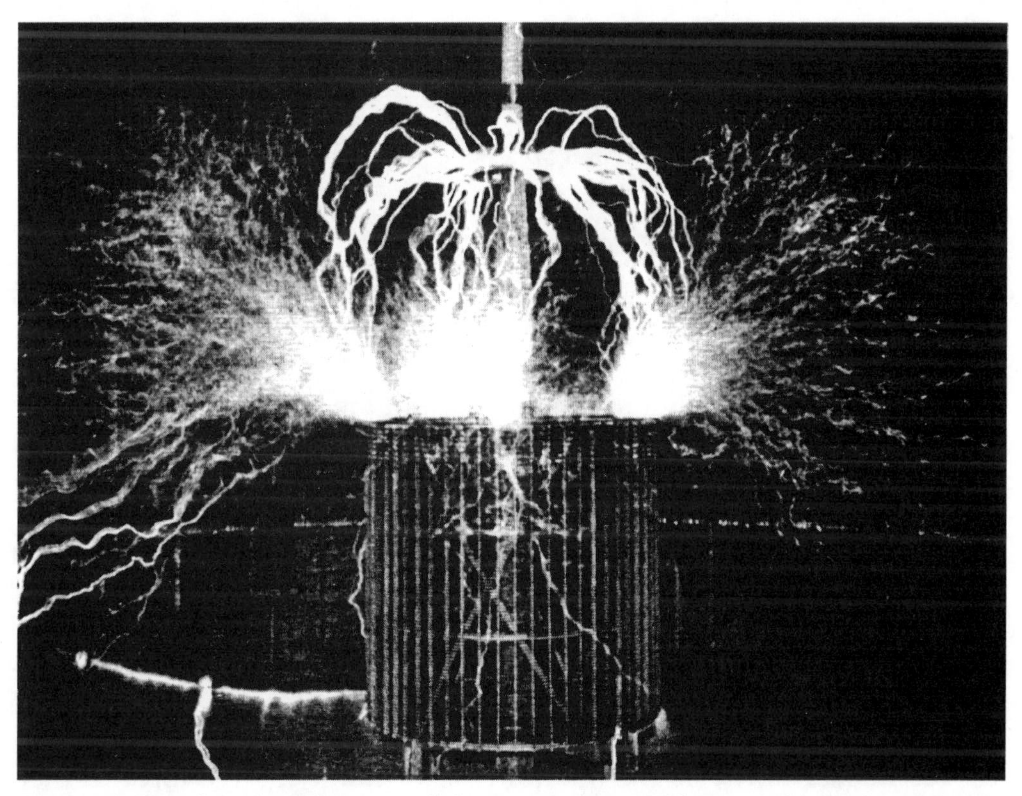

Figure 1. Burning the Nitrogen of the atmosphere. This result is produced by the discharge of an electrical oscillator giving twelve million volts. The electrical pressure, alternating one hundred thousand times per second, excites the normally inert nitrogen, causing it to combine with the oxygen. The flame-like discharge shown in the photograph measures sixty-five feet across.

ents of the atmosphere, and they combine readily, even if no further provision is made for intensifying the chemical action of the discharge. In the manufacture of nitrogen compounds by this method, of course, every possible means bearing upon the intensity of this action and the efficiency of the process will be taken advantage of, and, besides, special arrangements will be provided for the fixation of the compounds formed, as they are generally unstable, the nitrogen becoming again inert after a little lapse of time. Steam is a simple and effective means for fixing permanently the compounds. The result illustrated makes it practicable to oxidize the atmospheric nitrogen in unlimited quantities, merely by the use of cheap mechanical power and simple electrical apparatus. In this manner many compounds of nitrogen may be manufactured all over the world, at a small cost, and in any desired amount, and by means of these compounds the soil can be fertilized and its productiveness indefinitely increased. An abundance of cheap and healthful food, not artificial, but such as we are accustomed to, may thus be obtained. This new and inexhaustible source of food supply will be of incalculable benefit to mankind, for it will enormously contribute to the increase of the human mass, and thus add immensely to human energy. Soon, I hope, the world will see the beginning of an industry which, in time to come, will, I believe, be in importance next to that of iron.

THE SECOND PROBLEM: HOW TO REDUCE THE FORCE RETARDING THE HUMAN MASS—THE ART OF TELAUTOMATICS.

As before stated, the force which retards the onward movement of man is partly frictional and partly negative. To illustrate this distinction I may name, for example, ignorance, stupidity, and imbecility as some of the purely frictional forces, or resistances devoid of any directive tendency. On the other hand, visionariness, insanity, self-destructive tendency, religious fanaticism, and the like, are all forces of a negative character, acting in definite directions. To reduce or entirely to overcome these dissimilar retarding forces, radically different methods must be employed. One knows, for instance, what a fanatic may do, and one can take preventive measures, can enlighten, convince, and possibly direct him, turn his vice into virtue; but one does not know, and never can know, what a brute or an imbecile may do, and one must deal with him as with a mass, inert, without mind, let loose by the mad elements. A negative force always implies some quality, not infrequently a high one, though badly directed, which it is possible to turn to good advantage; but a directionless, frictional force involves unavoidable loss. Evidently, then, the first and general answer to the above question is: turn all negative force in the right direction and reduce all frictional force.

There can be no doubt that, of all the frictional resistances, the one that most retards human movement is ignorance. Not without reason said that man of wisdom, Buddha: "Ignorance is the greatest evil in the world." The friction which results from ignorance, and which is greatly increased owing to the numerous languages and nationalities, can be reduced only by the spread of knowledge and the unification of the heterogeneous elements of humanity. No effort could be better spent. But however ignorance may have retarded the onward movement of man in times past, it is certain that, nowadays, negative forces have become of greater importance. Among these there is one of far greater moment than any other. It is called organized warfare. When we consider the millions of individuals, often the ablest in mind and body, the flower of humanity, who are compelled to a life of inactivity

Figure 2. The first practical Telautomaton. A machine having all its bodily or translatory movements and the operation of the interior mechanism controlled from a distance without wires. The crewless boat shown in the photograph contains its own motive power, propelling and steering machinery, and numerous other accessories, all of which are controlled by transmitting from a distance, without wires, electrical oscillations to a circuit carried by the boat and adjusted to respond only to these oscillations.

and unproductiveness, the immense sums of money daily required for the maintenance of armies and war apparatus, representing ever so much of human energy, all the effort uselessly spent in the production of arms and implements of destruction, the loss of life and the fostering of a barbarous spirit, we are appalled at the inestimable loss to mankind which the existence of these deplorable conditions must involve. What can we do to combat best this great evil?

Law and order absolutely require the maintenance of organized force. No community can exist and prosper without rigid discipline. Every country must be able to defend itself, should the necessity arise. The conditions of today are not the result of yesterday, and a radical change cannot be effected tomorrow. If the nations would at once disarm, it is more than likely that a state of things worse than war itself would follow. Universal peace is a beautiful peace is a beautiful dream, but not at once realizable. We have seen recently that even the noble effort of the man invested with the greatest worldly power has been virtually without effect. And no wonder, for the establishment of universal peace is, for the time being, a physical impossibility. War is a negative force, and cannot be turned in a positive direction without passing through the intermediate phases. It is the problem of making a wheel, rotating one way, turn in the opposite direction without slowing it down, stopping it, and speeding it up again the other way.

It has been argued that the perfection of guns of great destructive power Will stop warfare. So I myself thought for a long time, but now I believe this to be a profound mistake. Such developments will greatly modify, but not arrest it. On the contrary, I think that every new arm that is invented, every new departure that is made in this direction, merely invites new talent and skill, engages new effort, offers a new incentive, and so only gives a fresh impetus to further development. Think of the discovery of gun powder. Can we conceive of any more radical departure than was effected by this innovation? Let us imagine ourselves living in that period: would we not have thought then that warfare was at an end, when the armor of the knight became an object of ridicule, when bodily strength and skill, meaning so much before, became of comparatively little value? Yet gunpowder did not stop warfare; quite the opposite—it acted as a most powerful incentive. Nor do I believe that warfare can ever be arrested by any scientific or ideal development, so long as similar conditions to those now prevailing exist, because war has itself become a science, and because war involves some of the most sacred sentiments of which man is capable. In fact, it is doubtful whether men who would not be ready to fight for a high principle would be good for anything at all. It is not the mind which makes man, nor is it the body; it is mind and body. Our virtues and our failings are inseparable, like force and matter. When they separate, man is no more.

Another argument, which carries considerable force, is frequently made, namely, that war must soon become impossible because the means of defense are outstripping the means of attack. This is only in accordance with a fundamental law which may be expressed by the statement that it is easier to destroy than to build. This law defines human capacities and human conditions. Were these such that it would be easier to build than to destroy, man would go on unresisted, creating and accumulating without limit. Such conditions are not of this earth. A being which could do this would not be a man; it might be a god. Defense will always have the advantage over attack, but this alone, it seems to me, can never stop war. By the use of new principles of defense we can render harbors impregnable against attack, but we cannot by such means prevent two war-ships meeting in battle on the high sea. And then, if we follow this idea to its ultimate development, we are led to the conclusion that it would be better for mankind if attack and defense were just oppositely related; for if every country, even the smallest, could surround itself with a wall absolutely impenetrable, and could defy the rest of the world, a state of things would surely be brought on which would be extremely unfavorable to human progress. It is by abolishing all barriers which separate nations and countries that civilization is best furthered.

Again, it is contended by some that the advent of the flying-machine must bring on universal peace. This, too, I believe to be an entirely erroneous view. The flying-machine is certainly coming, and very soon, but the conditions will remain the same as before. In fact, I see no reason why a ruling power, like Great Britain, might not govern the air as well as the sea. Without wishing to put myself on record as a prophet, I do not hesitate to say that the next years will see the establishment of an "air-power," and its center may not be far from New York. But, for all that, men will fight on merrily.

The ideal development of the war principle would ultimately lead to the transformation of the whole energy of war into purely potential, explosive energy, like that of an electrical condenser. In this form the war energy could be maintained without effort; it would need to be much smaller in amount, while incomparably more effective.

As regards the security of a country against foreign invasion, it is interesting to note that it depends only on the relative, and not on the absolute, number of the individuals or magnitude of the forces, and that, if every country should reduce the war-force in the same ratio, the security would remain unaltered. An international agreement with the object of reducing to a minimum the war-force which, in view of the present still imperfect education of the masses, is absolutely indispensable, would, therefore, seem to be the first rational step to take toward diminishing the force retarding human movement.

Fortunately, the existing conditions cannot continue indefinitely, for a new element is beginning to assert itself. A change for the better is imminent, and I shall now endeavor to show what, according to my ideas, win be the first advance toward the establishment of peaceful relations between nations, and by what means it will eventually be accomplished.

Let us go back to the early beginning, when the law of the stronger was the only law. The light of reason was not yet kindled, and the weak was entirely at the mercy of the strong. The weak individual then began to learn how to defend him self. He made use of a club, stone, spear, sling, or bow and arrow, and in the course of time, instead of physical strength, intelligence became the chief deciding factor in the battle. The wild character was gradually softened by the awakening of noble sentiments, and so Imperceptibly, after ages of continued progress, we have come from the brutal fight of the unreasoning animal to what we call the "civilized warfare" of today, in which the combatants shake hands, talk in a friendly way, and smoke cigars in the *entr'actes,* ready to engage again in deadly conflict at a signal. Let pessimists say what they like, here is an absolute evidence of great and gratifying advance.

But now, what is the next phase in this evolution? Not peace as yet, by any means. The next change which should naturally follow from modern developments should be the continuous diminution of the number of individuals engaged in battle. The apparatus will be one of specifically great power, but only a few individuals will be required to operate it. This evolution will bring more and more into prominence a machine or mechanism with the fewest individuals as an element of warfare, and the absolutely unavoidable consequence of this will be the abandonment of large, clumsy, slowly moving, and unmanageable units. Greatest possible speed and maximum rate of energy-delivery by the war apparatus will be the main object. The loss of life will become smaller and smaller, and finally, the number of the individuals continuously diminishing, merely machines will meet in a contest without bloodshed, the nations being simply interested, ambitious spectators. When this happy condition is realized, peace will be assured. But, no matter to what degree of perfection rapid fire guns, high power cannon, explosive projectiles, torpedo-boats, or other implements of war may be brought, no matter how destructive they may be made, that condition can never be reached through any such development. All such implements require men for their operation; men are indispensable parts of the machinery. Their object is to kill and to destroy. Their power resides in their capacity for doing evil. So long as men meet in battle, there will be bloodshed. Bloodshed will ever keep up barbarous passion. To break this fierce spirit, a radical departure must be made, an entirely new principle must be introduced, something that never existed before in warfare—a principle which will forcibly, unavoidably, turn the battle into a mere spectacle, a play, a contest without loss of blood. To bring on this result men must be dispensed with: machine must fight machine. But how accomplish that which seems impossible? The answer is simple enough: produce a machine capable of acting as though it were part of a human being—no mere mechanical contrivance, comprising levers, screws, wheels, clutches, and nothing more, but a machine embodying a higher principle, which will enable it to perform its duties as though it had intelligence, experience, reason, judgment, a mind! This conclusion is the result of my thoughts and observations which have extended through virtually my whole life, and I shall now briefly describe how I came to accomplish that which at first seemed an unrealizable dream.

A long time ago, when I was a boy, I was afflicted with a singular trouble, which seems to have been due to an extraordinary excitability of the retina. It was the appearance of images which, by their persistence, marred the vision of real objects and interfered with thought. When a word was said to me, the image of the object which it designated would appear vividly before my eyes, and many times it was impossible for me to tell whether the object I saw was real or not. This caused me great discomfort and anxiety, and I tried hard to free myself of the spell. But for a long time I tried in vain, and it was not, as I still clearly recollect, until I was about twelve years old that I succeeded for the first time, by an effort of the will, in banishing an image which presented itself. My happiness will never be as complete as it was then, but, unfortunately (as I thought at that time), the old trouble returned, and with it my anxiety. Here it was that the observations to which I refer began. I noted, namely, that whenever the image of an object appeared before my eyes I had seen something which reminded me of it. In the first instances I thought this to be purely accidental, but soon I convinced myself that it was not so. A visual impression, consciously or unconsciously received, invariably preceded the appearance of the image. Gradually the desire arose in me to find out, every time, what caused the images to appear, and the satisfaction of this desire soon became a necessity. The next observation I made was that, just as these images followed as a result of something I had seen, so also the thoughts which I conceived were suggested in like manner. Again, I experienced the same desire to locate the image which caused the thought, and this search for the original visual impression soon grew to be a second nature. My mind became automatic, as it were, and in the course of years of continued, almost unconscious performance, I acquired the ability of locating every time and, as a rule, instantly the visual impression which started the thought. Nor is this all. It was not long before I was aware that also all my movements were prompted in the same way, and so, searching, observing, and verifying continuously, year after year, I have, by every thought and every act of mine, demonstrated, and do so daily, to my absolute satisfaction, that I am an automaton endowed with power of movement, which merely responds to external stimuli beating upon my sense organs, and thinks and acts and moves accordingly. I remember only one or two cases in all my life in which I unable to locate the first impression which prompted a movement or a thought, or a dream.

With these experiences it was only natural that long ago, I conceived the idea of constructing an automaton which would mechanically represent me, and which would respond, as I do myself, but of course, in a much more primitive manner, to external influences. Such an automaton evidently had to have motive power, organs for locomotion, directive organs, and one or more sensitive organs so adapted as to be excited by external stimuli. This machine would, I reasoned, perform its movements in the manner of a living being, for it would have all the chief mechanical characteristics or elements of the same. There was still the capacity for growth, propagation, and, above all, the mind which would be wanting to make the model complete. But growth was not necessary in this case, since a machine could be manufactured full-grown, so to speak. As to the capacity for propagation, it could likewise be left out of consideration, for in the mechanical model it merely signified a process of manufacture. Whether the automaton be of flesh and bone, or wood and steel, it mattered little, provided it could perform all the duties required of it like an intelligent being. To do so, it had to have an element corresponding to the mind, which would effect the control of all its movements and operations, and cause it to act, in my unforeseen case that might present itself, with knowledge, reason, judgment, and experience. But this element I could easily embody in it by conveying to it my own intelligence, my own understanding. So this invention was evolved, and so a new art came into existence, for which the name "telautomatics" has been suggested, which means the art of controlling the movements and operations of distant automatons.

This principle evidently was applicable to any kind of machine that moves on land or in the water or in the air. In applying it practically for the first time, I selected a boat (see *Figure* 2). A storage battery placed within it furnished the motive power. The propeller, driven by a motor, represented the locomotive organs. The rudder, controlled by another motor likewise driven by the battery, took the place of the directive organs. As to the sensitive organ, obviously the first thought was to utilize a device responsive to rays of light, like a selenium cell, to represent the human eye. But upon closer inquiry I found that, owing to experimental and other difficulties, no thoroughly satisfactory control of the automaton could be effected by light, radiant heat, Hertzian radiations, or by rays in general, that is, disturbances which pass in straight lines through space. One of the reasons was that any obstacle coming between the operator and the distant automaton would place it beyond his control. Another reason was that the sensitive device representing the eye would have to be in a definite position with respect to the distant controlling apparatus, and this necessity would impose great limitations in the control. Still another and very important reason was that, in using rays, it would be difficult if not impossible, to give to the automaton individual features or characteristics distinguishing it from other machines of this kind. Evidently the automatons should respond only to an individual call, as a person responds to a name. Such considerations led me to conclude that the sensitive device of the machine should correspond to the ear rather than to the eye of a human being, for in this case its actions could be controlled irrespective of intervening obstacles, regardless of its position relative to the distant controlling apparatus, and last, but not least, it would remain deaf and unresponsive, like a faithful servant, to all calls but that of its master. These requirements made it imperative to use, in the control of the automaton, instead of light or other rays, waves or disturbances which propagate in all directions through space, like sound, or which follow a path of least resistance, however curved. I attained the result aimed by means of an electric circuit placed within the boat, and adjusted, or "tuned," exactly to electrical vibrations of the proper kind transmitted to it from a distant "electrical oscillator." This circuit in responding, however feebly, to the transmitted vibrations, affected magnets and other contrivances, through the medium of which were controlled the movements of the propeller and rudder, and also the operations of numerous other appliances.

By the simple means described the knowledge, experience, judgment—the mind, so to speak—of the distant operator were embodied in that machine, which was thus enabled to move and to perform all its operations with reason and intelligence. It behaved just like a blindfolded person obeying directions received through the ear.

The automatons so far constructed had "borrowed minds," so to speak, as each merely formed part of the distant operator who conveyed to it his intelligent orders; but this an is only in the beginning. I purpose to show that, however impossible it may now seem, an automaton may be contrived which will have its "own mind," and by this I mean that it will be able, independent of any operator, left entirely to itself, to perform, in response to external influences affecting its sensitive organs, a great variety of acts and operations as if it had intelligence. It will be able to follow a course laid out or to obey orders given far in advance; it will be capable of distinguishing between what it ought and what it ought not to do, and of making experiences or, otherwise stated, of recording impressions which will definitely affect its subsequent actions. In fact, I have already conceived such a plan.

Although I evolved this invention many years ago and explained it to my visitors very frequently in my laboratory demonstrations, it was not until much later, long after I had perfected it, that it became known, when, naturally enough, it gave rise to much discussion and to sensational reports. But the true significance of this new art was not grasped by the majority, nor was the great force of the underlying principle recognized. As nearly as I could judge from the numerous comments which then appeared, the results I had obtained were considered as entirely impossible. Even the few who were disposed to admit the practicability of the invention saw in it merely an automobile torpedo, which was to be used for the purpose of blowing up battle-ships, with doubtful success. The general impression was that I contemplated simply the steering of such a vessel by means of Hertzian or other rays, There are torpedoes steered electrically by wires, and there are means of communicating without wires, and the above was, of course, an obvious inference. Had I accomplished nothing more than this, I should have made a small advance indeed. But the art I have evolved does not contemplate merely the change of direction of a moving vessel; it affords a means of absolutely controlling, in every respect, all the innumerable translatory movements, as well as the operations of all the internal organs, no matter how many, of an individual automaton. Criticism to the effect that the control of the automaton could be interfered with were made by people who do not even dream of the wonderful results which can be accomplished by the use of electrical vibrations. The world moves slowly, and new truths are difficult to see. Certainly, by the use of this principle, an arm for attack as well as defense may be provided, of a destructiveness all the greater as the principle is applicable to submarine and aerial vessels. There is virtually no restriction as to the amount of explosive it can carry, or as to the distance at which it can strike, and failure is almost impossible. But the force of this new principle does not wholly reside in its destructiveness. Its advent introduces into warfare an element which never existed before—a fighting machine without men as a means of attack and defense. The continuous development in this direction must ultimately make war a mere contest of machines without men and without loss of life—a condition which would have been impossible without this new departure, and which, in my opinion, must be reached as preliminary to permanent peace. The future will either bear out or disprove these views. My ideas on this subject have been put forth with deep conviction, but in a humble spirit.

The establishment of permanent peaceful relations between nations would most effectively reduce the force retarding the human mass, and would be the best solution of this great human problem. But will the dream of universal peace ever be realized? Let us hope that it will. When all darkness shall be dissipated by the light of science, when all nations shall be merged into one, and patriotism shall be identical with religion, when there shall be one language, one country, one end, then the dream will have become reality.

THE THIRD PROBLEM: HOW TO INCREASE THE FORCE ACCELERATING THE HUMAN MASS—THE HARNESSING OF THE SUN'S ENERGY.

Of the three possible solutions of the main problem of increasing human energy, this is by far the most important to consider, not only because of its intrinsic significance, but also because of its intimate bearing on all the many elements and conditions which determine the movement of humanity. In order to proceed systematically, it would be necessary for me to dwell on all those considerations which have guided me from the outset in my efforts to arrive at a solution, and which have led me, step by step, to the results I shall now describe. As a preliminary study of the problem an analytical investigation, such as I have made, of the chief forces which determine the onward movement, would be of advantage, particularly in conveying an idea of that hypothetical "velocity" which, as explained in the beginning, is a measure of human energy; but to deal with this specifically here, as I would desire, would lead me far beyond the scope of the present subject. Suffice it to state that the resultant of all these forces is always in the direction of reason, which, therefore, determines, at any time, the direction of human movement. This is to say that every effort which is scientifically applied, rational, useful, or practical, must be in the direction in which the mass is moving. The practical, rational man, the observer, the man of business, he who reasons, calculates, or determines in advance, carefully applies his effort so that when coming into effect it will be in the direction of the movement, making it thus most efficient, and in this knowledge and ability lies the secret of his success. Every new fact discovered, every new experience or new element added to our knowledge and entering into the domain of reason, affects the same and, therefore, changes the direction of the movement, which, however, must always take place along the resultant of all those efforts which, at that time, we designate as reasonable, that is, self-preserving, useful, profitable, or practical. These efforts concern our daily life, our necessities and comforts, our work and business, and it is these which drive man onward.

But looking at all this busy world about us, on all this complex mass as it daily throbs and moves, what is it but an immense clock work driven by a spring? In the morning, when we rise, we cannot fail to note that all the objects about us are manufactured by machinery: the water we use is lifted by steam-power; the trains bring our breakfast from distant localities; the elevators in our dwelling and in our office building, the cars that carry us there, are all driven by power, in all our daily errands, and in our very life pursuit, we depend upon it; all the objects we see tell us of it; and when we return to our machine-made dwelling at night, lest we should forget it, all the material comforts of our home, our cheering stove and lamp, remind us how much we depend on power. And when there is an accidental stoppage of the machinery, when the city is snow-bound, or the life-sustaining movement otherwise temporarily arrested, we are affrighted to realize how impossible it would be for us to live the life we live without motive power. Motive power means work. To increase the force accelerating human movement means, therefore, to perform more work.

So we find that the three possible solutions of the great problem of increasing human energy are answered by the three words: food, peace, work. Many a year I have thought and pondered, lost myself in speculations and theories, considering man as a mass moved by force, viewing his inexplicable movement in the light of a mechanical one, and applying the simple principles of mechanics to the analysis of the same until I arrived at these solutions, only to realize that they were taught to me in my early childhood. These three words sound the key-notes of the Christian religion. Their scientific meaning and purpose are now clear to me: food to increase the mass, peace to diminish the retarding force, and work to increase the force accelerating human movement. These are the only three solutions which are possible of that great problem, and all of them have one object, one end, namely, to increase human energy. When we recognize this, we cannot help wondering how profoundly wise and scientific and how immensely practical the Christian religion is, and in what a marked contrast it stands in this respect to other religions. It is unmistakably the result of practical experiment and scientific observation which have extended through ages, while other religions seem to be the outcome of merely abstract reasoning. Work, untiring effort, useful and accumulative, with periods of rest and recuperation aiming at higher efficiency, is its chief and ever-recurring command. Thus we are inspired both by Christianity and Science to do our utmost toward increasing the performance of mankind. This most important of human problems I shall now specifically consider.

THE SOURCE OF HUMAN ENERGY—
THE THREE WAYS OF DRAWING ENERGY FROM THE SUN.

First let us ask: Whence comes all the motive power? What is the spring that drives all? We see the ocean rise and fall, the rivers flow, the wind, rain, hail, and snow beat on our windows, the trains and steamers come and go; we hear the rattling noise of carriages, the voices from the street; we feel, smell, and taste; and we think of all this. And all this movement, from the surging of the mighty ocean to that subtle movement concerned on our thought, has but one common cause. All this energy emanates from one single center, one single source—the sun. The sun is the spring that drives all. The sun maintains all human life and supplies all human energy. Another answer we have now found to the above great question: To increase the force accelerating human movement means to turn to the uses of man more of the sun's energy. We honor and revere those great men of bygone times whose names are linked with immortal achievements, who have proved themselves benefactors of humanity—the religious reformer with his wise maxims of life, the philosopher with his deep truths, the mathematician with his formulae, the physicist with his laws, the discoverer with his principles and secrets wrested from nature, the artist with his forms of the beautiful; but who honors him, the greatest of all,—who can tell the name of him,—who first turned to use the sun's energy to save the effort of a weak fellow-creature? That was man's first act of scientific philanthropy, and its consequences have been incalculable.

From the very beginning three ways of drawing energy from the sun were open to man. The savage, when he warmed his frozen limbs at a fire kindled in some way, availed himself of the energy of the sun stored in the burning material. When he carried a bundle of branches to his cave and burned them there, he made use if the sun's stored energy transported from one to another locality. When he set sail to his canoe, he utilized the energy of the sun supplied to the atmosphere or ambient medium. There can be no doubt that the first is the oldest way. A fire, found accidentally, taught the savage to appreciate its beneficial heat. He then very likely conceived the idea of carrying the glowing embers to his abode. Finally he learned to use the force of a swift current of water or air. It is characteristic of modern development that progress has been effected in the same order. The utilization of the energy stored in wood or coal, or, generally speaking, fuel, led to the steam engine. Next a great stride in advance was made in energy transportation by the use of electricity, which permitted the transfer of energy from one locality to another without transporting the material. But as to the utilization of the energy of the ambient medium, no radical step forward has yet been made known.

The ultimate results of development in these three directions are: first, the burning of coal by a cold process in a battery; second, the efficient utilization of the energy of the ambient medium; and, third, the transmission without wires of electrical energy to any distance. In whatever way these results may be arrived at, their practical application will necessarily involve an extensive use of iron, and this invaluable metal will undoubtedly be an essential element in the further development along these three lines. If we succeed in burning coal by a cold process and thus obtaining electrical energy in an efficient and inexpensive manner, we shall require in many practical uses of this energy electric motors—that is, iron. If we are successful in deriving energy from the ambient medium, we shall need, both in the obtainment and utilization of the energy, machinery—again, iron. If we realize the transmission of electrical energy without wires on an industrial scale, we shall be compelled to use extensively electric generators—once more, iron. Whatever we may do, iron will probably be the chief means of accomplishment in the near future, possibly more so than in the past. How long its reign will last is difficult to tell, for even now aluminum is looming up as a threatening competitor. But for the time being, next to providing new resources of energy, it is of the greatest importance to make improvements in the manufacture and utilization of iron. Great advances are possible in these latter directions, which, if brought about, would enormously increase the useful performance of mankind.

GREAT POSSIBILITIES OFFERED BY IRON FOR INCREASING HUMAN PERFORMANCE—
ENORMOUS WASTE IN IRON MANUFACTURE.

Iron is by far the most important factor in modern progress. Its contributes more than any other industrial product to the force accelerating human movement. So general is the use of this metal, and so intimately is it connected with all that concerns life, that it has become as indispensable to us as the very air we breathe. Its name is synonymous with usefulness, But, however great the influence of iron may be on the present human development, it does not add to the force urging man onward nearly as much as it might. First of all, its manufacture as now carried on is connected with an appalling waste of fuel—that is, waste of energy. Then, again, only a pan of all the iron produced is applied for useful purposes. A good part of it goes to create frictional resistances, while still another large part is the means of developing negative forces greatly retarding human movement. Thus the negative force of war is almost wholly represented in iron. It is impossible to estimate with any degree of accuracy the magnitude of this greatest of all retarding forces, but it is certainly very considerable. If the present positive impelling force due to all useful applications of iron be represented by ten, for instance, I should not think it exaggeration to estimate the negative force of war, with due consideration of all its retarding influences and results, at, say, six. On the basis of this estimate the effective impelling force of iron in the positive direction would be measured by the difference of these two numbers, which is four. But if, through the establishment of universal peace, the manufacture of war machinery should cease, and all struggle for supremacy between nations should be turned into healthful, ever active and productive commercial competition, then the positive impelling force due to iron would be measured by the sum of those two numbers, which is sixteen—that is, this force would have four times its present value. This example is, of course, merely intended to give an idea of the immense increase in the useful performance of mankind which would result from a radical reform of the iron industries supplying the implements of warfare.

A similar inestimable advantage in the saving of energy available to man would be secured by obviating the great waste of coal which is inseparably connected with the present methods of manufacturing iron. In some countries, as in Great Britain, the hurtful effects of this squandering of fuel are beginning to be felt. The price of coal is constantly rising, and the poor are made to suffer more and more. Though we are still far from the dreaded "exhaustion of the coal-fields," philanthropy commands us to invent novel methods of manufacturing iron, which will not involve such barbarous waste of this valuable material from which we derive at present most of our energy. It is our duty to coming generations to leave this store of energy intact for them, or at least not to touch it until we shall have perfected processes for burning coal more efficiently. Those who are to come after us will need fuel more than we do. We should be able to manufacture the iron we require by using the sun's energy, without wasting any coal at all. As an effort to this end the idea of smelting iron ores by electric currents obtained from the energy of falling water has naturally suggested itself to many. I have myself spent much time in endeavoring to evolve such a practical process, which would enable iron to be manufactured at small cost. After a prolonged investigation of the subject, finding that it was unprofitable to use the currents generated directly for smelting the ore, I devised a method which is far more economical.

ECONOMICAL PRODUCTION OF IRON BY A NEW PROCESS.

The industrial project, as I worked it out six years ago, contemplated the employment of the electric currents derived from the energy of a waterfall, not directly for smelting the ore, but for decomposing water, as a preliminary step. To lessen the cost of the plant, I proposed to generate the currents in exceptionally cheap and simple dynamos, which I designed for this sole purpose. The hydrogen liberated in the electrolytic decomposition was to be burned or recombined with oxygen, not with that from which it was separated, but with that of the atmosphere. Thus very nearly the total electrical energy used up in the decomposition of the water would be recovered in the form of heat resulting from the recombinations of the hydrogen. This heat was to be applied to the smelting of the ore. The oxygen gained as a by-product in the decomposition of the water I intended to use for certain other industrial purposes, which would probably yield good financial returns, inasmuch as this is the cheapest way of obtaining this gas in large quantities. In any event, it could be employed to burn all kinds of refuse, cheap hydrocarbon, or coal of the most inferior quality which could not be burned in air or be otherwise utilized to advantage, and thus again a considerable amount of heat would be made available for the smelting of the ore. To increase the economy of the process I contemplated, furthermore, using an arrangement such that the hot metal and the products of combustion, coming out of the furnace, would give up their heat upon the cold ore going into the furnace, so that comparatively little of the heat-energy would be lost in the smelting. I calculated that probably forty thousand pounds of iron could be produced per horse-power per annum by this method. Liberal allowances were made for those losses which are unavoidable, the above quantity being about half of that theoretically obtainable. Relying on this estimate and on practical data with reference to a certain kind of sand ore existing in abundance in the region of the Great Lakes, including cost of transportation and labor, I found that in some localities iron could be manufactured in this manner cheaper than by any of the adopted methods. This result would be attained all the more surely if the oxygen obtained from the water, instead of being used for smelting the ore, as assumed, should be more profitably employed. Any new demand for this gas would secure a higher revenue from the plant, thus cheapening the iron. This project was advanced merely in the interest of industry. Some day, I hope, a beautiful industrial butterfly will come out of the dusty and shriveled chrysalis.

The production of iron from sand ores by a process of magnetic separation is highly commendable in principle, since it involves no waste of coal; but the usefulness of this method is largely reduced by the necessity of melting the iron afterward. As to the crushing of iron ore, I would consider it rational only if done by water-power, or by energy otherwise obtained without consumption of fuel. An electrolytic cold process, which would make it possible to extract iron cheaply, and also to mold it into the required forms without any fuel consumption, would, in my opinion, be very great advance in iron manufacture. In common with some other metals, iron has so far resisted electrolytic treatment, but there can be no doubt that such a cold process will ultimately replace in metallurgy the present crude method of casting, and thus obviate the enormous waste of fuel necessitated by the repeated heating of metal in the foundries.

Up to a few decades ago the usefulness of iron was based almost wholly on its remarkable mechanical properties, but since the advent of the commercial dynamo and electric motor its value to mankind has been greatly increased by its unique magnetic qualities. As regards the latter, iron has been greatly improved of late. The signal progress began about thirteen years ago, when I discovered that in using soft Bessemer steel instead of wrought iron, as then customary, in an alternating motor, the performance of the machine was doubled. I brought this fact to the attention of Mr. Albert Schmid, to whose untiring efforts and ability is largely due the supremacy of American electrical machinery, and who was then superintendent of an industrial corporation engaged in this field. Following my suggestion, he constructed transformers of steel, and they showed the same marked improvement. The investigation was then systematically continued under Mr. Schmid's guidance, the impurities being gradually eliminated from the "steel" (which was only such in name, for in reality it was pure soft iron), and soon a product resulted which admitted of little further improvement.

THE COMING AGE OF ALUMINUM—
DOOM OF THE COPPER INDUSTRY—
THE GREAT CIVILIZING POTENCY OF THE NEW METAL.

With the advances made in iron of late years we have arrived virtually at the limits of improvement. We cannot hope to increase very materially its tensile strength, elasticity, hardness, or malleability, nor can we expect to make it much better as regards its magnetic qualities. More recently a notable gain was secured by the mixture of a small percentage of nickel with the iron, but there is not much room for further advance in this direction. New discoveries may be expected, but they cannot greatly add to the valuable properties of the metal, though they may considerably reduce the cost of manufacture. The immediate future of iron is as-

sured by its cheapness and its unrivaled mechanical and magnetic qualities. These are such that no other product can compete with it now. But there can be no doubt that, at a time not very distant, iron, in many of its now uncontested domains, will have to pass the scepter to another: the coming age will be the age of aluminum. It is only seventy years since this wonderful metal was discovered by Wochler, and the aluminum industry, scarcely forty years old, commands already the attention of the entire world. Such rapid growth has not been recorded in the history of civilization before. Not long ago aluminum was sold at the fanciful price of thirty or forty dollars per pound; today it can be had in any desired amount for as many cents. What is more, the time is not far off when this price, too, will be considered fanciful, for great improvements are possible in the methods of its manufacture. Most of the metal is now produced in the electric furnace by a process combining fusion and electrolysis, which offers a number of advantageous features, but involves naturally a great waste of the electrical energy of the current. My estimates show that the price of aluminum could be considerably reduced by adopting in its manufacture a method similar to that proposed by me for the production of iron. A pound of aluminum requires for fusion only about seventy per cent of the heat needed for melting a pound of iron, and inasmuch as its weight is only about one third of that of the latter, a volume of aluminum four times that of iron could be obtained from a given amount of heat-energy. But a cold electrolytic process of manufacture is the ideal solution, and on this I have placed my hope.

The absolutely unavoidable consequence of the advance of the aluminum industry will be the annihilation of the copper industry. They cannot exist and prosper together, and the latter is doomed beyond any hope of recovery. Even now it is cheaper to convey an electric current through aluminum wires than through copper wires; aluminum castings cost less, and in many domestic and other uses copper has no chance of successfully competing. A further material reduction of the price of aluminum cannot but be fatal to copper. But the progress of the former will not go on unchecked, for, as it ever happens in such cases, the larger industry will absorb the smaller, one: the giant copper interests will control the pygmy aluminum interests, and the slow-pacing copper will reduce the lively gait of aluminum. This will only delay, not avoid, the impending catastrophe.

Aluminum, however, will not stop at downing copper. Before many year's have passed it will be engaged in a fierce struggle with iron, and in the latter it will find an adversary not easy to conquer. The issue of the contest will largely depend on whether iron shall be indispensable electric machinery. This the future alone can decide. The magnetism as exhibited in iron is an isolated phenomenon in nature. What it is that makes this metal behave so radically different from all other materials in this respect has not yet been ascertained, though many theories have been suggested. As regards magnetism, the molecules of the various bodies behave like hollow beams partly filled with a heavy fluid and balanced in the middle in the manner of a see-saw. Evidently some disturbing influence exists in nature which causes each molecule, like such a beam, to tilt either one or the other way. If the molecules are tilted one way, the body is magnetic; if they are tilted the other way, the body is non-magnetic; but both positions are stable, as they would be in the case of the hollow beam, owing to the rushing of the fluid to the lower end. Now, the wonderful thing is that the molecules of all known bodies went one way, while those of iron went the other way. This metal, it would seem, has an origin entirely different from that of the rest of the globe. It is highly improbable that we shall discover some other and cheaper material which will equal or surpass iron in magnetic qualities.

Unless we should make a radical departure in the character of the electric currents employed, iron will be indispensable. Yet the advantages it offers are only apparent. So long as we use feeble magnetic forces it is by far superior to any other material; but if we find ways of producing great magnetic forces, then better results will be obtainable without it. In fact, I have already produced electric transformers in which no iron is employed, and which are capable of performing ten times as much work per pound of weight as those with iron. This result is attained by using electric currents of a very high rate of vibration, produced in novel ways, instead of the ordinary currents now employed in the industries. I have also succeeded in operating electric motors without iron by such rapidly vibrating currents, but the results, so far, have been inferior to those obtained with ordinary motors constructed of iron, although theoretically the former should be capable of performing incomparably more work per unit of weight than the latter. But the seemingly insuperable difficulties which arc now in the way may be overcome in the end, and then iron will be done away with, and all electric machinery will be manufactured of aluminum, in all probability, at prices ridiculously low. This would be a severe, if not a fatal, blow to iron. In many other branches of industry, as ship-building, or wherever lightness of structure is required, the progress of the new metal will be much quicker. For such uses it is eminently suitable, and is sure to supersede iron sooner or later. It is highly probable that in the course of time we shall be able to give it many of those qualities which make iron so valuable.

While it is impossible to tell when this industrial revolution will be consummated, there can be no doubt that the future belongs to aluminum, and that in times to come it will be the chief means of increasing human performance. It has in this respect capacities greater by far than those of any other metal. I should estimate its civilizing potency at fully one hundred times that of iron. This estimate, though it may astonish, is not at all exaggerated. First of all, we must remember that there is thirty times as much aluminum as iron in bulk, available for the use of man. This in itself offers great possibilities. Then, again, the new metal is much more easily workable, which adds to its value. In many of its properties it partakes of the character of a precious metal, which gives it additional worth. Its electric conductivity, which, for a given weight, is greater than that of any other metal, would be alone sufficient to make it on of the most important factors in future human progress, Its extreme lightness makes it far more easy to transport the objects manufactured. By virtue of this property it will revolutionize naval construction, and in facilitating transport and travel it will add enormously to the useful performance of mankind. But its greatest civilizing potency will be, I believe, in aerial travel, which is sure to be brought about by means of it. Telegraphic instruments will slowly enlighten the barbarian. Electric motors and lamps will do it more quickly, but quicker than anything else the flying-machine win do it. By rendering travel ideally easy it will be the best means for unifying the heterogeneous elements of humanity. As the first step toward this realization we should produce a lighter storage-battery or get more energy from coal.

EFFORTS TOWARD OBTAINING MORE ENERGY FROM COAL— THE ELECTRIC TRANSMISSION— THE GAS-ENGINE — THE COLD-COAL BATTERY.

I remember that at one time I considered the production of electricity by burning coal in a battery as the greatest achievement toward advancing civilization, and I am surprised to find how much the continuous study of these subjects has modified my views. It now seems to me that to burn coal, however efficiently, in a battery would be a mere makeshift, a phase in the evolution toward something much more perfect. After all, in generating electricity in this manner, we should be destroying material, and this would be a barbarous process. We ought to be able to obtain the energy we need without consumption of material. But I am far from underrating the value of such an efficient method of burning fuel. At the present time most motive power comes from coal, and, either directly or by its products, it adds vastly to human energy. Unfortunately, in all the processes now adopted, the larger portion of the energy of the coal is uselessly dissipated. The best steam-engines utilize only a small part of the total energy. Even in gas-engines, in which, particularly of late, better results are obtainable, there is still a barbarous waste going on. In our electric-lighting systems we scarcely utilize one third of one per cent, and in lighting by gas a much smaller fraction, of the total energy of the coal. Considering the various uses of coal throughout the world, we certainly do not utilize more than two per cent of its energy theoretically available. The man who should stop this senseless waste would be a great benefactor of humanity, though the solution he would offer could not be a permanent one, since it would ultimately lead to the exhaustion of the store of material. Efforts toward obtaining more energy from coal are now being made chiefly in two directions—by generating electricity and by producing gas for motive power purposes. In both of these lines notable success has already been achieved.

The advent of the alternating-current system of electric power-transmission marks an epoch in the economy of energy available to man from coal. Evidently all electrical energy obtained from a waterfall, saving so much fuel, is a net gain to mankind, which is all the more effective as it is secured with little expenditure of human effort, and as this most perfect of all known methods of deriving energy from the sun contributes in many ways to the advancement of civilization. But electricity enables us also to get from coal much more energy than was practicable in the old ways. Instead of transporting the coal to distant places of consumption, we burn it near the mine, develop electricity in the dynamos, and transmit the current to remote localities, thus effecting a considerable saving. Instead of driving the machinery in a factory in the old wasteful way by belts and shafting, we generate electricity by steam-power and operate electric motors. In this manner it is not uncommon to obtain two or three times as much effective motive power from the fuel, besides securing many other important advantages. It is in this field as much as in the transmission of energy to great distances that the alternating system, with its ideally simple machinery, is bringing about an industrial revolution. But in many lines this progress has not yet been felt. For example, steamers and trains are still being propelled by the direct application of steam-power to shafts or axles. A much greater percentage of the heat-energy of the fuel could be transformed into motive energy by using, in place of the adopted marine engines and locomotives, dynamos driven by specially designed high-pressure steam or gas-engines and by utilizing the electricity generated for the propulsion. A gain of fifty to one hundred per cent in the effective energy derived from the coal could be secured in this manner. It is difficult to understand why a fact so plain and obvious is not receiving more attention from engineers. In ocean steamers such an improvement would be particularly desirable, as it would do away with noise and increase materially the speed and the carrying capacity of the liners.

Still more energy is now being obtained from coal by the latest improved gas-engine, the economy of which is, on the average, probably twice that of the best steam-engine. The introduction of the gas-engine is very much facilitated by the importance of the gas industry. With the increasing use of the electric light more and more of the gas is utilized for heating and motive-power purposes. In many instances gas is manufactured close to the coal-mine and conveyed to distant places of consumption, a considerable saving both in the cost of transportation and in utilization of the energy of the fuel being thus effected. In the present state of the mechanical and electrical arts the most rational way of deriving energy from coal is evidently to manufacture gas close to the coal store, and to utilize it, either on the spot or elsewhere, to generate electricity for industrial uses in dynamos driven by gas-engines. The commercial success of such a plant is largely dependent upon the production of gas-engines of great nominal horse-power, which, judging from the keen activity in this field, will soon be forthcoming. Instead of consuming coal directly, as usual, gas should be manufactured from it and burned to economize energy.

But all such improvements cannot be more than passing phases in the evolution toward something far more perfect, for ultimately we must succeed in obtaining electricity from coal in a more direct way, involving no great loss of its heat-energy. Whether coal can be oxidized by a cold process is still a question. Its combination with oxygen always evolves heat, and whether the energy of the combination of the carbon with another element can be turned directly into electrical energy has not yet been determined. Under certain conditions nitric acid will burn the carbon, generating an electric current, but the solution does not remain cold. Other means of oxidizing coal have been proposed, but they have offered no promise of leading to an efficient process. My own lack of success has been complete, though perhaps not quite so complete as that of some who have "perfected" the cold-coal battery. This problem is essentially one for the chemist to solve. It is not for the physicist, who determines all his results in advance, so that, when the experiment is tried, it cannot fail. Chemistry, though a positive science, does not yet admit of a solution by such positive methods as those which are available in the treatment of many physical problems. The result, if possible, will be arrived at through patient trying rather than through deduction or calculation. The time will soon come, however, when the chemist will be able to follow a course clearly mapped out beforehand, and when the process of his arriving at a desired result will be purely constructive. The cold-coal battery would give a great impetus to electrical development; it would lead very shortly to a practical flying machine, and would enormously enhance the introduction of the automobile. But these and many other problems will be better solved, and in a more scientific manner, by a light-storage battery.

ENERGY FROM THE MEDIUM—
THE WINDMILL AND THE SOLAR ENGINE—
MOTIVE POWER FROM TERRESTRIAL HEAT—
ELECTRICITY FROM NATURAL SOURCES.

Besides fuel, there is abundant material from which we might eventually derive power. An immense amount of energy is locked up in limestone, for instance, and machines can be driven by liberating the carbonic acid through sulfuric acid or otherwise. I once constructed such an engine, and it operated satisfactorily.

But, whatever our resources of primary energy may be in the future, we must, to be rational, obtain it without consumption of any material. Long ago I came to this conclusion, and to arrive at this result only two ways, as before indicated, appeared possible—either to try to use the energy of the sun, stored in the ambient medium, or to transmit, through the medium, the sun's energy to distant places from some locality where it was obtainable without consumption of material. At that time I at once rejected the latter method as entirely impracticable, and turned to examine the possibilities of the former.

It is difficult to believe, but it is, nevertheless, a fact, that since time immemorial man has had at his disposal a fairly good machine which has enabled him to utilize the energy of the ambient medium. This machine is the windmill. Contrary to popular belief, the power obtainable from wind is very considerable. Many a deluded inventor has spent years of his life in endeavoring to "harness the tides," and some have even proposed to compress air by tide or wave power for supplying energy, never understanding the signs of the old windmill on the hill, as it sorrowfully waved its arms about and bade them stop. The fact is that a wave or tide motor would have, as a rule, but a small chance of competing commercially with the windmill, which is by far the better machine, allowing a much greater amount of energy to be obtained in a simpler way. Wind-power has been, in old times, of inestimable value to man, if for nothing else but for enabling him to cross the seas, and it is even now a very important factor in travel and transportation. But there arc great limitations in this ideally simple method of utilizing the sun's energy. The machines are large for a given output, and the power is intermittent, thus necessitating the storage of energy and increasing the cost of the plant.

A far better way, however, to obtain power would be to avail ourselves of the sun's rays, which beat the earth incessantly and supply energy at a maximum rate of over four million horse-power per square mile. Although the average energy received per square mile in any locality during the year is only a small fraction of that amount, yet an inexhaustible source of power would be opened up by the discovery of some efficient method of utilizing the energy of the rays. The only rational way known to me at the time when I began the study of this subject was to employ some kind of heat or thermodynamic engine, driven by a volatile fluid evaporated in a boiler by the heat of the rays. But closer investigation of this method, and calculations, showed that, notwithstanding the apparently vast amount of energy received from the sun's rays, only a small fraction of that energy could be actually utilized in this manner. Furthermore, the energy supplied through the sun's radiations is periodical, and the same limitations as in use of the windmill I found to exist here also. After along study of this mode of obtaining motive power from the sun, taking into account the necessarily large bulk of the boiler, the low efficiency of the heat-engine, the additional cost of storing the energy, and other drawbacks, I came to the conclusion that the "solar engine," a few instances excepted, could not be industrially exploited with success.

Another way of getting motive power from the medium without consuming any material would be to utilize the heat contained in the earth, the water, or the air for driving an engine. It is a well-known fact that the interior portions of the globe are very hot, the temperature rising, as observations show, with the approach to the center at the rate of approximately 1°C. for every hundred feet of depth. The difficulties of sinking shafts and placing boilers at depths of, say, twelve thousand feet, corresponding to an increase in temperature of about 120°C., are not insuperable, and we could certainly avail ourselves in this way of the internal heat of the globe. In fact, it would not be necessary to go to any depth at all in order to derive energy from the stored terrestrial heat. The superficial layers of the earth and the air strata close to the same are at a temperature sufficiently high to evaporate some extremely volatile substances, which we might use in our boilers instead of water. There is no doubt that a vessel might be propelled on the ocean by an engine driven by such a volatile fluid, no other energy being used but the heat abstracted from the water. But the amount of power which could be obtained in this manner would be, without further provision, very small.

Electricity produced by natural causes is another source of energy which might be rendered available. Lightning discharges involve great amounts of electrical energy, which we could utilize by transforming and storing it. Some years ago I made known a method of electrical transformation which renders the first part of this task easy, but the storing of the energy of lightning discharges will be difficult to accomplish. It is well known, furthermore, that electric currents circulate constantly through the earth, and that there exists between the earth and any air stratum a difference of electrical pressure, which varies in proportion to the height.

In recent experiments I have discovered two novel facts of importance in this connection. One of these facts is that an electric current is generated in a wire extending from the ground to a great height by the axial, and probably also by the translatory, movement of the earth. No appreciable current, however, will flow continuously in the wire unless the electricity is allowed to leak out into the air. Its escape is greatly facilitated by providing at the elevated end of the wire a conducting terminal of great surface, with many sharp edges or points. We are thus enabled to get a continuous supply of electrical energy by merely supporting a wire at a height, but, unfortunately, the amount of electricity which can be so obtained is small.

The second fact which I have ascertained is that the upper air strata are permanently charged with electricity opposite to that of the earth. So, at least, I have interpreted my observations, from which it appears that the earth, with its adjacent insulating and outer conducting envelope, constitutes a highly charged electrical condenser containing, in all probability, a great amount of electrical energy which might be turned to the uses of man, if it were possible to reach with a wire to great altitudes.

It is possible, and even probable, that there will be, in time, other resources of energy opened up, of which we have no knowledge now. We may even find ways of applying forces such as magnetism or gravity for driving machinery without using any other means. Such realizations, though highly improbable, are not impossible. An example will best convey an idea of what we can hope to attain and what we can never attain. Imagine a disk of some homogeneous material turned perfectly true and arranged to turn in frictionless bearings on a horizontal shaft above the ground. This disk, being under the above conditions perfectly balanced, would rest in any position. Now, it is possible that we may learn how to make such a disk rotate continuously and perform work by the force of gravity without any further effort on our pan; but it is perfectly impossible for the disk to turn and to do work without any force from the outside. If it could do so, it would be what is designated scientifically as a "perpetuum mobile," a machine creating its own motive power. To make the disk rotate by the force of gravity we have only to invent a screen against this force. By such a screen we could prevent this force from acting on one half of the disk, and the rotation of the latter would follow. At least, we cannot deny such a possibility until we know exactly the nature of the force of gravity. Suppose that this force were due to a movement comparable to that of a stream of air passing from above toward the center of the earth. The effect of such a stream upon both halves of the disk would be equal, and the latter would not rotate ordinarily; but if one half should be guarded by a plate arresting the movement, then it would turn.

A DEPARTURE FROM KNOWN METHODS—
POSSIBILITY OF A "SELF-ACTING" ENGINE OR MACHINE, INANIMATE, YET CAPABLE, LIKE A LIVING BEING, OF DERIVING ENERGY FROM THE MEDIUM—
THE IDEAL WAY OF OBTAINING MOTIVE POWER.

When I began the investigation of the subject under consideration, and when the preceding or similar ideas presented themselves to me for the first time, though I was then unacquainted with a number of the facts mentioned, a survey of the various ways of utilizing the energy of the medium convinced me, nevertheless, that to arrive at a thoroughly satisfactory practical solution a radical departure from the methods then known had to be made. The windmill, the solar engine, the engine driven by terrestrial heat, had their limitations in the amount of power obtainable. Some new way had to be discovered which would enable us to get more energy. There was enough heat-energy in the medium, but only a small part of it was available for the operation of an engine in the ways then known. Besides, the energy was obtainable only at a very slow rate. Clearly, then, the problem was to discover some new method which would make it possible both to utilize more of the heat-energy of the medium and also to draw it away from the same at a more rapid rate. I was vainly endeavoring to form an idea of how this might be accomplished, when I read some statements from Camot and Lord Kelvin (then Sir William Thomson) which meant virtually that it is impossible for an inanimate mechanism or self-acting machine to cool a portion of the medium below the temperature of the surrounding, and operate by heat abstracted. These statements interested me intensely. Evidently a living being could do this very thing, and since the experiences of my early life which I have related had convinced me that a living being is only an automaton, or, otherwise stated, a "self-acting engine," I came to the conclusion that it was possible to construct a machine which would do the same. As the first step toward this realization I conceived the following mechanism. Imagine a thermopile consisting of a number of bars of metal extending from the earth to the outer space beyond the atmosphere. The heat from below, conducted upward along these metal bars, would cool the earth or the sea or the air, according to the location of the lower parts of the bars, and the result, as is well known, would be an electric current circulating in these bars. The two terminals of the thermopile could now be joined through an electric motor, and, theoretically, this motor would run on and on, until the media below would be cooled down to the temperature of the outer space. This would be an inanimate engine which, to all evidence, would be cooling a portion of the medium below the temperature of the surrounding, and operating by the heat abstracted.

But was it not possible to realize a similar condition without necessarily going to a height? Conceive, for the sake of illustration, an enclosure T, as illustrated in *diagram b*, such that energy could not be transferred across it except through a channel or path O, and that, by some means or other, in this enclosure a medium were maintained which would have little energy, and that on the outer side of the same there would be the ordinary ambient medium with much energy. Under these assumptions the energy would flow through the path O, as indicated by the arrow, and might then be converted on its passage into some other form of energy. The question was, Could such a condition be attained? Could we produce artificially such a "sink" for the energy of the ambient medium to flow in? Suppose that an extremely low temperature could be maintained by some process in a given space; the surrounding medium would then be compelled to give off heat, which could be converted into mechanical or other form of energy, and utilized. By realizing such a plan, we

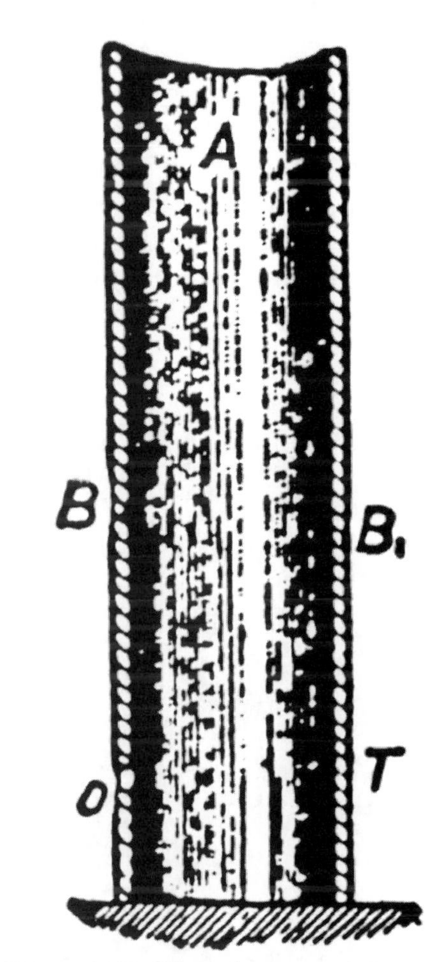

Diagram B. Obtaining energy from the ambient medium. A: Medium with little energy; B and B_1: ambient medum with much energy; O: path of energy.

should be enabled to get at any point of the globe a continuous supply of energy, day and night. More than this, reasoning in the abstract, it would seem possible to cause a quick circulation of the medium, and thus draw the energy at a very rapid rate.

Here, then, was an idea which, if realizable, afforded a happy solution of the problem of getting energy from the medium. But was it realizable? I convinced myself that it was so in a number of ways, of which one is the following. As regards heat, we are at a high level, which may be represented by the surface of a mountain lake considerably above the sea, the level of which may mark the absolute zero of temperature existing in the interstellar space. Heat, like water, flows from high to low level, and, consequently, just as we can let the water of the lake run down to the sea, so we are able to let heat from the earth's surface travel up into the cold region above. Heat, like water, can perform work in flowing down, and if we had any doubt as to whether we could derive energy from the medium by means of a thermopile, as before described, it would be dispelled by this analog. But can we produce cold in a given portion of the space and cause that the heat to flow in continually? To create such a "sink," or "cold hole," as we might say, in the medium, would be equivalent to producing in the lake a space either empty or filled with something much lighter than water. This we could do by placing in the lake a tank, and pumping all the water out of the latter. We know, then, that the water, if allowed to flow back into the tank, would, theoretically, be able to perform exactly the same amount of work which was used in pumping it out, but not a bit more. Consequently noth-

The Columbian Exposition in Chicago, 1893, featuring Tesla's 3-phase alternating current motor.

ing could be gained in this double operation of first raising the water and then letting it fall down. This would mean that it is possible to create such a sink in the medium. But let us reflect a moment. Heat, though following certain general laws of mechanics, like a fluid, is not such; it is energy which may be converted into other forms of energy as it passes from a high to a low level. To make our mechanical analogy complete and true, we must, therefore, assume that the water, in its passage into the tank, is convened into something else, which may be taken out of it without using any, or by using very little, power.

For example, if heat be represented in this analog by the water of the lake, the oxygen and hydrogen composing the water may illustrate other forms of energy into which the heat is transformed in passing from hot to cold. If the process of heat transformation were absolutely perfect, no heat at all would arrive at the low level, since all of it would be converted into other forms of energy. Corresponding to this ideal case, all the water into the tank would be decomposed into oxygen and hydrogen before reaching the bottom, and the result would be that water would continually flow in, and yet the tank would remain entirely empty, the gases formed escaping. We would thus produce, by expending initially a certain amount of work to create a sink for the heat or, respectively, the water to flow in, a condition enabling us to get any amount of energy without further effort. This would be an ideal way of obtaining motive power. We do not know of any such absolutely perfect process of heat-conversion, and consequently some heat will generally reach the low level, which means to say, in our mechanical analog, that some water will arrive at the bottom of the tank, and a gradual and slow filling of the latter will take place, necessitating continuous pumping out. But evidently there will be less to pump out than flows in, or, in other words, less energy will be needed to maintain the initial condition than is developed by the fall, and this is to say that some energy will be gained from the medium. What is not convened in flowing down can just be raised up with its own energy, and what is converted is clear gain. Thus the virtue of the principle I have discovered resides wholly in the conversion of the energy on the downward flow.

FIRST EFFORTS TO PRODUCE THE SELF-ACTING ENGINE—THE MECHANICAL OSCILLATOR WORK OF DEWAR AND LINDE—LIQUID AIR.

Having recognized this truth, I began to devise means for carrying out my idea, and, after long thought, I finally conceived a combination of apparatus which should make possible the obtaining of power from the medium by a process of continuous cooling of atmospheric air. This apparatus, by continually transforming heat into mechanical work, tended to become colder and colder, and if it only were practicable to reach a very low temperature in this manner, then a sink for the heat could be produced, and energy could be derived from the medium. This seemed to be contrary to the statements of Carnot and Lord Kelvin before referred to, but I concluded from the theory of the process that such a result could be attained. This conclusion I reached, I think, in the latter part of 1883, when I was in Paris, and it was at a time when my mind was being more and more dominated by an invention which I had evolved during the preceding year, and which has since become known under the name of the "rotating magnetic field." During the few years which followed I elaborated further the plan I had imagined, and studied the working conditions, but made little headway. The commercial introduction in this country of the invention before referred to required most of my energies until 1889, when I again took up the idea of the self-acting machine. A closer investigation of the principles involved, and calculation, now showed that the result I aimed at could not be reached in a practical manner by ordinary machinery, as I had in the beginning expected. This led me, as a next step, to the study of a type of engine generally designated as "turbine," which at first seemed to offer better chances for a realization of the idea. Soon I found, however, that the turbine, too, was unsuitable. But my conclusions showed that if an engine of a peculiar kind could be brought to a high degree of perfection, the plan I had conceived was realizable, and I resolved to proceed with the development of such an engine, the primary object of which was to secure the greatest economy of transformation of heat into mechanical energy. A characteristic feature of the engine was that the work-performing piston was not connected with anything else, but was perfectly free to vibrate at an enormous rate. The mechanical difficulties encountered in the construction of this engine were greater than I had anticipated, and I made slow progress. This work was continued until early in 1892, when I went to London, where I saw Professor Dewar's admirable experiments with liquefied gases. Others had liquefied gases before, and notably Ozlewski and Pictet had performed creditable early experiments in this line, but there was such a vigor about the work of Dewar that even the old appeared new. His experiments showed, though in a way different from that I had imagined, that it was possible to reach a very low temperature by transforming heat into mechanical work, and I returned, deeply impressed with what I had seen, and more than ever convinced that my plan was practicable. The work temporarily interrupted was taken up anew, and soon I had in a fair state of perfection the engine which I have named "the mechanical oscillator." In this machine I succeeded in doing away with all packings, valves, and lubrication, and in producing so rapid a vibration of the piston that shafts of tough steel, fastened to the same and vibrated longitudinally, were torn asunder. By combining this engine with a dynamo of special design I produced a highly efficient electrical generator, invaluable in measurements and determinations of physical quantities on account of the unvarying rate of oscillation obtainable by its means. I exhibited several types of this machine, named "mechanical and electrical oscillator," before the Electrical Congress at the World's Fair in Chicago during the summer of 1893, in a lecture which, on account of other pressing work, I was unable to prepare for publication. On that occasion I exposed the principles of the mechanical oscillator, but the original purpose of this machine is explained here for the first time.

In the process, as I had primarily conceived it, for the utilization of the energy of the ambient medium, there were five essential elements in combination, and each of these had to be newly designed and perfected, as no such machines existed. The mechanical oscillator was the first element of this combination, and having perfected this, I turned to the next, which was an air-compressor of a design in certain respects resembling that of the mechanical oscillator. Similar difficulties in the construction were again encountered, but the work was pushed vigorously, and at the close of 1894 I had completed these two elements of the combination, and thus produced an apparatus for compressing air, virtually to any desired pressure, incomparably simpler, smaller, and more efficient than the ordinary. I was just beginning work on the third clement, which together with the first two would give a refrigerating machine of exceptional efficiency and simplicity, when a misfortune befell me in the burning of my laboratory, which crippled my labors and delayed me. Shortly afterward Dr. Carl Linde announced the liquefaction of air by a self-cooling process, demonstrating that it was practicable to proceed with the cooling until liquefaction of the air took place. This was the only experimental proof which I was still wanting that energy was obtainable from the medium in the manner contemplated by me.

The liquefaction of air by a self-cooling process was not, as popularly believed, an accidental discovery, but a scientific result which could not have been delayed much longer, and which, in all probability, could not have escaped Dewar. This fascinating advance, I believe, is largely due to the powerful work of this great Scotchman. Nevertheless, Linde's is an immortal achievement. The manufacture of liquid air has been carried on for four years in Germany, on a scale much larger than in any other country, and this strange product has been applied for a variety of purposes. Much was expected of it in the beginning, but so far it has been an industrial *ignis fatuus.* By the use of such machinery as I am perfecting, its cost will probably be greatly lessened, but even then its commercial success will be questionable. When used as a refrigerant it is uneconomical, as its temperature is unnecessarily low. It is as expensive to maintain a body at a very low temperature as it is to keep it very hot; it takes coal to keep air cold. In oxygen manufacture it cannot yet compete with the electrolytic method. For use as an explosive it is unsuitable, because its low temperature again condemns it to a small efficiency, and for motive-power purposes its cost is still by far too high. It is of interest to note, however, that in driving an engine by liquid air a certain amount of energy may be gained from the engine, or, stated otherwise, from the ambient medium which keeps the engine warm, each two hundred pounds of iron casting of the latter contributing energy at the rate of about one effective horse-power during one hour. But this gain of the consumer is offset by an equal loss of the producer.

Much of this task on which I have labored so long remains to be done. A number of mechanical details are still to be perfected and some difficulties of a different nature to be mastered, and I cannot hope to produce a self-acting machine deriving energy from the ambient medium for a long time yet, even if all my expectations should materialize. Many circumstances have occurred which have retarded my work of late, but for several reasons the delay was beneficial.

One of these reasons was that I had ample time to consider what the ultimate possibilities of this development might be. I worked for a long time fully convinced that the practical realization of this method of obtaining energy from the sun would be of incalculable industrial value, but the continued study of the subject revealed the fact that while it will be commercially profitable if my expectations are well founded, it will not be so to an extraordinary degree.

DISCOVERY OF UNEXPECTED PROPERTIES OF THE ATMOSPHERE—STRANGE EXPERIMENTS—TRANSMISSION OF ELECTRICAL ENERGY THROUGH ONE WIRE WITHOUT RETURN—TRANSMISSION THROUGH THE EARTH WITHOUT ANY WIRE.

Another of these reasons was that I was led to recognize the transmission of electrical energy to any distance through the media as by far the best solution of the great problem of harnessing the sun's energy for the use of man. For a long time I was convinced that such a transmission on an industrial scale could never be realized, but a discovery which I made changed my view. I observed that under certain conditions the atmosphere, which is normally a high insulator, assumes conducting properties, and so becomes capable of conveying any amount of electrical energy. But the difficulties in the way of a practical utilization of this discovery for the purpose of transmitting electrical energy without wires were seemingly insuperable. Electrical pressures of many millions of volts had to be produced and handled; generating apparatus of a novel kind, capable of withstanding the immense electrical stresses, had to be invented and perfected, and a complete safety against the dangers of the high-tension currents had to be attained in the system before its practical introduction could be even thought of. All this could not be done in a few weeks or months, or even years. The work required patience and constant application, but the improvements came, though slowly. Other valuable results were, however, arrived at in the course of this long continued work, of which I shall endeavor to give a brief account, enumerating the chief advances as they were successively effected.

The discovery of the conducting properties of the air, though unexpected, was only a natural result of experiments in a special field which I had carried on for some years before. It was, I believe, during 1889 that certain possibilities offered by extremely rapid electrical oscillations determined me to design a number of special machines adapted for their investigation. Owing to the peculiar requirements, the

Figure 3. Experiment to illustrate the supplying of electrical energy through a single wire without return. An ordinary incandescent lamp, connected with one or both of its terminals to the wire forming the upper free end of the coil shown in the photograph, is lighted by electrical vibrations conveyed to it through the coil from an electrical oscillator, which is worked only to one fifth of one per cent, of its full capacity.

construction of these machines was very difficult, and consumed much time and effort; but my work was generously rewarded, for I reached by their means several novel and important results. One of the earliest observations I made with these new machines was that electrical oscillations of an extremely high rate act in an extraordinary manner upon the human organism. Thus, for instance, I demonstrated that powerful electrical discharges of several hundred thousand volts, which at that time were considered absolutely deadly, could be passed through the body without inconvenience or hurtful consequences. These oscillations produced other specific physiological effects, which, upon my announcement, were eagerly taken up by skilled physicians and further investigated. This new field has proved itself fruitful beyond expectation and in the few years which have passed since, it has been developed to such an extent that it now forms a legitimate and important department of medical science. Many results, thought impossible at that time, are now readily obtainable with these oscillations, and many experiments undreamed of then can now be readily performed by their means. I still remember with pleasure how, nine years ago, I passed the discharge of a powerful induction-coil through my body to demonstrate before a scientific society the comparative harmlessness of very rapidly vibrating electric currents, and I can still recall the astonishment of my audience. I would now undertake, with much less apprehension than I had in that experiment, to transmit through my body with such currents the entire electrical energy of the dynamos now working at Niagara—forty or fifty thousand horsepower. I have produced electrical oscillations which were of such intensity that when circulating through my arms and chest they have melted wires which joined my hands, and still I felt no inconvenience, I have energized with such oscillations a loop of heavy copper wire so powerfully that masses of metal, and even objects of an elec-

Figure 4. Experiment to illustrate the transmission of electrical energy through the earth without wire. The coil shown in the photograph has its lower end of terminal connected to the ground, and is exactly attuned to the vibrations of a distant electrical oscillator. The lamp lighted is in an independent wire loop, energized by induction from the coil excited by the electrical vibrations transmitted to it through the ground from the oscillator, which is worked only to five per cent of its full capacity.

trical resistance specifically greater that of human tissue, brought close to or placed within the loop, were heated to a high temperature and melted, often with the violence of an explosion, and yet into this very space in which this terribly destructive turmoil was going on I have repeatedly thrust my head without feeling anything or experiencing injurious after-effects.

Another observation was that by means of such oscillations light could be produced in a novel and more economical manner, which promised to lead to an ideal system of electric illumination by vacuum-tubes, dispensing with the necessity of renewal of lamps or incandescent filaments, and possibly also with the use of wires in the interior of buildings. The efficiency of this light increases in proportion to the rate of the oscillations, and its commercial success is, therefore, dependent on the economical production of electrical vibrations of transcending rates. In this direction I have met with gratifying success of late, and the practical introduction of this new system of illumination is not far off.

The investigations led to many other valuable observations and results, one of the more important of which was the demonstration of the practicability of supplying electrical energy through one wire without return. At first I was able to transmit in this novel manner only very small amounts of electrical energy, but in this line also my efforts have been rewarded with similar success.

The photograph shown in *Figure 3* illustrates, as its title explains, an actual transmission of this kind effected with apparatus used in other experiments here described. To what a degree the appliances have been perfected since my first demonstrations early in 1891 before a scientific society, when my apparatus was barely capable of lighting one lamp (which result was considered wonderful), will appear when l state that l have now no difficulty in lighting in this manner four or five hundred lamps, and could light many more. In fact, there is no limit to the amount of energy which may in this way be supplied to operate any kind of electrical device.

After demonstrating the practicability of this method of transmission, the thought naturally occurred to me to use the earth as a conductor, thus dispensing with all wires. Whatever electricity may be, it is a fact that it behaves like an incompressible fluid, and the earth may be looked upon as an immense reservoir of electricity, which, I thought, could be disturbed effectively by a properly designed electrical machine. Accordingly, my next efforts were directed toward perfecting a special apparatus which would be highly effective in creating a disturbance of electricity in the earth.

The progress in this new direction was necessarily very slow and the work discouraging, until I finally succeeded in perfecting a novel kind of transformer or induction-coil, particularly suited for this special purpose. That it is practicable, in this manner, not only to transmit minute amounts of electrical energy for operating delicate electrical devices, as I contemplated at first, but also electrical energy in appreciable quantities will appear from inspection of *Figure 4,* which illustrates an actual experiment of this kind performed with the same apparatus. The result obtained was all the more remarkable as the top end of the coil was not connected to a wire or plate for magnifying the effect.

"WIRELESS" TELEGRAPHY— THE SECRET OF TUNING— ERRORS IN THE HERTZIAN INVESTIGATION— A RECEIVER OF WONDERFUL SENSITIVENESS.

As the first valuable result of my experiments in this latterline a system of telegraphy without wires resulted, which I described in two scientific lectures in February and March, 1893. It is mechanically illustrated in *diagram c,* the upper part of which shows the electrical arrangement as I described it then, while the lower part illustrates its mechanical analog. The system is extremely simple in principle. Imagine two tuning forks F, F_1, one at the sending and the other at the receiving station respectively, each having attached to its lower prong a minute piston p, fitting in a cylinder. Both the cylinders communicate with a large reservoir R, with elastic walls, which is supposed to be closed and filled with a light and incompressible fluid. By striking repeatedly one of the prongs of the tuning-fork F, the small piston p below would be vibrated, and its vibrations, transmitted through the fluid, would reach the distant fork F_1, which is "tuned" to the fork F, or, stated otherwise, of exactly the same note as the latter. The fork F_1 would now be set vibrating, and its vibration would be intensified by the continued action of the distant fork F until its upper prong, swinging far out, would make an electrical connection with a stationary contact c", starting in this manner some electrical or other appliances which may be used for recording the signals. In this simple way messages could be exchanged between the two stations, a similar contact c' being provided for this purpose, close to the upper prong of the fork F, so that the apparatus at each station could be employed in turn as receiver and transmitter.

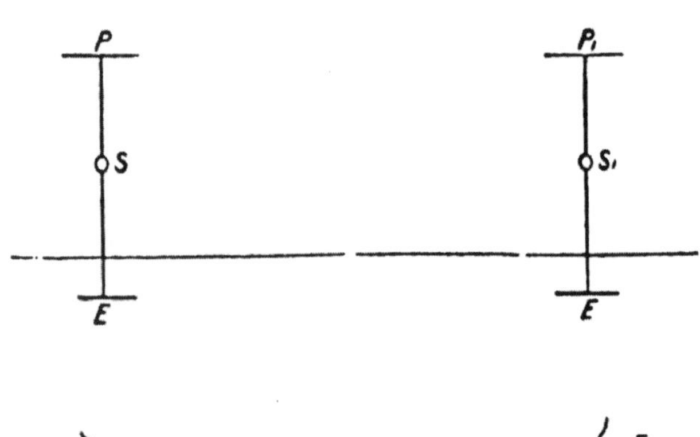

Diagram C. "Wireless" telegraphy mechanically illustrated.

The electrical system illustrated in the upper figure of *diagram c* is exactly the same in principle, the two wires or circuits ESP and $E_1S_1P_1$, which extend vertically to a height, representing the two tuning-forks with the pistons attached to them. These circuits are connected with the ground by plates E, E_1, and to two elevated metal sheets P, P_1, which store electricity and thus magnify considerably the effect. The closed reservoir R, with elastic walls, is in this case replaced by the earth, and the fluid by electricity. Both of these circuits are "tuned" and operate just like the two tuning-forks. Instead of striking the fork F at the sending-station, electrical oscillations are produced in the vertical sending or transmitting-wire ESP, as by the action of a source S, included in this wire, which spread through the ground and reach the distant vertical receiving-wire $E_1S_1P_1$, exciting corresponding electrical oscillations in the same. In the latter wire or circuit is included a sensitive device or receiver S_1, which is thus set in action and made to operate a relay or other appliance. Each station is, of course, provided both with a source of electrical oscillations S and a sensitive receiver S_1, and a simple provision is made for using each of the two wires alternately to send and to receive the messages.

The exact attunement of the two circuits secures great advantages, and, in fact, it is essential in the practical use of the system. In this respect many popular errors exist, and, as a rule, in the technical reports on this subject circuits and appliances are described as affording these advantages when from their very nature it is evident that this is impossible. In order to attain the best results it is essential that the length of each wire or circuit, from the ground connection to the top, should be equal to one quarter of the wavelength of the electrical vibration in the wire, or else equal to that length multiplied by an odd number. Without the observation of this rule it is virtually impossible to prevent the interference and insure the privacy of messages. Therein lies the secret of tuning. To obtain the most satisfactory results it is, however, necessary to resort to electrical vibrations of low pitch. The Hertzian spark apparatus, used generally by experimenters, which produces oscillations of a very high rate, permits no effective tuning, and slight disturbances are sufficient to render an exchange of messages impracticable. But scientifically designed, efficient appliances allow nearly perfect adjustment. An experiment performed with the improved apparatus repeatedly referred to, and intended to convey an idea of this feature, is illustrated in *Figure 5,* which is sufficiently explained by its note.

Since I described these simple principles of telegraphy without wires I have had frequent occasion to note that the identical features and elements have been used, in the evident belief that the signals are being transmitted to considerable distances by "Hertzian" radiations. This is only one of many misapprehensions to which the investigations of the lamented physicist have given rise. About thirty-three years ago Maxwell, following up a suggestive experiment made by Faraday in 1845, evolved an ideally simple theory which intimately connected light, radiant heat, and electrical phenomena, interpreting them as being all due to vibrations of a hypothetical fluid of inconceivable tenuity, called the ether. No experimental verification was arrived at until Hertz, at the suggestion of Helmholtz, undertook a series

Figure 5. Photographic view of coil responding to electrical oscillations. The picture shows a number of coils, differently attuned and responding to the vibrations transmitted to them through the earth from an electrical oscillator. The large coil on the right, discharging strongly, is tuned to the fundamental vibration, which is fifty thousand per second; the two larger vertical coils to twice that number; the smaller white coil to four times that number, and the remaining small coils to higher tones. The vibrations produced by the oscillator were so intense that they affected perceptibly a small coil tuned to the twenty-sixth higher tone.

of experiments to this effect. Hertz proceeded with extraordinary ingenuity and insight, but devoted little energy to the perfection of his old-fashioned apparatus. The consequence was that he failed to observe the important function which the air played in his experiments, and which I subsequently discovered. Repeating his experiments and reaching different results, I ventured to point out this oversight. The strength of the proofs brought forward by Hertz in support of Maxwell's theory resided in the correct estimate of the rates of vibration of the circuits he used. But I ascertained that he could not have obtained the rates he thought he was getting. The vibrations with identical apparatus he employed are, as a rule, much slower, this being due to the presence of air, which produces a dampening effect upon a rapidly vibrating electric circuit of high pressure, as a fluid does upon a vibrating tuning-fork. I have, however, discovered since that time other causes of error, and I have long ago ceased to look upon his results as being an experimental verification of the poetical conceptions of Maxwell. The work of the great German physicist has acted as an immense stimulus to contemporary electrical research, but it has likewise, in a measure, by its fascination, paralyzed the scientific mind, and thus hampered independent inquiry. Every new phenomenon which was discovered was made to fit the theory, and so very often the truth has been unconsciously distorted.

When I advanced this system of telegraphy, my mind was dominated by the idea of effecting communication to any distance through the earth or environing medium, the practical consummation of which I considered of transcendent importance, chiefly on account of the moral effect which it could not fail to produce universally. As the first effort to this end I proposed, at that time, to employ relay-stations with tuned circuits, in the hope of making thus practicable signaling over vast distances, even with apparatus of very moderate power then at my command.

Figure 6. Photographic view of the essential parts of electrical oscillator used in the experiments described.

I was confident, however, that with properly designed machinery signals could be transmitted to any point of the globe, no matter what the distance, without the necessity of using such intermediate stations. I gained this conviction through the discovery of a singular electrical phenomenon, which I described early in 1892, in lectures delivered before some scientific societies abroad, and which I have called a "rotating brush." This is a bundle of light which is formed, under certain conditions, in a vacuum-bulb, and which is of a sensitiveness to magnetic and electric influences bordering, so to speak, on the supernatural. This light bundle is rapidly rotated by the earth's magnetism as many as twenty thousand times per second, the rotation in these parts being opposite to what it would be in the southern hemisphere, while in the region of the magnetic equator it should not rotate at all. In its most sensitive state, which is difficult to attain, it is responsive to electric or magnetic influences to an incredible degree. The mere stiffening of the muscles of the arm and consequent slight electrical change in the body of an observer standing at some distance from it, will perceptibly affect it. When in this highly sensitive state it is capable of indicating the slightest magnetic and electric changes taking place in the earth. The observation of this wonderful phenomenon impressed me strongly that communication at any distance could be easily effected by its means, provided that apparatus could be perfected capable of producing an electric or magnetic change of state, however small, in the terrestrial globe or environing medium.

DEVELOPMENT OF A NEW PRINCIPLE—
THE ELECTRICAL OSCILLATOR—
PRODUCTION OF IMMENSE ELECTRICAL MOVEMENTS—
THE EARTH RESPONDS TO MAN—
INTERPLANETARY COMMUNICATION NOW PROBABLE.

I resolved to concentrate my efforts upon this venturesome task, though it involved great sacrifice, for the difficulties to be mastered were such that I could hope to consummate it only after years of labor. It meant delay of other work to which I would have preferred to devote myself, but I gained the conviction that my energies could not be more usefully employed; for I recognized that an efficient apparatus for the production of powerful electrical oscillations, as was needed for that specific purpose, was the key to the solution of other most important electrical and, in fact, human problems. Not only was communication, to any distance, without wires possible by its means, but, likewise, the transmission of energy in great amounts, the burning of the atmospheric nitrogen, the production of an efficient illuminant, and many other results of inestimable scientific and industrial value. Finally, however, I had the satisfaction of accomplishing the task undertaken by the use of anew principle, the virtue of which is based on the marvelous properties of the electrical condenser. One of these is that it can discharge or explode its stored energy in an inconceivably short time. Owing to this it is unequaled in explosive violence. The explosion of dynamite is only the breath of a consumptive compared with its discharge. It is the means of producing the strongest current, the highest electrical pressure, the greatest commotion in the medium. Another of its properties, equally valuable, is that its discharge may vibrate at any rate desired up to many millions per second.

I had arrived at the limit of rates obtainable in other ways when the happy idea presented itself to me to resort to the condenser. I arranged such an instrument so as to be charged and discharged alternately in rapid succession through a coil with a few turns of stout wire, forming the primary of a transformer or induction-coil. Each time the condenser was discharged the current would quiver in the primary wire and induce corresponding oscillations in the secondary. Thus a transformer or induction-coil on new principles was evolved, which I have called "the electrical oscillator," partaking of those unique qualities which characterize the condenser, and enabling results to be attained impossible by other means. Electrical effects of

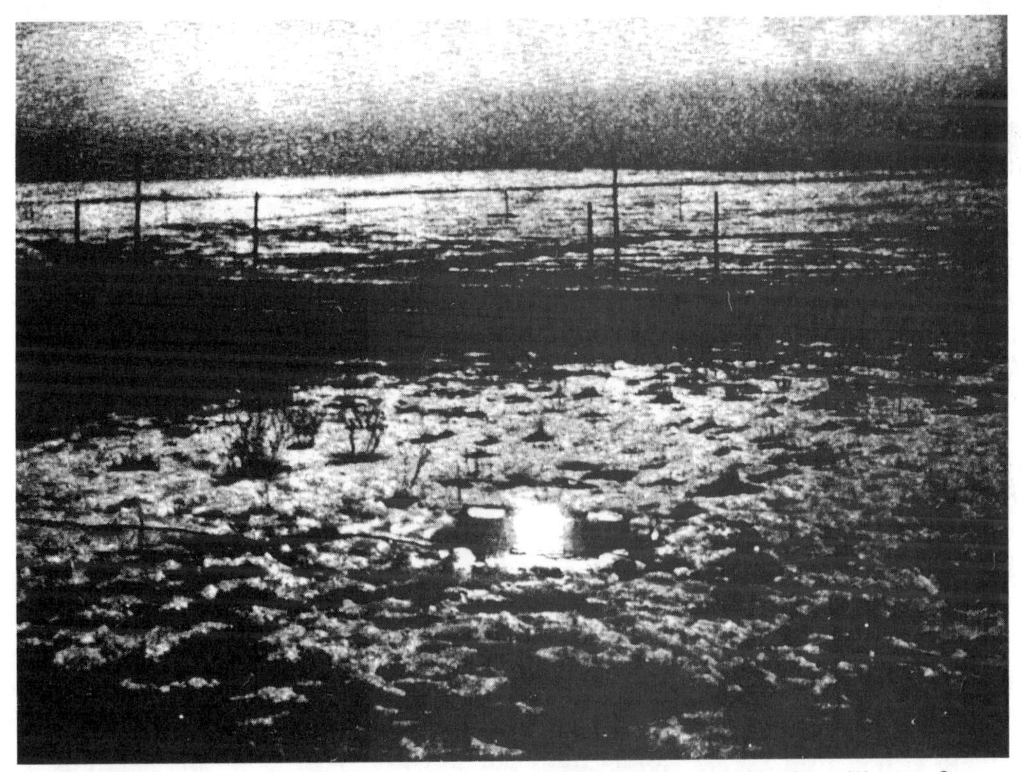

Figure 7. Experiment to illustrate an inductive effect of an electrical oscillator of great power. The photograph shows three ordinary incandescent lamps lighted to full candle-power by currents induced in a local loop consisting of a single wire forming a square of fifty feet each side, which includes the lamps, and which is at a distance of one hundred feet from the primary circuit energized by the oscillator. The loop likewise includes an electrical condenser, and is exactly attuned to the vibrations of the oscillator, which is worked at less than five per cent of its total capacity.

Figure 8. Experiment to illustrate the capacity of the oscillator for producing electrical explosions of great power. The coil, partly shown in the photograph, creates an alternative movement of electricity from the earth into a large reservoir and back at the rate of one hundred thousand alternations per second. The adjustments are such that the reservoir is filled full and bursts at each alternation just at the moment when the electrical pressure reaches the maximum. The discharge escapes with a deafening noise, striking an unconnected coil twenty-two feet away, and creating such a commotion of electricity in the earth that sparks an inch long can be drawn from a water-main at a distance of three hundred feet from the laboratory.

any desired character and of intensities undreamed of before are now easily producible by perfected apparatus of this kind, to which frequent reference has been made, and the essential parts of which are shown in *Figure 6.* For certain purposes a strong inductive effect is required; for others the greatest possible suddenness; for others again, an exceptionally high rate of vibration or extreme pressure; while for certain other objects immense electrical movements are necessary. The photographs in *Figures 7, 8, 9,* and *10,* of experiments performed with such an oscillator, may serve to illustrate some of these features and convey an idea of the magnitude of the effects actually produced. The completeness of the titles of the figures referred to makes a further description of them unnecessary.

However extraordinary the results shown may appear, they are but trifling compared with those which arc attainable by apparatus designed on these same principles. I have produced electrical discharges the actual path of which, from end to end, was probably more than one hundred feet long; but it would not be difficult to reach lengths one hundred times as great. I have produced electrical movements occurring at the rate of approximately one hundred thousand horse-power, but rates of one, five, or ten million horsepower are easily practicable. In these experiments effects were developed incomparably greater than any ever produced by human agencies, and yet these results are but an embryo of what is to be.

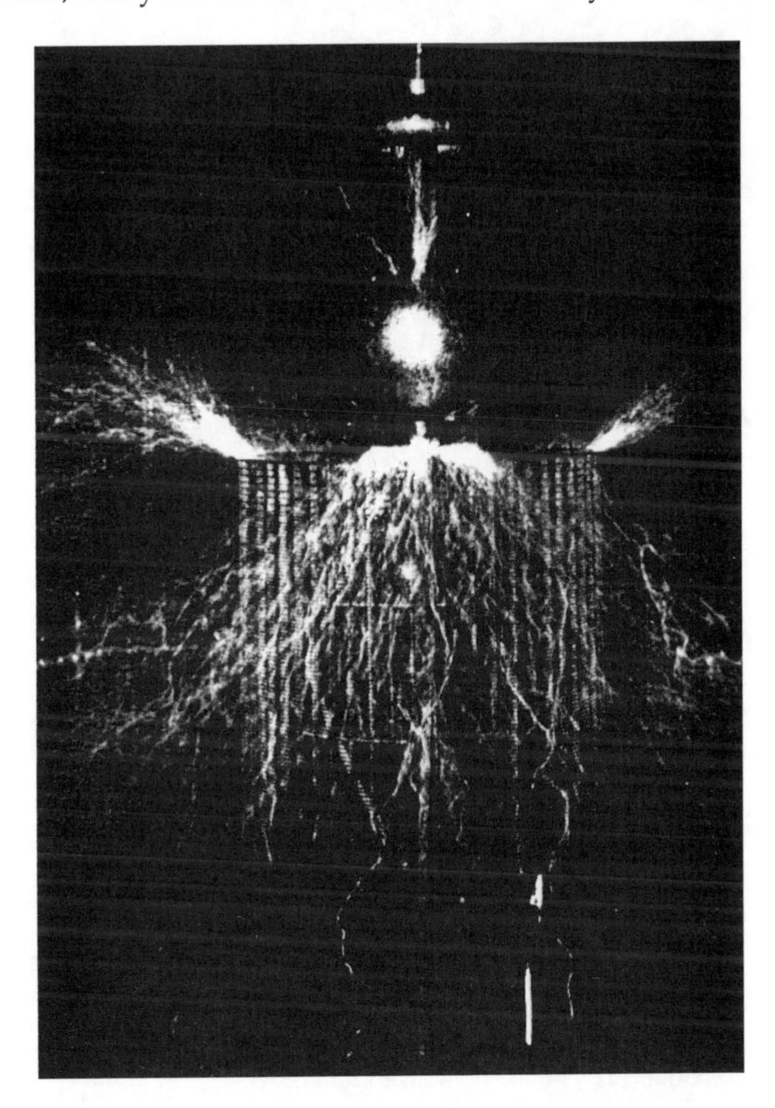

Figure 9. Experiment to illustrate the capacity of the oscillator for creating great electrical movement. The ball shown in the photograph, covered with a polished metallic coating of twenty square feet of surface, represents a large reservoir of electricity, and the inverted tin pan underneath, with a sharp rim, a big opening through which the electricity can escape before filling the reservoir. The quantity of electricity set in movement is so great that, although most of it escapes through the rim of the pan or opening provided, the ball or reservoir is nevertheless alternately emptied and filled to overflowing (as is evident from the discharge escaping on the top of the ball) one hundred and fifty thousand times per second.

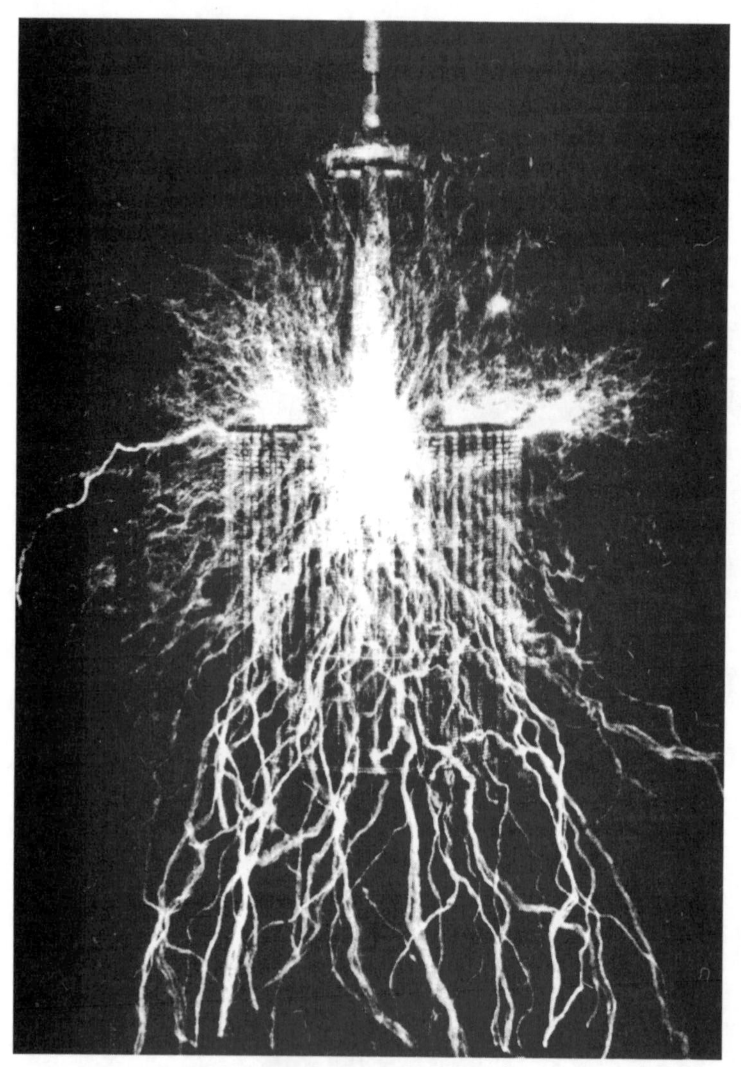

Figure 10. Photographic view of an experiment to illustrate an effect of an electrical oscillator delivering energy at a rate of 75,000 horse-power. The discharge, creating a strong draft owing to the heating of the air, is carried upward through the open roof of the building. The greatest width across is nearly seventy feet. The pressure is over twelve million volts, and the current alternates one hundred and thirty thousand times per second.

That communication without wires to any point of the globe is practicable with such apparatus would need no demonstration, but through a discovery which I made I obtained absolute certitude. Popularly explained, it is exactly this: When we raise the voice and hear an echo in reply, we know that the sound of the voice must have reached a distant wall, or boundary, and must have been reflected from the same. Exactly as the sound, so an electrical wave is reflected, and the same evidence which is afforded by an echo is offered by an electrical phenomenon known as a "stationary" wave—that is, a wave with fixed nodal and ventral regions. Instead of sending sound vibrations toward a distant wall, I have sent electrical vibrations toward the remote boundaries of the earth, and instead of the wall the earth has replied. In place of an echo I have obtained a stationary electrical wave, a wave reflected from afar.

Stationary waves in the earth mean something more than mere telegraphy without wires to any distance. They will enable us to attain many important specific results impossible otherwise. For instance, by their use we may produce at will, from a sending-station, an electrical effect in any particular region of the globe; we may determine the relative position or course of a moving object, such as a vessel at sea, the distance traversed by the same, or its speed; or we may send over the earth a wave of electricity traveling at any rate we desire, from the pace of a turtle up to lightning speed.

With these developments we have every reason to anticipate that in a time not very distant most telegraphic messages across the oceans will be transmitted without cables. For short distances we need a "wireless" telephone, which requires no expert operators. The greater the spaces to be bridged, the more rational becomes communication without wires. The cable is not only an easily damaged and costly instrument, but it limits us in the speed of transmission by reason of a certain electrical property inseparable from its construction. A properly designed plant for effecting communication without wires ought to have many times the working capacity of a cable, while it will involve incomparably less expense. Not a long time will pass, I believe, before communication by cable will become obsolete, for not only will signaling by this new method be quicker and cheaper, but also much safer. By using some new means for isolating the messages which I have contrived, an almost perfect privacy can be secured.

I have observed the above effects so far only up to a limited distance of about six hundred miles, but inasmuch as there is virtually no limit to the power of the vibrations producible with such an oscillator, I feel quite confident of the success of such a plant for effecting transoceanic communication. Nor is this all. My measurements and calculations have shown that it is perfectly practicable to produce on our globe, by the use of these principles, an electrical movement of such magnitude that, without the slightest doubt, its effect will be perceptible on some of our nearer planets, such as Venus and Mars. Thus from mere possibility interplanetary communication has entered the stage of probability. In fact, that we can produce a distinct effect on one of these planets in this novel manner, namely, by disturbing the electrical condition of the earth, is beyond any doubt. This way of effecting such communication is, however, essentially different from all others which have so far been proposed by scientific men. In all the previous instances only a minute fraction of the total energy reaching the planet—as much as it would be possible to concentrate in a reflector-could be utilized by the supposed observer in his instrument. But by the means I have developed he would be enabled to concentrate the larger portion of the entire energy transmitted to the planet in his instrument, and the chances of affecting the latter are thereby increased many million fold.

Besides machinery for producing vibrations of the required power, we must have delicate means capable of revealing the effects of feeble influences exerted upon the earth. For such purposes, too, I have perfected new methods. By their use we shall likewise be able, among other things, to detect at considerable distance the presence of an iceberg or other object at sea. By their use, also, I have discovered some terrestrial phenomena still unexplained. That we can send a message to a planet is certain, that we can get an answer is probable: man is not the only being in the Infinite gifted with a mind.

TRANSMISSION OF ELECTRICAL ENERGY TO ANY DISTANCE WITHOUT WIRES— NOW PRACTICABLE— THE BEST MEANS OF INCREASING THE FORCE ACCELERATING THE HUMAN MASS.

The most valuable observation made in the course of these investigations was the extraordinary behavior of the atmosphere toward electric impulses of excessive electromotive force. The experiments showed that the air at the ordinary pressure became distinctly conducting, and this opened up the wonderful prospect of transmitting large amounts of electrical energy for industrial purposes to great distances without wires, a possibility which, up to that time, was thought of only as a scientific dream. Further investigation revealed the important fact that the conductivity imparted to the air by these electrical impulses of many millions of volts increased very rapidly with the degree of rarefacation, so that air strata at very moderate altitudes, which are easily accessible, offer, to all experimental evidence, a perfect conducting path, better than a copper wire, for currents of this character.

Thus the discovery of these new properties of the atmosphere not only opened up the possibility of transmitting, without wires, energy in large amounts, but, what was still more significant, it afforded the certitude that energy could be transmitted in this manner economically. In this new system it matters little—in fact, almost nothing—whether the transmission is effected at a distance of a few miles or of a few thousand miles. While I have not, as yet, actually effected a transmission of a considerable amount of energy, such as would be of industrial importance, to a great distance by this new method, I have operated several model plants under exactly the same conditions which will exist in a large plant of this kind, and the practicability of the system is thoroughly demonstrated. The experiments have shown conclusively that, with two terminals maintained at an elevation of not more than thirty thousand to thirty-five thousand feet above sea-level, and with an electrical pressure of fifteen to twenty million volts, the energy of thousands of horsepower can be transmitted over distances which may be hundreds and, if necessary, thousands of miles. I am hopeful, however, that I may be able to reduce very considerably the elevation of the terminals now required, and with this object I am following up an idea which promises such a realization. There is, of course, a popular prejudice against using an electrical pressure of millions of volts, which may cause sparks to fly at distances of hundreds of feet, but, paradoxical as it may seem, the system, as I have described it in a technical publication, offers greater personal safety than most of the ordinary distribution circuits now used in the cities. This is, in a measure, borne out by the fact that, although I have carried on such experiments for a number of years, no injury has been sustained either by me or any of my assistants.

But to enable a practical introduction of the system, a number of essential requirements are still to be fulfilled. It is not enough to develop appliances by means of which such a transmission can be effected. The machinery must be such as to allow the transformation and transmission of electrical energy under highly economical and practical conditions. Furthermore, an inducement must be offered to those who are engaged in the industrial exploitation of natural sources of power, as waterfalls, by guaranteeing greater returns on the capital invested than they can secure by local development of the property.

From that moment when it was observed that, contrary to the established opinion, low and easily accessible strata of the atmosphere are capable of conducting electricity, the transmission of electrical energy without wires has become a rational task of the engineer, and one surpassing all others in importance. Its practical consummation would mean that energy would be available for the uses of man at any point of the globe, not in small amounts such as might be derived from the ambient medium by suitable machinery, but in quantities virtually unlimited, from waterfalls. Export of power would then become the chief source of income for many happily situated countries, as the United States, Canada, Central and South America, Switzerland, and Sweden. Men could settle down everywhere, fertilize and irrigate the soil with little effort, and convert barren deserts into gardens, and thus the entire globe could be transformed and made a fitter abode for mankind. It is highly probable that if there are intelligent beings on Mars they have long ago realized this very idea, which would explain the changes on its surface noted by astronomers. The atmosphere on that planet, being of considerably smaller density than that of the earth, would make the task much more easy.

It is probable that we shall soon have a self-acting heat-engine capable of deriving moderate amounts of energy from the ambient medium. There is also a possibility—though a small one—that we may obtain electrical energy direct from the sun. This might be the case if the Maxwellian theory is true, according to which electrical vibrations of all rates should emanate from the sun. I am still investigating this subject. Sir William Crookes has shown in his beautiful invention known as the "radiometer" that rays may produce by impact a mechanical effect, and this may lead to some important revelation as to the utilization of the sun's rays in novel ways. Other sources of energy may be opened up, and new methods of deriving energy from the sun discovered, but none of these or similar achievements would equal in importance the transmission of power to any distance through the medium. I can conceive of no technical advance which would tend to unite the various elements of humanity more effectively than this one, or of one which would more add to and more economize human energy. It would be the best means of increasing the force accelerating the human mass. The mere moral influence of such a radical departure would be incalculable. On the other hand if at any point of the globe energy can be obtained in limited quantities from the ambient medium by means of a self-acting heat-engine or otherwise, the conditions will remain the same as before. Human performance will be increased, but men will remain strangers as they were.

I anticipate that many, unprepared for these results, which, through long familiarity, appear to me simple and obvious, will consider them still far from practical application. Such reserve, and even opposition, of some is as useful a quality and as others. Thus, a mass which resists the force at first, once set in movement, adds to the energy. The scientific man does not aim at an immediate result. He does not expect that his advanced ideas will be readily taken up. His work is like that of the planter—for the future. His duty is to lay the foundation for those who are to come, and point the way. He lives and labors and hopes with the poet who says:

Schaff', das Tagwerk meiner Hande,
Hohes Gluck, dass ich's vollende!
Lass, o lass mich nicht ermatten!
Nein, es sind nicht leere Traume:
Jetzt nur Stangen, diese Baume
Geben einst noch Frucht und Schatten.

Daily work—my hands' employment,
To complete is pure enjoyment!
Let, oh, let me never falter!
No! there is no empty dreaming:
Lo! these trees, but bare poles seeming,
Yet will yield both fruit and shelter!
Goethe's "Hope,"
Translated by William Gibson, Com. U.S.N.

Part Three
The Wireless Transmission of Electricity

WIRELESS TRANSMISSION OF POWER
Resonating Planet Earth
by Toby Grotz

Many researchers have speculated on the meaning of the phrase "non- Hertzian waves" as used by Dr. Nikola Tesla. Dr. Tesla first began to use this term in the mid 1890's in order to explain his proposed system for the wireless transmission of electrical power. In fact, it was not until the distinction between the method that Heinrich Hertz was using and the system Dr. Tesla had designed, that Dr. Tesla was able to receive the endorsement of the renowned physicist, Lord Kelvin.1

To this day, however, there exists a confusion amongs researchers, experimentalists, popular authors and laymen as to the meaning of non- Hertzian waves and the method Dr. Tesla was promoting for the wireless transmission of power. In this paper, the terms pertinent to wireless transmission of power will be explained and the methods being used by present researchers in a recreation of the Tesla's 1899 Colorado Springs experiments will be defined.

Early Theories of Electromagnetic Propagation

In pre-World War I physics, scientists postulated a number of theories to explain the propagation of electromagnetic energy through the ether. There were three popular theories present in the literature of the late 1800's and early 1900's. They were:

1. Transmission through or along the Earth,
2. Propagation as a result of terrestrial resonances,
3. Coupling to the ionosphere using propagation through electrified gases.

We shall concern our examination at this time to the latter two theories as they were both used by Dr. Tesla at various times to explain his system of wireless transmission of power. It should be noted, however, that the first theory was supported by Fritz Lowenstein, the first vice-president of the Institute of Radio Engineers, a man who had the enviable experience of assisting Dr. Tesla during the Colorado Springs experiments of 1899. Lowenstein presented what came to be known as the "gliding wave" theory of electromagnetic radiation and propagation during a lecture before the IRE in 1915. (Fig. 1)

Dr. Tesla delivered lectures to the Franklin Institute at Philadelphia, in February, 1983, and to the National Electric Light Association in St. Louis, in March, 1983, concerning electromagnetic wave propagation. The theory presented in those lectures proposed that the Earth could be considered as a conducting sphere and that it could support a large electrical charge. Dr. Tesla proposed to disturb the charge distribution on the surface of the Earth and record the period of the resulting oscillations as the charge returned to its state of equilibrium. The problem of a single charged sphere had been analyzed at that time by J.J. Thompson and A.G. Webster in a treatise entitled "The Spherical Oscillator." This was the beginning of an examination of what we may call the science of terrestrial resonances, culminating in the 1950's and 60's with the engineering of VLF radio systems and the research and discoveries of W.O. Schumann and J.R. Waite.

The second method of energy propagation proposed by Dr. Tesla was that of the propagation of electrical energy through electrified gases. Dr. Tesla experimented with the use of high frequency RF currents to examine the properties of gases over a wide range of pressures. It was determined by Dr. Tesla that air under a partial vacuum could conduct high frequency electrical currents as well or better than copper wires. If a transmitter could be elevated to a level where the air pressure was on the order of 75 to 130 millimeters in pressure and an excitation of megavolts was applied, it was theorized that;

"...the air will serve as a conductor for the current produced, and the latter will be transmitted through the air with, it may be, even less resistance than through an ordinary copper wire".2 (Fig. 2) Resonating Planet Earth

Dr. James T. Corum and Kenneth L. Corum, in chapter two of their soon to be published book, *A Tesla Primer,* point out a number of statements made by Dr. Tesla which indicate that he was using resonator fields and transmission line modes.

1. When he speaks of tuning his apparatus until Hertzian radiations have been eliminated, he is referring to using ELF vibrations: "...the Hertzian effect has gradually been reduced through the lowering of frequency."3

2. "...the energy received does not diminish with the square of the distance, as it should, since the Hertzian radiation propagates in a hemisphere."3

3. He apparently detected resonator or standing wave modes: "...my discovery of the wonderful law governing the movement of electricity through the globe...the projection of the wavelengths (measured along the surface) on the earth's diameter or axis of symmetry...are all equal."3

4. "We are living on a conducting globe surrounded by a thin layer of insulating air, above which is a rarefied and conducting atmosphere...The Hertz waves represent energy which is radiated and unrecoverable. The current energy, on the other hand, is preserved and can be recovered, theoretically at least, in its entirety."4

As Dr. Corum points out, "The last sentence seems to indicate that Tesla's Colorado Springs experiments could be properly interpreted as characteristic of a wave-guide probe in a cavity resonator."5 This was in fact what led Dr. Tesla to report a measurement which to this day is not understood and has led many to erroneously assume that he was dealing with faster than light velocities.

The Controversial Measurement: It does not indicate faster than light velocity

The mathematical models and experimental data used by Schumann and Waite to describe ELF transmission and propagation are complex and beyond the scope of this paper. Dr. James F. Corum, Kenneth L. Corum and Dr. A-Hamid Aidinejad have, however, in a series of papers presented at the 1984 Tesla Centennial Symposium and the 1986 International Tesla Symposium, applied the experimental values obtained by Dr. Tesla during his Colorado Springs experiments to the models and equations used by Schumann and Waite.

The results of this exercise have proved that the Earth and the surrounding atmosphere can be used as a cavity resonator for the wireless transmission of electrical power. (Fig. 3)

Dr. Tesla reported that .08484 seconds was the time that a pulse emitted from his laboratory took to propagate to the opposite side of the planet and to return. From this statement many have assumed that his transmissions exceeded the speed of light and many esoteric and fallacious theories and publications have been generated. As Corum and Aidinejad point out, in their 1986 paper, "The Transient Propagation of ELF Pulses in the Earth Ionosphere Cavity", this measurement represents the coherence time of the Earth cavity resonator system. This is also known to students of radar systems as a determination of the range dependent parameter. The accompanying diagrams from Corum's and Aidinejad's paper graphically illustrate the point. (Fig. 3 & Fig. 4)

We now turn to a description of the methods to be used to build, as Dr. Tesla did in 1899, a cavity resonator for the wireless transmission of electrical power.

PROJECT TESLA:
The Wireless Transmission of Electrical Energy Using Schumann Resonance

It has been proven that electrical energy can be propagated around the world between the surface of the Earth and the ionosphere at extreme low frequencies in what is known as the Schumann Cavity. The Schumann cavity surrounds the Earth between ground level and extends upward to a maximum 80 kilometers. Experiments to date have shown that electromagnetic waves of extreme low frequencies in the range of 8 Hz, the fundamental Schumann Resonance frequency, propagate with little attenuation around the planet within the Schumann Cavity. Knowing that a resonant cavity can be excited and that power can be delivered to that cavity similar to the methods used in microwave ovens for home use, it should be possible to resonate and deliver power via the Schumann Cavity to any point on Earth. This will result in practical wireless transmission of electrical power.

Background

Although it was not until 1954-1959 when experimental measurements were made of the frequency that is propagated in the resonant cavity surrounding the Earth, recent analysis shows that it was Nikola Tesla who, in 1899, first noticed the existence of stationary waves in the Schumann cavity. Tesla's experimental measurements of the wave length and frequency involved closely match Schumann's theoretical calculations. Some of these observations were made in 1899 while Tesla was monitoring the electromagnetic radiations due to lightning discharges in a thunderstorm which passed over his Colorado Springs laboratory and then moved more than 200 miles eastward across the plains. In his Colorado Springs Notes, Tesla noted that these stationary waves "... can be produced with an oscillator," and added in parenthesis, "This is of immense importance."6 The importance of his observations is due to the support they lend to the prime objective of the Colorado Springs laboratory. The intent of the experiments and the laboratory Tesla had constructed was to prove that wireless transmission of electrical power was possible.

Schumann Resonance is analogous to pushing a pendulum. The intent of Project Tesla is to create pulses or electrical disturbances that would travel in all directions around the Earth in the thin membrane of non- conductive air between the ground and the ionosphere. The pulses or waves would follow the surface of the Earth in all directions expanding outward to the maximum circumference of the Earth and contracting inward until meeting at a point opposite to that of the transmitter. This point is called the anti-pode. The traveling waves would be reflected back from the anti-pode to the transmitter to be reinforced and sent out again.

At the time of his measurements Tesla was experimenting with and researching methods for "...power transmission and transmission of intelligible messages to any point on the globe." Although Tesla was not able to commercially market a system to transmit power around the globe, modern scientific theory and mathematical calculations support his contention that the wireless propagation of electrical power is possible and a feasible alternative to the extensive and costly grid of electrical transmission lines used today for electrical power distribution.

The Need for a Wireless System of Energy Transmission

A great concern has been voiced in recent years over the extensive use of energy, the limited supply of resources, and the pollution of the environment from the use of present energy conversion systems. Electrical power accounts for much of the energy consumed. Much of this power is wasted during transmission from power plant generators to the consumer. The resistance of the wire used in the electrical grid distribution system causes a loss of 26-30% of the energy generated. This loss implies that our present system of electrical distribution is only 70-74% efficient.

A system of power distribution with little or no loss would conserve energy. It would reduce pollution and expenses resulting from the need to generate power to overcome and compensate for losses in the present grid system.

The proposed project would demonstrate a method of energy distribution calculated to be 90-94% efficient. An electrical distribution system, based on this method would eliminate the need for an inefficient, costly, and capital intensive grid of cables, towers, and

substations. The system would reduce the cost of electrical energy used by the consumer and rid the landscape of wires, cables, and transmission towers.

There are areas of the world where the need for electrical power exists, yet there is no method for delivering power. Africa is in need of power to run pumps to tap into the vast resources of water under the Sahara Desert. Rural areas, such as those in China, require the electrical power necessary to bring them into the 20th century and to equal standing with western nations.

As first proposed by Buckminster Fuller, wireless transmission of power would enable world wide distribution of off peak demand capacity. This concept is based on the fact that some nations, especially the United States, have the capacity to generate much more power than is needed. This situation is accentuated at night. The greatest amount of power used, the peak demand, is during the day. The extra power available during the night could be sold to the side of the planet where it is day time. Considering the huge capacity of power plants in the United States, this system would provide a saleable product which could do much to aid our balance of payments.

MARKET ANALYSIS

Of the 56 billion dollars spent for research by the the U.S government in 1987, 64% was for military purposes, only 8% was spent on energy related research. More efficient energy distribution systems and sources are needed by both developed and under developed nations. In regards to Project Tesla, the market for wireless power transmission systems is enormous. It has the potential to become a multi-billion dollar per year market.

Market Size

The increasing demand for electrical energy in industrial nations is well documented. If we include the demand of third world nations, pushed by their increasing rate of growth, we could expect an even faster rise in the demand for electrical power in the near future.

In 1971, nine industrialized nations, (with 25 percent of the world's population), used 690 million kilowatts, 76 percent of all power generated. The rest of the world used only 218 million kilowatts. By comparison, China generated only 17 million kilowatts and India generated only 15 million kilowatts (less than two percent each).7 If a conservative assumption was made that the three-quarters of the world which is only using one-quarter of the current power production were to eventually consume as much as the first quarter, then an additional 908 million kilowatts will be needed. The demand for electrical power will continue to increase with the industrialization of the world.

Market Projections

The Energy Information Agency (EIA), based in Washington, D.C., reported the 1985 net generation of electric power to be 2,489 billion kilowatt hours. At a conservative sale price of $.04 per kilowatt hour that results in a yearly income of 100 billion dollars. The EIA also reported that the 1985 capacity according to generator name plates to be 656,118 million watts. This would result in a yearly output of 5,740 billion kilowatt hours at 100%

utilization. What this means is that we use only about 40% of the power we can generate (an excess capability of 3,251 billion kilowatt hours).

Allowing for down time and maintenance and the fact that the night time off peak load is available, it is possible that half of the excess power generation capability could be utilized. If 1,625 billion kilowatt hours were sold yearly at $.06/kilowatt, income would total 9.7 billion dollars.

Project Tesla: Objectives

The objectives of Project Tesla are divided into three areas of investigation.

1. Demonstration that the Schumann Cavity can be resonated with an open air, vertical dipole antenna; 2. Measurement of power insertion losses; 3. Measurement of power retrieval losses, locally and at a distance.

Methods

A full size, 51 foot diameter, air core, radio frequency resonating coil and a unique 130 foot tower, insulated 30 feet above ground, have been constructed and are operational at an elevation of approximately 11,000 feet. This system was originally built by Robert Golka in 1973- 1974 and used until 1982 by the United States Air Force at Wendover AFB in Wendover, Utah. The USAF used the coil for simulating natural lightning for testing and hardening fighter aircraft. The system has a capacity of over 600 kilowatts. The coil, which is the largest part of the system, has already been built, tested, and is operational.

A location at a high altitude is initially advantageous for reducing atmospheric losses which work against an efficient coupling to the Schumann Cavity. The high frequency, high voltage output of the coil will be half wave rectified using a uniquely designed single electrode X-ray tube. The X-ray tube will be used to charge a 130 ft. tall, vertical tower which will function to provide a vertical current moment. The mast is topped by a metal sphere 30 inches in diameter. X-rays emitted from the tube will ionize the atmosphere between the Tesla coil and the tower. This will result in a low resistance path causing all discharges to flow from the coil to the tower. A circulating current of 1,000 amperes in the system will create an ionization and corona causing a large virtual electrical capacitance in the medium surrounding the sphere. The total charge around the tower will be in the range of between 200-600 coulombs. Discharging the tower 7-8 times per second through a fixed or rotary spark gap will create electrical disturbances, which will resonantly excite the Schumann Cavity, and propagate around the entire Earth.

The propagated wave front will be reflected from the antipode back to the transmitter site. The reflected wave will be reinforced and again radiated when it returns to the transmitter. As a result, an oscillation will be established and maintained in the Schumann Cavity. The loss of power in the cavity has been estimated to be about 6% per round trip. If the same amount of power is delivered to the cavity on each

Tesla's concept of an anti-gravity airship with no wings or propeller, from *Electrical Experimenter,* Oct. 1919.

cycle of oscillation of the transmitter, there will be a net energy gain which will result in a net voltage, or amplitude increase. This will result in reactive energy storage in the cavity. As long as energy is delivered to the cavity, the process will continue until the energy is removed by heating, lightning discharges, or as is proposed by this project, loading by tuned circuits at distant locations for power distribution.

The resonating cavity field will be detected by stations both in the United States and overseas. These will be staffed by engineers and scientists who have agreed to participate in the experiment.

Measurement of power insertion and retrieval losses will be made at the transmitter site and at distant receiving locations. Equipment constructed especially for measurement of low frequency electromagnetic waves will be employed to measure the effectiveness of using the Schumann Cavity as a means of electrical power distribution. The detection equipment used by project personnel will consist of a pick up coil and industry standard low noise, high gain operational amplifiers and active band pass filters.

In addition to project detection there will be a record of the experiment recorded by a network of monitoring stations that have been set up specifically to monitor electromagnetic activity in the Schumann Cavity.

Evaluation Procedure

The project will be evaluated by an analysis of the data provided by local and distant measurement stations. The output of the transmitter will produce a 7-8 Hz sine wave as a result of the discharges from the antenna. The recordings made by distant stations will be time synchronized to ensure that the data received is a result of the operation of the transmitter.

Power insertion and retrieval losses will be analyzed after the measurements taken during the transmission are recorded. Attenuation, field strength, and cavity Q will be calculated using the equations presented in Dr. Corum's papers. These papers are noted in the references. If recorded results indicate power can be efficiently coupled into or transmitted in the Schumann Cavity, a second phase of research involving power reception will be initiated.

Environmental Considerations

The extreme low frequencies (ELF), present in the environment have several origins.

The time varying magnetic fields produced as a result of solar and lunar influences on ionospheric currents are on the order of 30 nanoteslas. The largest time varying fields are those generated by solar activity and thunderstorms. These magnetic fields reach a maximum of 0.5 microteslas (uT) The magnetic fields produced as a result of lightning discharges in the Schumann Cavity peak at 7, 14, 20 and 26 Hz. The magnetic flux densities associated with these resonant frequencies vary from 0.25 to 3.6 picoteslas. per root hertz (pT/Hz1/2).

Exposure to man made sources of ELF can be up to 1 billion (1000 million or 1 x 109) times stronger than that of naturally occurring fields. Household appliances operated at 60 Hz can produce fields as high as 2.5 mT. The field under a 765 kV, 60 Hz power line carrying 1 amp per phase is 15 uT. ELF antennae systems that are used for submarine communication produce fields of 20 uT. Video display terminals produce fields of 2 uT, 1,000,000 times the strength of the Schumann Resonance frequencies.9

Project Tesla will use a 150 kw generator to excite the Schumann cavity. Calculations predict that the field strength due to this excitation at 7.8 Hz will be on the order of 46 picoteslas.

Future Objectives

The successful resonating of the Schumann Cavity and wireless transmission of power on a small scale resulting in proof of principle will require a second phase of engineering, the design of receiving stations. On completion of the second phase, the third and fourth phases of the project involving further tests and improvements and a large scale demonstration project will be pursued to prove commercial feasibility. Total cost from proof of principle to commercial prototype is expected to total $3 million. Interest in participation in this project may be directed to the author.

REFERENCES

The following four papers were presented at the 1984 Tesla Centennial Symposium and the 1986 International Tesla Symposium.

"The Transient Propagation of ELF Pulses in the Earth-Ionosphere Cavity", by A-Ahamid Aidinejad and James F. Corum.

"Disclosures Concerning the Operation of an ELF Oscillator", by James F. Corum and Kenneth L. Corum.

"A Physical Interpretation of the Colorado Springs Data", by James F. Corum and Kenneth L. Corum.

"Critical Speculations Concerning Tesla's Invention and Applications of Single Electrode X-Ray Directed Discharges for Power Processing, Terrestrial Resonances and Particle Beam Weapons" by James F. Corum and Kenneth L. Corum.

FOOTNOTES

1. *Tesla Said,* Compiled by John T. Ratzlaff, Tesla Book Company, Millbrae, CA, 1984.
2. *Dr. Nikola Tesla: Selected Patent Wrappers,* compiled by John T. Ratzlaff, Tesla Book Company, 1980, Vol. I, Pg. 128.
3. "The Disturbing Influence of Solar Radiation on the Wireless Transmission of Energy", by Nikola Tesla, *Electrical Review,* July 6, 1912, PP. 34, 35.
4. "The Effect of Static on Wireless Transmission", by Nikola Tesla, *Electrical Experimenter,* January 1919, PP. 627, 658.
5. *Tesla Primer and Handbook,* Dr. James T. Corum and Kenneth L. Corum, unpublished. Corum and Associates, 8551 ST Rt 534, Windsor, Ohio 44099

6. *Colorado Springs Notes, 1899-1900,* Nikola Tesla, Nikola Tesla Museum, Beograd, Yugoslavia, 1978, Pg. 62.
7. *Van Nostrands Scientific Encylopedia, Fith Edition,* Page. 899.
8. "PC Monitors Lightning Worldwide", Davis D. Sentman, *Computers in Science,* Premiere Issue, 1987.
9. "Artificially Stimulated Resonance of the Earth's Schumann Cavity Waveguide", Toby Grotz, *Proceedings of the Third International New Energy Technology Symposium/Exhibition,* June 25th-28th, 1988, Hull, Quebec, Planetary Association for Clean Energy, 191 Promenade du Portage/600, Hull, Quebec J8X 2K6 Canada

ABOUT THE AUTHOR

Mr. Grotz, is an electrical engineer and has 15 years experience in the field of geophysics, aerospace and industrial research and design. While working for the Geophysical Services Division of Texas Instruments and at the University of Texas at Dallas, Mr. Grotz was introduced to and worked with the geophysical concepts which are of importance to the proposed project. As a Senior Engineer at Martin Marietta, Mr. Grotz designed and supervised the construction of industrial process control systems and designed and built devices and equipment for use in research and development and for testing space flight hardware. Mr. Grotz organized and chaired the 1984 Tesla Centennial Symposium and the 1986 International Tesla Symposium and was President of the International Tesla Society, a not for profit corporation formed as a result the first symposium. As Project Manager for Project Tesla, Mr. Grotz aided in the design and construction of a recreation of the equipment Nikola Tesla used for wireless transmission of power experiments in 1899 in Colorado Springs. Mr. Grotz received his B.S.E.E. from the University of Connecticut in 1973.

Tesla's wireless transmission tower in action, sending power to electrical airships in this illustration for *Radio News,* December, 1925.

Part Four
Tesla's Electric Car

Tesla's Electric Car

From the *Dallas Morning News* Texas Sketches column January 24th, 1993—

The Electric Auto that almost triumphed Power Source of '31 car still a mystery by A.C. Greene Not long ago, Texas Sketches told the story of Henry "Dad" Garrett and his son C.H.'s water-fueled automobile, which was successfully demonstrated in 1935 at White Rock Lake in Dallas.

Eugene Langkop of Dallas (a Packard lover, like so many of us) notes that the "wonder car" of the future may be a resurrection of the electric car. It uses no gasoline, no oil -just some grease fittings—has no radiator to fill or freeze, no carburetor problems, no muffler to replace and gives off no pollutants.

Famous former electrics include Columbia, Rauch & Lang and Detroit Electric. Dallas had electric delivery trucks in the 1920s and 30s. Many electric delivery vehicles were used in big cities into the 1960s.

The problem with electrics was slow speed and short range.

Within the past decade two Richardson men, George Thiess and Jack Hooker, claimed to have used batteries operating on magnesium from seawater to increase the range of their electric automobile from 100 miles to 400 or 500 miles.

But it is a mystery car once demonstrated by Nikola Tesla, developer of alternating current, that might have made electrics triumphant.

Supported by the Pierce-Arrow Co. and General Electric in 1931, he took the gasoline engine from a new Pierce-Arrow and replaced it with an 80-horsepower alternating-current electric motor with no external power source.

At a local radio shop he bought 12 vacuum tubes, some wires and assorted resistors, and assembled them in a circuit box 24 inches long, 12 inches wide and 6 inches high, with a pair of 3-inch rods sticking out. Getting into the car with the circuit box in the front seat beside him, he pushed the rods in, announced, "We now have power," and proceeded to test drive the car for a week, often at speeds of up to 90 mph.

As it was an alternating-current motor and there were no batteries involved, where did the power come from?

Popular responses included charges of "black magic," and the sensitive genius didn't like the skeptical comments of the press. He removed his mysterious box, returned to his laboratory in New York—and the secret of his power source died with him.

A.C. Greene is an author and Texas historian who lives in Salado.

The original article from which Mr. Greene gleaned the above info was from a Packard Newsletter. Mr. Gene Langkopf kindly sent us a copy of that article which now follows:

The Forgotten Art of Electric—Powered Automobiles
by Arthur Abrom

Electric powered automobiles were one of the earliest considerations and this mode of propulsion enjoyed a brief but short reign. The development of electricity as a workable source of power for mankind has been studded with great controversy.

Thomas A. Edison was the first to start to market systems (i.e. electric generators) of any commercial value. His research and developmental skills were utilized to market a "direct current" system of electricity. Ships were equipped with D.C. systems and municipalities began lighting their streets with this revolutionary D.C. electric system. (At that time) Edison

was the sole source of electricity!

While in the process of commercializing electricity, Thomas Edison hired men who knew of the new scientific gift to the world and were capable of new applications for electricity. One such man was a foreigner named Nikola Tesla. This man, although not known to many of us today, was without a doubt the greatest scientific mind that has ever lived. His accomplishments dwarfed even Thomas Edison's! Whereas Mr. Edison was a great experimenter, Mr. Tesla was a great theoretician. Nikola Tesla became frustrated and very much annoyed at the procedures Edison followed.

Tesla would rather calculate the possibility of something working (i.e. mathematical investigation) than the hit and miss technique of constant experimentation. So in the heat of an argument, he quit one day and stormed out of Edison's laboratory in West Orange, New Jersey.

Working on his own, Tesla conceived and built the first working alternating current generator. He, and he alone, is responsible for all of the advantages we enjoy today because of A.C. electric power.

Angered by Edison, Tesla sold his new patents to George Westinghouse for 15 million dollars in the very early 1900's. Tesla became totally independent and proceeded to carry on his investigative research in his laboratory on 5th Avenue in New York City.

George Westinghouse began to market this new system of electric generators and was in competition with Edison. Westinghouse prevailed because of the greater superiority of the A.C. generators over the less efficient D.C. power supplies of Thomas Edison. Today, A.C. power is the only source of electricity the world uses. And, please remember, Nikola Tesla is the man who developed it.

Now specifically dealing with automobiles in the infant days of their development, electric propulsion was considered and used. An electric powered automobile possessed many advantages that the noisy, cantankerous, smoke-belching gasoline cars could not offer.

First and foremost is the absolute silence one experiences when riding in an electrically powered vehicle. There is not even a hint of noise. One simply turns a key and steps on the accelerator—the vehicle moves instantly! No cranking from the start, no crank to turn (this was before electric starters), no pumping of the accelerator, no spark control to advance and no throttle linkage to pre-set before starting. One simply turned the ignition switch to on!

Second, is a sense of power. If one wants to increase speed, you simply depress the accelerator further—there is never any hesitation. Releasing the accelerator causes the vehicle to slow down immediately—you are always in complete control. It is not difficult to understand why these vehicles were so very popular around the turn of the century and until 1912 or so.

The big disadvantage to these cars was their range and need for re-charging every single night. All of these electric vehicles used a series of batteries and a D.C. motor to move itself about. The batteries require recharging every night and the range of travel was restricted to about 100 miles. Understand that this restriction was not a serious one in the early part of this century. Doctors began making house calls with electric cars (do you remember doctors making house calls?) because he no longer needed to tend to the horse at night time—

just plug the car into an electric socket! No feeding, no rub-down and no mess to clean up!

Many of the large department stores in metropolitan areas began purchasing delivery trucks that were electrically powered. They were silent and emitted no pollutants. And, maintenance was a minimum on electrically powered vehicles. There were few mechanics and garages in operation in the early 1900's. So city life and travel appeared to be willing to embrace the electric automobile. Remember, these masterfully built vehicles all ran on D.C. current.

Two things happened to dampen the popularity of the electric automobile. One was the subconscious craving for speed that gripped all auto enthusisasts of this era. Each manufacturer was eager to show how far his car could travel (i.e. the transcontinental races) and what was its top speed!

Col. Vanderbilt constructed the first all concrete race track in Long Island and racing became the passion for the well-to-do. Newspapers constantly record new records of speed achieved by so-in-so. And, of course, the automobile manufacturers were quick to capitalize on the advertising effect of these new peaks of speed. Both of these events made the electrically powered vehicles appear to only belong to the "little old lady" down the street or the old retired gentleman who talked about the "good old days".

Electric vehicles could not reach speeds of 45 or 50 m.p.h. for this would have destroyed the batteries in moments. Bursts of speeds of 25 to 35 m.p.h. could be maintained for a moment or so. Normal driving speed-depending upon traffic conditions, was 15 to 20 m.p.h. by 1900 to 1910 standards, this was an acceptable speed limit to obtain from your electric vehicle.

Please note that none of the manufacturers of electric cars ever installed a D.C. generator. This would have put a small charge back into the batteries as the car moved about and would have thereby increased its operating range. This was considered by some to be approaching perpetual motion—and that, of course, was utterly impossible! Actually, D.C. generators would have worked and helped the electric car cause.

As mentioned earlier, Mr. Westinghouse's A.C. current generating equipment was being sold and installed about the country. The earlier D.C. equipment was being retired and disregarded. As a side note, Consolidated Edison Power Company of New York City still has one of Thomas Edison's D.C. generators installed in its 14th St. powerhouse—it still works! About this time, another giant corporation was formed and entered the A.C. generating equipment field—General Electric. This spelled the absolute end for Edison's D.C. power supply systems as a commercial means of generating and distributing electric power.

The electric automobile could not be adapted to accomodate and utilize a polyphase motor (i.e. A.C. power). Since they used batteries as a source of power, their extinction was sealed. No battery can put out an A.C. signal. True, a converter could be utilized (i.e. convert the D.C. signal from the battery to an A.C. signal), but the size of the equipment at this time was too large to fit in an automobile—even one with the generous dimensions of this era.

So, somewhere around 1915 or so, the electric automobile became a memory. True, United Parcel Service still utilizes several electric trucks in New York City today but the bulk of their fleet of vehicles utilizes gasoline or diesel fuel. For all intensive purposes, the electrically powered automobile is dead—they are considered dinosaurs of the past.

But, let us stop a moment and consider the advantages of utilizing electric power as a means of propelling vehicles. Maintenance is absolutely minimal for the only oil required is for the two bearings in the motor and the necessary grease fittings. There is no oil to change, no radiator to clean and fill, no transmission to foul up, no fuel pump, no water pump, no carburetion problems, no muffler to rot out or replace and no pollutants emitted into the atmosphere. It appears as though it might be the answer we have been searching for!

Therefore, the two problems facing us become top speed and range of driving—providing, of course, the A.C. and D.C. problems could be worked out. With today's technology this does not seem to be insurmountable. In fact, the entire problem has already been solved—in the past, the distant past and the not so distant! Stop! Re-read the last sentence again. Ponder it for a few moments before going on.

Several times earlier in this article, I mentioned the man, Nikola Tesla and stated that he was the greatest mind that ever lived. The U.S. Patent Office has 1,200 patents registered in the name of Nikola Tesla and it is estimated that he could have patented an additional 1,000 or so from memory!

But, back to our electric automobiles—in 1931, under the financing of Pierce-Arrow and George Westinghouse, a 1931 Pierce-Arrow was selected to be tested at the factory grounds in Buffalo, N.Y. The standard internal combustion engine was removed and an 80-H.P. 1800 r.p.m electric motor installed to the clutch and transmission. The A.C. motor measured 40 inches long and 30 inches in diameter and the power leads were left standing in the air—no external power source!

At the appointed time, Nikola Tesla arrived from New York City and inspected the Pierce-Arrow automobile. He then went to a local radio store and purchased a handful of tubes (12), wires and assorted resistors. A box measuring 24 inches long, 12 inches wide and 6 inches high was assembled housing the circuit. The box was placed on the front seat and had its wires connected to the air-cooled, brushless motor. Two rods 1/4" in diameter stuck out of the box about 3" in length.

Mr. Tesla got into the driver's seat, pushed the two rods in and stated, "We now have power". He put the car into gear and it moved forward! This vehicle, powered by an A.C. motor, was driven to speeds of 90 m.p.h. and performed better than any internal combustion engine of its day! One week was spent testing the vehicle. Several newspapers in Buffalo reported this test. When asked where the power came from, Tesla replied, "From the ethers all around us". Several people suggested that Tesla was mad and somehow in league with sinister forces of the universe. He became incensed, removed his mysterious box from the vehicle and returned to his laboratory in New York City. His secret died with him!

It is speculated that Nikola Tesla was able to somehow harness the earth's magnetic field that encompasses our planet. And, he somehow was able to draw tremendous amounts of power by cutting these lines of force or causing them to be multiplied together. The exact nature of his device remains a mystery but it did actually function by powering the 80 h.p. A.C. motor in the Pierce-Arrow at speeds up to 90 m.p.h. and no recharging was ever necessary!

In 1969, Joseph R. Zubris took his 1961 Mercury and pulled out the Detroit internal combustion engine. He then installed an electric motor as a source of power. His unique wiring system cuts the energy drain at starting to 75% of normal and doubles the electrical efficiency of the electric motor when it is operating! The U.S. Patent Office issued him a patent No. 3,809,978. Although he approached many concerns for marketing, no one really seemed to be interested. And, his unique system is still not on the market.

In the 1970's, an inventor used an Ev-Gray generator, which intensified battery current, the voltage being induced to the field coils by a simple programmer (sequencer). By allowing the motor to charge separate batteries as the device ran, phenomenally tiny currents were needed. The device was tested at the Crosby Research Institute of Beverly Hills, Ca., a 10-horepower EMA motor ran for over a week (9 days) on four standard automobile batteries.

The inventors estimated that a 50-horsepower electric motor could traverse 300 miles at 50 m.p.h. before needing a re-charge. Dr. Keith E. Kenyon, the inventor of Van Nuys, California discovered a discrepancy in the normal and long accepted laws relating to electric motor magnets. Dr. Kenyon demonstrated his invention for many scientists and engineers in 1976 but their reaction was astounding. Although admitting Dr. Kenyon's device worked, they saw little or no practical application for it!

So the ultimate source for our electrically powered automobile would be to have an electric motor that required no outside source of power. Sounds impossible because it violates all scientific thought! But it has been invented and H.R. Johnson has been issued a patent No. 4,151,431 on April 24, 1979 on such a device!

This new design although originally suggested by Nikola Tesla in 1905, is a permanent magnet motor. Mr. Johnson has arranged a series of permanent magnets on the rotor and a corresponding series—with different spacing—on the stator. One simply has to move the stator into position and rotation of the rotor begins immediately.

Howard Johnson Permanent Magnet Motor

His patent states, "The invention is directed to the method of utilizing the unpaired electron spins in ferro magnetic and other materials as a source of magnetic fields for producing power without any electron flow as occurs in normal conductors and to permanent magnet motors for utilization of this method to produce a power source.

In the practice of this invention, the unpaired electron spins occurring within permanent magnets are utilized to produce a motive power source solely through the super-conducting characteristics of a permanent magnet and the magnetic flux created by the magnets are controlled and concentrated to orient the magnetic forces generated in such a manner to do useful continuous work such as the displacement of a rotor with respect to a stator.

The timing and orientation of magnetic forces at the rotor and stator components produced by permanent magnets to produce a motor is accomplished with the proper geometrical relationship of these components".

Now before you dismiss the idea of a magnetically run motor—a free energy source, consider the following : Engineers of Hitachi Magnetics Corp. of California have stated that a motor run solely by magnets is feasible and logical but the politics of the matter make it impossible for them to pursue developing a magnet motor or any device that would compete with the energy cartels.

In a book entitled, *Keely and His Discoveries* by Clara B. Moore published in 1893, we find the following statemtents, "The magnet that lifts a pound today if the load is gradually increased day by day will lift double that amount in time. Whence comes this energy? Keely teaches that it comes from sympathetic association with one of the currents of the polar stream and that its energy increases as long as the sympathetic flow lasts, which is through eternity".

Now consider some basic observations concerning magnets:

1) Two permanent magnets can either attract or repel depending on the arrangement of the magnetic poles.

2) Two magnets repel further than they attract because of friction and inertia forces.

3) Most of our energy comes directly or indirectly from electromagnetic energy of the sun, e.g. photosynthesis and watercycle of ocean to water vapor to rain or snow to ocean.

4) Magnetic energy "travels" between poles at the speed of light.

5) Permanent magnets on both sides of an iron shield are attracted to the shield and only weakly to each other at close proximity to the shield.

6) Permanent magnets are ferrous metals and are attractive only. Attraction is an inverse square force.

7) Magnetic energy can be shielded.

8) The sliding or perpendicular force of a keeper is much less than the force in the direction of the field to remove the keeper.

9) Most of the magnetic energy is concentrated at the poles of the magnet.

10) A permanent magnet loses little strength unless dropped or heated. Heating misaligns the magnetic elements within the magnet.

11) If a weight lifted by a permanet magnet is slowly increased, the lifting power of the magnet can be increased until all the magnetic domains in the magnet are aligned in the same direction. This becomes the limit.

12) Using magnets to repel tends to weaken them as it causes more misalignment of the domains.

13) A magnetic material placed between two magnets will always be attracted to the stronger magnet.

So, our ultimate motor becomes a permanent magnet motor of proper size with speed being controlled through the automobiles transmission. And, here is the biggest plus, permanent magnets keep their strength for a minimum of 95 years! So here we have a fuel-less automobile that would last us our lifetime.

There is only one drawback to an automobile powered by a permanent magnet motor—if the vehicle gets involved in an accident, the shock of the crash could jar the magnets and cause them to lose power! But this seems to be a small price to pay for an automobile that could run all day at 60 m.p.h.—use no fuel—and never need a recharge!

Now the only question left to be answered is, "Where do you buy one?" or perhaps, "When will we be able to buy one?" At present there are several companies offering interim solutions. Some offer electric powered designs—but this is strictly batteries, while others offer a hybrid combination of batteries and small gasoline engines. All of these so-called "modern alternatives" suffer from the same lack of accessories we've become accustomed to.

They do not, or cannot offer power steering, brakes or windows or air- conditioning, etc. Since they are small aerodynamically shaped packages holding only two people, their appeal is distinctly limited.

When someone constructs an automobile run by a permanent magnet motor attached to the differential thus eliminating the transmission, the world will beat a path to his door—providing the energy cartel doesn't find him first!

In Richardson, Texas last year, two men—George Thiess and Jack Hooker have advanced the storage battery to a new level. Their new batteries will operate on magnesium made from seawater.

Thiess/Hooker Advanced Storage Battery

The magnesium is used to charge the battery while in an electrolene solution and the range of their auto is increased by replacing the magnesium rods every 400 to 500 miles. Their studies are being officially watched by the Department of Energy. Perhaps an all new era of electrically powered automobiles may be on its way to reality.

This subject is intensely interesting to many researchers so if you have any suggestions or comments, we here at KeelyNet would greatly appreciate your sharing with us.

Tesla's Electric Car—

by Gerry Vassilatos

Tesla had already considered the condition of charged particles, each representing a tightly constricted whorl of aether. The force necessarily exerted at close distances by such aetheric constrictions was incalculably large. Aetheric ponderance maintained particulate stability.

Crystalline lattices were therefore places within which one could expect to find unexpected voltages. Indeed, the high voltages inherent in certain metallic lattices, intra-atomic field energies, are enormous. The close Coulomb gradient between atomic centers are electrostatic potentials reaching humanly unattainable levels.

By comparison, the voltages which Tesla once succeeded in releasing were quite insignificant. In these balanced lattices, Tesla sought the voltages needed to initiate directed aetheric streams in matter.

Once such a flow began, one could simply tap the stream for power. In certain materials, these ether streams might automatically produce the contaminating electrons, a source of energy for existing appliances. One could theoretically then "tailor" the materials needed to produce unexpected aetheric power with or without the attendant detrimental particles.

Tesla did mention the latent aetheric power of charged forces, the explosive potentials of bound Ether, and the aetheric power inherent in matter.

By these studies, Tesla sought replacement for the 100,000,000 volt initiating pulses which natural law required for the implementation of space *Ether.* Tesla had long been forced to abandon those gigantic means by other, less natural laws.

Thereafter, Tesla shifted his attentions from the appreciation of the gigantic to an appreciation of the miniature. He sought a means for proliferating an immense number of small and compact aether power receivers.

With one such device, Tesla succeeded in obtaining power to drive am electric car. But for the exceptional account which follows, we would have little information on this last period in Tesla's productive life, one which very apparently did not cease its prolific streams of creativity to his last breath.

The information comes through an unlikely source, one rarely mentioned by Tesla biographers. It chanced that an aeronautical engineer, Derek Ahlers, met with one of Tesla's nephews then living in New York. Theirs was an acquaintance lasting some 10 years, consisting largely of anecdotal commentaries on Dr. Tesla. Mr. Savo provided an enormous fund of knowledge concerning many episodes in Tesla's last years.

Himself an Austrian military man and a trained aviator, Mr. Savo was extremely open about certain long-cherished incidents in which his uncle's genius was consistency made manifest. Mr. Savo reported that in 1931, he participated in an experiment involving aetheric power. Unexpectedly, almost inappropriately, he was asked to accompany his uncle on a long train ride to Buffalo.

A few times in this journey, Mr. Savo asked the nature of their journey. Dr. Tesla remained unwilling to disclose any information, speaking rather directly to this issue. Taken into a small garage, Dr. Tesla walked directly to a Pierce Arrow, opened the hood and began making a few adjustments. In place of the engine, there was an AC motor.

This measured a little more than 3 feet long, and a little more than 2 feet in diameter. From it trailed two very thick cables which connected with the dashboard. In addition, there was an ordinary 12 volt storage battery. The motor was rated at 80 horsepower.

Maximum rotor speed was stated to be 30 turns per second. A 6 foot antenna rod was fitted into the rear section of the car.

Dr. Tesla stepped into the passenger side and began making adjustments on a "power receiver" which had been built directly into the dashboard.

The receiver, no larger than a short-wave radio of the day, used 12 special tubes which Dr. Tesla brought with him in a boxlike case.

The device had been prefitted into the dashboard, no larger than a short-wave receiver. Mr. Savo told Mr. Ahler that Dr. Tesla built the receiver in his hotel room, a device 2 feet in length, nearly 1 foot wide, a 1/2 foot high.

These curiously constructed tubes having been properly installed in their sockets, Dr. Tesla pushed in 2 contact rods and informed Peter that power was now available to drive.

Several additional meters read values which Dr. Tesla would not explain. Not sound was heard. Dr. Tesla handed Mr. Savo the ignition key and told him to start the engine, which he promptly did. Yet hearing nothing, the accelerator was applied, and the car instantly moved. Tesla's nephew drove this vehicle without other fuel for an undetermined long interval.

Mr. Savo drove a distance of 50 miles through the city and out to the surrounding countryside. The car was tested to speeds of 90 mph, with the speedometer rated to 120.

After a time, and with increasing distance from the city itself, Dr. Tesla felt free enough to speak. Having now become sufficiently impressed with the performance of both his device and the automobile.

Dr. Tesla informed his nephew that the device could not only supply the needs of the car forever, but could also supply the needs of a household—with power to spare. When originally asked how the device worked, Tesla was initially adamant and refused to speak.

Many who have read this "apocryphal account" have stated it to be the result of an "energy broadcast". This misinterpretation has simply caused further confusions concerning this stage of Tesla's work. He had very obviously succeeded in performing, with this small and compact device, what he had learned in Colorado and Shoreham.

As soon as they were on the country roads, clear of the more congested areas, Tesla began to lecture on the subject. Of the motive source he referred to "a mysterious radiation which comes out of the aether". The small device very obviously and effectively appropriated this energy.

Tesla also spoke very glowingly of this providence, saying of the energy itself that "it is available in limitless quantities".

Dr. Tesla stated that although "he did not know where it came from, mankind should be very grateful for its presence".

The two remained in Buffalo for 8 days, rigorously testing the car in the city and countryside. Dr. Tesla also told Mr. Savo that the device would soon be used to drive boats, planes, trains, and other automobiles. Once, just before leaving the city limits, they stopped at a streetlight and a bystander joyfully commented concerning their lack of exhaust fumes.

Mr. Savo spoke up whimsically, saying that they had "no engine". They left Buffalo and traveled to a predetermined location which Dr. Tesla knew, an old farmhouse barn some 20 miles from Buffalo. Dr. Tesla and Mr. Savo left the car in this barn, took the 12 tubes and the ignition key, and departed.

Later on, Mr. Savo heard a rumor that a secretary had spoken candidly about both the receiver and the test run, being promptly fired for the security breach. About a month after the incident, Mr. Savo received a call from a man who identified himself as Lee De Forest, who asked how he enjoyed the car.

Mr. Savo expressed his joy over the mysterious affair, and Mr. de Forest declared Tesla the greatest living scientist in the world. Later, Mr. Savo asked his uncle whether or not the power receiver was being used in other applications.

He was informed that Dr. Tesla had been negotiating with a major shipbuilding company to build a boat with a similarly outfitted engine. Asked additional questions, Dr. Tesla became annoyed. Highly concerned and personally strained over the security of this design, it seems obvious that Tesla was performing these tests in a desperate degree of secrecy for good reasons.

Tesla had already been the victim of several manipulations, deadly actions entirely sourced in a single financial house. For this reason, secrecy and care had become his only recent excess.

Tesla's Electric Car —The Moray Version
by Jerry Decker
This file was originally posted on the KeelyNet BBS on January 31, 1993.

In Tesla's electric car, the standard internal combustion engine was removed and an 80-H.P. 1800 r.p.m electric motor installed to the clutch and transmission.

The A.C. motor measured 40 inches long and 30 inches in diameter and the power leads were left standing in the air—no external power source!

He then went to a local radio store and purchased a handful of tubes (12), wires and assorted resistors. A box measuring 24 inches long, 12 inches wide and 6 inches high was assembled housing the circuit.

The box was placed on the front seat and had its wires connected to the air-cooled, brushless motor. Two rods 1/4" in diameter stuck out of the box about 3" in length." The mention of this experiment in a local paper kind of blew me away but it did give "some" detail of what was in this mysterious power box.

We know that T.H. Moray had probably the best known version of such a device. In his case he used a special "valve" which appeared to be basically a diode. Except this diode worked more like a Triac. That is, any electrical wave, both positive AND negative going currents, was picked up by an antenna and passed through this diode with minimal loss of energy. As far as we know, this valve was based on a composite substance with GERMANIUM as the host material.

From there it went through a tuned circuit based on vacuum tubes and capacitors to build and discharge the energy as demanded by the load. The tuned circuits were resonant with one or more earth or cosmic frequencies and the vacuum tubes acted as harmonic constructive interference amplifiers of the input signals.

We will note that Moray's resonant circuits used CAPACITORS, COILS and RESISTORS. Experiments done during Moray's heyday showed an output up to 50,000 Watts of high frequency energy. It is believed that the energy was high frequency because 100 watt light bulbs burned cool to the touch. One other CRITICAL POINT about Moray's converter was that it would ONLY energize RESISTIVE loads and NOT INDUCTIVE loads. This is because inductive loads imply coils of wire which are heated more so by HYSTERESIS (interferring electro-magnetic fields) rather than simple resistance from the flow of current through molecular/atomic patterns.

This type of interferring field caused an energy backup and subsequent de-tuning of Moray's generator. Since it was essentially a TUNED device, it could not compensate for any frequency changes or distortions ONCE TUNED. As a result, any attempt to hook up an inductive load would cause the device to stop generating electrical energy. To restart it, all inductive loading must be removed, the device re-tuned and restarted.

Moray also used an unusual mode of operation for a vacuum tube in that he operated with a "cold cathode." This did not require a heated plate for the "thermionic emissions" deemed necessary to successful vacuum tube operation.

There is also mention of radioactive elements in the antenna circuit which leads one to think he might have been tuning into the continual radioactive decay processes of nature, rather than cosmic or earth energies.

Now to the Tesla Power Box We will first of all note the use of an AC coil motor. This alone tells us that the Tesla device was superior and not so dependent on tuning as was Moray's machine which could only power RESISTIVE loads. All universal energy moves in WAVES and so is essentially for alternating current (AC). That is why Moray called his book "THE SEA OF ENERGY IN WHICH THE EARTH FLOATS". The entire universe is continually bathed in these AC energies and they cover the entire frequency spectrum.

What intrigues the hell out of me was how Tesla could use "off-the- shelf" vacuum tubes and other components, put them together in the correct configuration and make it work.

Another point we should note is the list of components :

1) 12 Vacuum Tubes (70L7-GT rectifier beam power tubes)
2) Wires
3) Assorted Resistors
4) 1/4" diameter rods 3" in length

NOTE, NO CAPACITORS! The wires could have been simply for connection or wound as coils. The 1/4" rods were either BUS BARS for power output taps OR more likely ANTENNAS! Resonant circuits can be constructed using several techniques. You can achieve the same effect from :

1) Resistors AND capacitors
2) Capacitors AND coils
3) Coils AND resistors

So, in the case of the Tesla Power Box, he either wound his own coils or simply used the wire to connect the resistors with the vacuum tubes. I am of the opinion that he used the wire ONLY for connection and DID NOT USE COILS! I also think he used a DIODE somewhere in the circuit in order to tap ONLY one polarity.

We have no specifications for the AC motor that Tesla used in the auto, so we have no idea if it was single or polyphase. In the case of a single phase motor, it only requires a single winding which projects a magnetic field that rotates according to the increase or decrease of the alternating current.

A polyphase (poly = two or more) motor uses multiple windings which are fed by phased input currents that alternate in such a manner as to reinforce each other. In the case of a 3 phase motor, the currents are phased 120 degrees apart. This gives much greater torque to the motor but requires 3 times the current because it uses 3 times the input energy.

Since the box powered an AC (coil) motor, it is probable it was TUNED to one or more frequencies, most likely polyphased frequencies.

So, if the 3" long rods were in fact ANTENNAS, we can calculate their frequency by using the following :

(I cannot express Lambda here so we will use w for wavelength)

w = wavelength

v = velocity of propagation

f = signal frequency

a short example : w = v / f = wavelength in feet

w = 984,000,000/1,500,000 = 656 feet

f = 984,000,000/656 = 1,500,000 or 1.5 MHZ

3 inches * 4 = 1 foot

984,000,000/1 = 984,000,000

984,000,000/4 = 246,000,000 or 246 MHZ

This would indicate the 3" rods (if they were truly 3" in length and functioning as antennas) would resonate at 246 MHZ.

Because of the parts list description, I am of the opinion that it was a DUAL circuit. That is, 6 vacuum tubes and one 1/4" diameter 3" rod along with assorted resistors were to pick up and "pump" ONLY the positive going signals, while the other 6 vacuum tubes, rod and resistors did the same for the negative going signals. Such a scheme could either use PARALLEL or SERIAL connections of the vacuum tubes. Since current conduction is proportional to surface area, one would think that a parallel arrangement of the 70L7-GT rectifier beam power tubes with all INPUTS connected to one antenna source and all OUTPUTS connected to a common terminal attached to the load, would provide for the MAXIMUM current flow from incoming energy waves.

The nature of these "energy waves" is the question here. Are they cosmic rays, electrostatic, Schumann peaks, magnetic force, something "other" or Aether flow into the neutral centers of mass as per Keely.

Vacuum tube construction takes several forms. Of these, the simplest is two plates separated by a grid wire. When the bottom plate is heated, thermally induced ions (thermionic emissions) are emitted by the bottom plate. The grid can be biased by the application of voltage to increase, decrease or halt the flow of these ions to the upper plate.

Other forms include more plates with more grids to allow better control of the ion flow. By proper biasing, vacuum tubes can be operated as switches, modulators or amplifiers among other uses.

Vacuum tubes operate primarily with high voltages that control the ion flows. Modern transistors are equivalent to vacuum tubes except that they operate using CURRENT instead of voltage. Transistors equate to Vacuum tubes by the following comparisons :

Vacuum Tube Transistor Polarity

Operates from Voltage Current

lower plate emitter negative—cathode

grid base neutral

upper plate collector positive—anode

In the case of the Tesla Power Box, the vacuum tube appears to function as a "pump," collecting incoming current in the form of ion intensification. Once this "compressed" ion field reaches a certain density, the pump allows it to be released into the next stage of the circuit, be it the actual load or another vacuum tube.

So if the circuit is 6 vacuum tubes in parallel, all fed from a common antenna, outputting to a common load terminal, then the common antenna input would feed all vacuum tubes with the same wave. This would give the greatest CURRENT accumulation because of the EXPANDED SURFACE AREA of the paralleled tubes.

Note, these vacuum tubes most likely operate in the "cold cathode" mode since the heaters of the vacuum tubes were not fed by any outside voltage to provide the heat for the more orthodox therionic emission.

If the vacuum tubes are hooked in series, then one "pump" would feed another "pump" to get successively higher densities of electrons. This would give higher VOLTAGES because of increased PRESSURE.

Keep in mind that electricity is much like air or water. We can think of voltage as pounds per square inch (PSI) and current as cubic feet per minute (CFM). That is PSI is pressure, CFM is flow.

Another analog is comparing a river to electricity. In such a comparison, the speed of the river is the VOLTAGE or pressure while the width of the river is the CURRENT or rate of flow.

Such a comparison shows WHY current requires THE GREATEST SURFACE AREA for the maximum flow. Fuses function on just this principle, when the current flows over the surface of the fuse, it creates heat. If too much current flows, it creates too much heat causing the fuse to melt and separate. The more surface area the fuse, the greater the amount of current can flow, another reason to not place a penny in a fuse socket.

So we have two antennas (1/4" diameter, 3" long rods), two sets of 6 vacuum tubes connected together by wire and assorted resistors. As the waves of energy are collected by the 3" rods, positive on one, negative on another, the energy builds up in the form of increased ions in each of the paralleled vacuum tubes. As in Moray's generator, the circuit will feed whatever load is attached as long as it does not EXCEED the current carrying capacity of the circuit components. What we have is an energy pumping system.

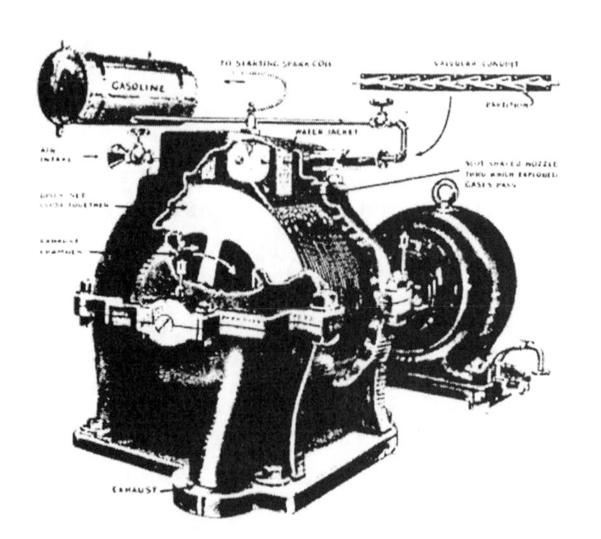

The Tesla Gasoline Turbine - a simple and practical gas engine.

Part Five
The Tesla Papers

THE FREE ENERGY RECIEVER
THE LOST INVENTIONS OF NIKOLA TESLA

Anonymous

For starters, think of this as a solar-electric panel. Tesla's invention is very different, but the closest thing to it in conventional technology is in photovoltaics. One radical difference is that conventional solar-electric panels consist of a substrate coated with crystalline silicon; the latest use amorphous silicon. Conventional solar panels are expensive, and, whatever the coating, they are manufactured by esoteric processes. But Tesla's "solar panel" is just a shiny metal plate with a transparent coating of some insulating material which today could be a spray plastic. Stick one of these antenna-like panels up in the air, the higher the better, and wire it to one side of a capacitor, the other going to a good earth ground. Now the energy from the sun is charging that capacitor. Connect across the capacitor some sort of switching device so that it can be discharged at rhythmic intervals, and you have an electric output. Tesla's patent is telling us that it is that simple to get electric energy. The bigger the area of the insulated plate, the more energy you get.

But this is more than a "solar panel" because it does not necessarily need sunshine to operate. It also produces power at night

Of course, this is impossible according to official science. For this reason, you could not get a patent on such an invention today. Many an inventor has learned this the hard way. Tesla had his problems with the patent examiners, but today's free-energy inventor has it much tougher. At the time of this writing, the U. S. Patent Office is headed by a Reagan appointee who came to the office straight from a top executive position with Phillips Petroleum.

Tesla's free-energy receiver was patented in 1901 as An Apparatus for the Utilization of Radiant Energy. The patent refers to "the sun, as well as other sources of radiant energy, like cosmic rays." That the device works at night is explained in terms of the night-time availability of cosmic rays. Tesla also refers to the ground as "a vast reservoir of negative electricity."

Tesla was fascinated by radiant energy and its free-energy possibilities. He called the Crooke's radiometer (a device which has vanes that spin in a vacuum when exposed to radiant energy) "a beautiful invention." He believed that it would become possible to harness energy directly by "connecting to the very wheelwork of nature." His free-energy receiver is as close as he ever came to such a device in his patented work. But on his 76th birthday at the ritual press conference, Tesla (who was without the financial wherewithal to patent but went on inventing in his head) announced a "cosmic-ray motor." When asked if it was more powerful than the Crooke's radiometer, he answered, "thousands of times more powerful."

How it works

From the electric Potential that exists between the elevated plate (plus) and the ground (minus), energy builds in the capacitor, and, after "a suitable time interval," the accumulated energy will "manifest itself in a powerful discharge" which can do work. The capacitor, says Tesla, should be "of considerable electrostatic capacity," and its dielectric made of "the best quality mica,' for it has to withstand potentials that could rupture a weaker dielectrictric.

Tesla gives various options for the switching device. One is a rotary switch that resembles a Tesia circuit controller. Another is an electrostatic device consisting of two very light, membranous conductors suspended in a vacuum. These sense the energy build-up in the capacitor, one going positive, the other negative, and, at a certain charge level, are attracted, touch, and thus fire the capacitor. Tesla also mentions another switching device consisting of a minute air gap or weak dielectric film which breaks down suddenly when a certain potential is reached.

The above is about all the technical detail you get in the patent. Although I've seen a few cursory references to Tesla's invention in my sampling of the literature of free-energy, I am not aware of any attempts to verify it experimentally.

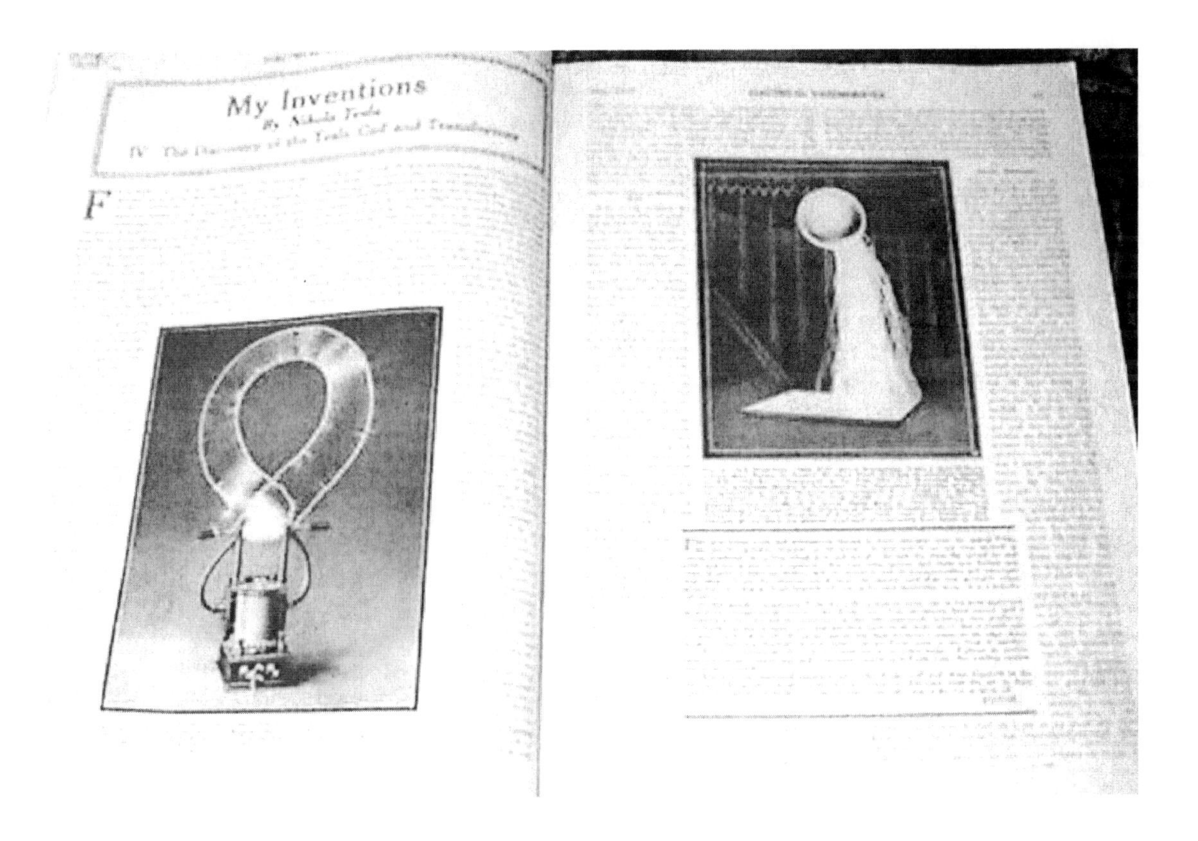
My Inventions

The Tesla Howitzer

(anonymous, from the Internet)

Before the turn of the century, Nikola Tesla had discovered and was utilizing a new type of electric wave. Tesla repeatedly stated his waves were non-Hertzian, and his wireless transmissions did not fall off as the square of the distance.

His discovery was apparently so fundamental (and his intent to provide free energy to all humankind was so clear) that it was responsible for the withdrawal of his financial backing, his deliberate isolation, and the gradual removal of his name from the history books.

By 1914 or so, Tesla had been successfully isolated and was already nearly a "nonperson." Thereafter Tesla lived in nearly total seclusion, occasionally surfacing (at his annual birthday party for members of the press) to announce the discovery of an enormous new source of free energy, the perfection of wireless transmission of energy without losses, fireball weapons to destroy whole armies and thousands of airplanes at hundreds of miles distance, and a weapon (the "Tesla Shield," I've dubbed it) that could provide an impenetrable defense and thus render war obsolete.

In my pursuit of Tesla's secret, it gradually became apparent to me that present orthodox electromagnetic theory is seriously flawed in some fundamental respects. One of these is in the definition and use of THETA, the scalar electrostatic potential. It is this error which has hidden the long-sought Unified Field Theory from the theorists.

In the theory of the scalar electrostatic potential (SEP), the idea is introduced of work accomplished on a charge brought in from a distance against the scalar field.

The SEP is not a vector field, but is a scalar field. Indeed, scalar potential cannot of itself perform work on a charged mass due to the extremely high SEP of the vacuum itself.

Only a differential of SEP between two spatial points can produce force or accomplish work. (Rigorously, a differential of scalar potential between two spatial points constitutes a vector. Only a vector can produce force and do work.)

Also, work can only be done on a mass. Further, it takes TIME to move an electron or other charged mass between two spatial points, and so the work performed by a spatial differential of the THETA-FIELD requires TIME. Rigorously, the delta SEP is voltage, not SEP per se, and is directly related to the voltage or "E" field.

The entire voltage concept depends on the work performed in moving a mass, after that mass has moved. The idea of "voltage" always implies the existence of a steady differential of THETA between two spatial points for a finite length of time, and it also involves the assumption of a flow of actual mass having occurred.

SEP, on the one hand, is always a single-point function; on the other hand, difference in potential (i.e., V) is always a two point function, as is any vector.

Yet many graduate level physics and electromagnetics papers and texts erroneously confuse THETA and V in the static case! Such an interpretation is of course quite incorrect.

Another common assumption in present EM theory—that the electrostatic potential (0,O) of the normal vacuum is zero—has no legitimate basis.

In fact, we know (0,O) is nonzero because the vacuum is filled with enormous amounts of fluctuating virtual state activity, including incredible charge fluctuations. And by virtue of its point definition, (0,O) must be the "instantaneous stress" on spacetime itself, and a measure of the intensity of the virtual state flux though a 4-dimensional spacetime point.

Potential theory was largely developed in the 1800's before the theory of relativity. Time flowrate was then regarded as immutable.

Accordingly, electrostatic "intensity" was chosen as "spatial intensity," with the connotation of "spatial flux density." This assumes a constant, immutable rate of flow of time, which need not be true at all if we believe relativity.

Such a spatial "point" intensity is actually a "line" in 4-space, and not a 4-dimensional "point" at all. Thus the spatial potential—0, 3—is a very special case of the real spacetime potential—0,4, or charge—and electromagnetic theory today is accordingly a special case of the real 4-space electromagnetism that actually exists! Note also that charge is a 4-dimensional concept.

Now mass is a spatial, 3-dimensional concept. Rigorously, mass does not exist in time—masstime exists in time. Mass and charge are thus of differing dimensionalities!

Also, according to quantum mechanics, the charge of a particle—e.g., of an electron—is due to the continual flux of virtual particles given off and absorbed by the observable particle of mass.

Thus charge also is conceptually a measure of the virtual flux density, and directly related to THETA. Further, since the charge exists in time, it is the charge of a particle of spatial mass that gives it the property of masstime, or existing in time.

Here a great confusion and fundamental error has been thrown into the present EM theory by the equating of "charge" and "charged mass." As we have seen, the two things are really very different indeed.

To speak of a spatial "amount" of charge erroneously limits the basic EM theory to a fixed time flowrate condition (which of course it was considered to be, prior to Einstein's development of relativity).

Thus when the limited present theory encounters a "relativistic" case (where the time flowrate changes), all sorts of extraordinary corrections must be introduced.

The real problem, of course, is with the fundamental definitions of electrostatic potential and charge. The spatial "amount" of charge (i.e., the coulomb), as we presently erroneously use the term, is actually the spatial amount of observable "charged mass."

To correct the theory, one must introduce the true 4-space SEP and separate the definitions of charge and charged mass.

Only when a mass is moved does one have work—and voltage or vector fields. (The reason one has voltage and E field connected to a normal electrostatically charged object in the laboratory is because an excess of charged-particle masses are assembled on the object, and these masses are in violent motion! A true static charge would have no E field at all.)

The THETA field need not involve observable mass accumulation, but only charge (virtual flowrate intensity) accumulation.

Accumulated masses are like so many gallons of water; accumulated charge is like so much pressure on both the water (space) and the time in which the water is existing.

Now, if one varies the SEP solely as a point function, one would have a purely scalar complex longitudinal wave, and not a vector wave at all. This is the fundamentally new electrical wave that Tesla discovered in 1899.

Rigorously, all vector fields are two-point functions and thus decomposable into two scalar fields, as Whittaker showed in 1903.

It follows that any vector wave can be decomposed into two scalar waves. By implication, therefore, a normal transverse EM vector wave, e.g., must simply be two coupled scalar (Tesla) waves—and these scalars independently would be longitudinal if uncoupled.

An ordinary transverse EM vector wave is thus two pair-coupled Tesla scalar longitudinal waves, and only a single special case of the much more fundamental electromagnetics discovered by Nikola Tesla.

A Tesla (scalar potential) wave—i.e., a massless wave in pure 0,O, the stress of the

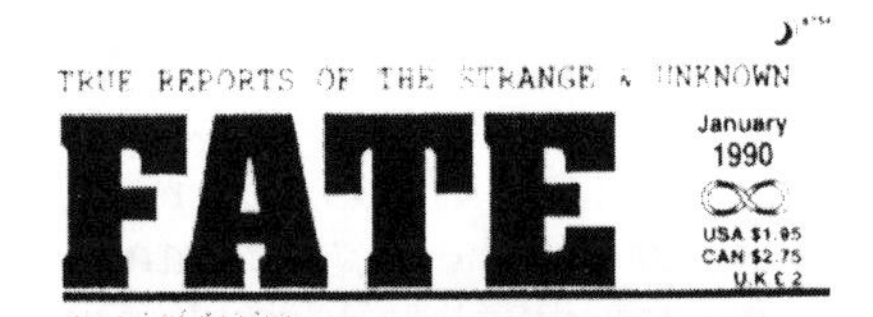

spacetime medium—would have very strange characteristics indeed.

For one thing, since it moves in a complex 4-space, it has many more modes of movement than does a simple wave in 3-space.

And for another thing, it need not be bound at all by the speed of (vector) light. In current theory, one 0,3-field does not directly interact or couple with other existing 0,3-fields except by simple superposition.

Therefore presently the THETA-field is considered to have no drag limitation at all, hence infinite velocity. (E.g., as stated in Jackson's, *Classical Electrodynamics,* 2nd edition, page 223.)

Actually, a 0,4-wave can and will interact with some of the other existing 0,4-waves in the medium transversed, and this interaction can involve pair-coupling into EM vector fields and waves, an interaction not presently in the electrodynamics theory.

The result of scalar pair-coupling creates a finite amount of vector "drag" on the 0,4-wave, so it then has less than infinite velocity.

However, if this drag is small due to limited pair coupling, the scalar wave's velocity through the slightly dragging medium still may be far greater than the speed of vector EM waves (light) in vacuum.

On the other hand, if the pair-coupling is made severe, the THETA-wave may move at a speed considerably below the speed of vector light waves in vacuum.

The velocity of the 0,4-wave is thus both variable and controllable or adjustable (e.g., simply by varying its initial amplitude which through a given medium changes the percentage of pair-coupling and hence the degree of drag on the scalar wave.) The Tesla scalar wave thus can have either subluminal or superluminal velocity, in contradiction to present theory.

Note that the scalar wave also violates one of Einstein's fundamental postulates—for the speed of our "new kind of light" wave is not limited to c, and need not be the same to every observer. Thus Tesla scalar waves lead to a new "super-relativity" of which the present Einstein relativity is only a highly special case!

But let us now look for some subtle but real examples of scalar waves and scalar pair-coupling in nature.

As is well known, a tectonic fault zone can provide anomalous lights, sounds, etc. from stresses, piezoelectrical activity, and telluric currents in the earth and through the fault zone.

In examining the fault zone phenomena, I finally realized that a fault zone was literally a scalar interferometer—i.e., if one can have scalar PHI-waves, they can interfere either constructively or destructively.

Their interference, however, produces scalar pair-coupling into vector EM waves. This coupling may be at a distance from the interferometer itself, and thus the interferometer can produce energy directly at a distance, without vector transmission through the intervening space.

Coupling of THETA waves with the paired scalars comprising ordinary EM vector waves can also occur. If this triplex coupling forms additional EM vector waves 180 degrees out of phase, the ordinary EM wave is diminished or extinguished.

If the scalar triplex coupling occurs so as to create vector EM waves, the amplitude of the ordinary vector wave is increased.

Scalar potential waves can thus augment or diminish, or create or destroy, ordinary EM waves at a distance by pair-coupling interference under appropriate conditions, and this is in consonance with the implications of Whittaker's fundamental 1903 work.

An earthquake fault zone is such a scalar interferometer. Stresses and charge pileups exist in the plates on each side adjacent to the fault, with stress relief existing in the middle in the fault fracture itself.

Since the rock is locally nonlinear, the mechanical stresses and electrical currents in it are also locally nonlinear. This results in the generation of multiple frequencies of THETA-4-waves from each side of the fault interferometer, yielding two complex Fourier expansion patterns of scalar potential waves.

On occasion these two Fourier-transformed scalar wave patterns couple at a distance to produce stable ordinary electromagnetic fields in a 3-dimensional spatial pattern—e.g., a stress light such as the Vestigia light covered in Part I of *The Excalibur Briefing.*

Driven by the erratic to scalar Fourier expansion patterns of the scalar interferometer (whose input stresses normally slowly change), an erratic, darting, hovering "spooklight" of the variety studied by Vestigia produced.

As the stresses change in each side of the interferometer, the distant scalar coupling zone is affected. Thus the stresslight moves and its form changes, but it may be relatively stable in form for seconds or minutes.

Since the stresses in the rock may be intense, the stress light may involve an intense pair or THETA-patterns coupling into the sphere or ball of vector EM energy. The atoms and molecules of the air in the region of the coupled stresslight ball thus become highly excited, giving off radiant energy as the excited states decay.

Since much of the piezoelectric material in the stressed rocks is quartz, the features of quartz are of particular interest. Each little quartz is itself highly stressed, and has stress cracks.

It is therefore a little scalar interferometer.

Further, quartz is transparent to infrared and ultraviolet; and the random orientation of all the quartz scalar interferometers may also form a Prigogine system far from thermodynamic equilibrium.

If so, this system can tap into highly energetic microscopic electromagnetic fluctuations to produce large-scale, ordered, relatively stable patterns of electromagnetic energy at a distance.

In short, all of this lends support to the formation of relatively stable but somewhat erratic patterns of electromagnetic energy at a distance from the fault itself.

In the atmosphere, such scalar interferometers could form in clouds or even in the air or between clouds and earth.

If so, such rare but occasional "weather" scalar interferometers could account for the rare phenomenon of ball lightening. The intense energy of the ball lightening, as compared to the lesser energy of an earth stress light, could well be due to the enormous electrical charges between clouds or between cloud and earth, available to fuel the scalar interferometer. Very probably it is this phenomenon which gave Tesla the clue to

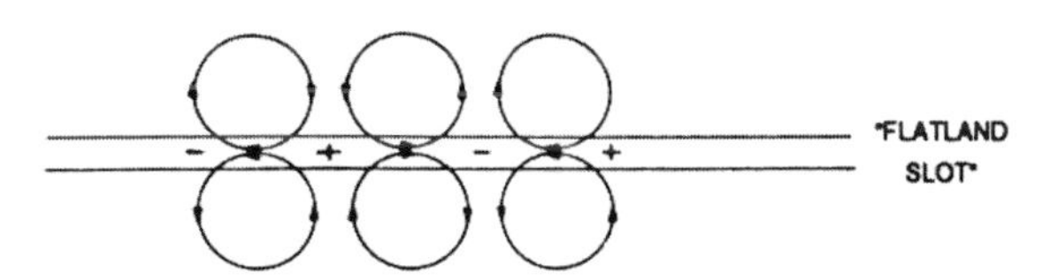

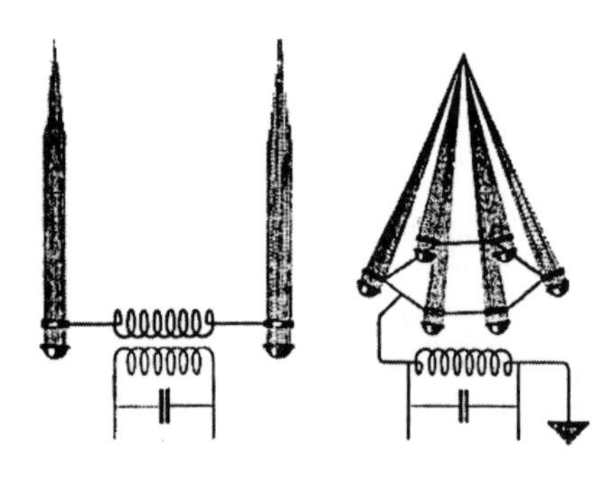

scalar wave interferometry.

Thus such phenomena as earth stress lights, ball lightening, and the Tesla system of wireless transmission of energy at a distance with negligible lasses and at speeds exceeding the speed of light may be explained. They are complex, however, and involve fundamental changes to present electromagnetic theory.

These changes include utilizing 4-space scalar electrostatic potentials, scalar waves, pair coupling, ordinary 3-dimensional Fourier expansion, the Prigogine effect, and the properties of piezoelectric materials in rocks.

Since the scalar potential also stresses time, it can change the rate of flow of time itself. Thus it affects anything which exists in time—including the mind, both of the individual and at various levels of unconsciousness.

Therefore the same functions that result in earth stress lights also affect mind and thought, and are in turn affected by mind and thought. This is the missing ingredient in Persinger's theory that UFO's are correlated with, and a result of, fault zones and earth stresses.

While Persinger seems to feel this is a "normal physics" explanation, it indeed involves a paranormal explanation.

The time-stressing ability of the true THETA scalar wave also explains the interaction of such earth stress lights with humans and human intent, as noted by other researchers. (E.g., the lights that repeatedly seemed to react to the observers, as detailed by Dr. Harley Rutledge in his epoch-making *Project Identification*, Prentice-Hall, 1981.)

These ideas in condensed form comprise the concepts required to violate the speed of light and produce an ordinary electromagnetic field at a distance, using scalar interferometry, without losses—as Tesla had done in his wireless transmission system which he had tested prior to 1900 and had perfected by the 1930's.

Scalar interferometry can give stable regions of EM or "light energy" at a distance without losses, particularly as detailed in the beautiful Vestigia experiments, and it is within our grasp to utilize the new effects.

Indeed, any stress crack in a material can result in the scalar potential interferometer effect. Exophoton and exoelectron emission—poorly understood but already known in fatiguing of materials—must be at least partly due to the scalar interferometer effect.

However, one additional caution should be advanced. Normal movement of electrons allows so much "sideplay" movement of the electrons—and there is so much such sideplay electron motion in the surrounding vicinity—that pair coupling is almost instantaneous for small waves.

Thus orbital electrons in atoms seem to absorb and emit vector EM photons. Actually they also emit some percentage of scalar waves as well.

Since a scalar wave is comprised exclusively of disturbance in the virtual state, it need not obey the conservation of energy law.

Further, a scalar wave of itself does not "push electrons" or other charges; hence it is nearly indetectable by present detectors.

Ionization detectors such as a Geiger counter tube, e.g., are exceptions if the scalar wave encountered is fairly strong.

In that case sufficient triplex coupling with the ionized gas occurs to produce additional ionization or charge, breaching the tube's cutoff threshold and producing a cascade discharge of electrons and voltage which is detected.

But weak scalar waves are presently indetectable by ordinary instruments. However, these small scalar waves are detectable by sensitive interferometry techniques—e.g., such as an electron interferometer. Since the use of such instruments is quite rare, then indeed we have been living immersed in a sea of scalar waves without knowing it.

Finally, the percentage of scalar waves produced by changes in charged mass pileups can be increased by utilizing charged mass streaming. Essentially the charged masses must be moved suddenly, as quickly as possible, at or near the complete breakdown of the medium.

For this reason, Tesla utilized sparkgaps in his early transmission systems, but also found that he could induce ionized media to "breakdown" in such fashion by a slow growth process.

One of his early patented atmospheric wireless transmission systems is based on this fact. However, it was necessary to use a very high voltage, insuring extreme stress on the medium and hence some spillover stress onto time itself.

In other words, THETA-3 is always an approximation; at sufficiently high spatial stress, sufficient spillover THETA-4 exists to give Tesla scalar waves.

For this reason, Tesla used very high voltages and extremely sharp discharges to give "streaming" of the charged masses and thus high percentages of THETA-4 waves. This suggests that the breakdown of dielectrics is a much richer phenomenon than is presently allowed for in the conventional theory.

To summarize, electrostatic potential—THETA field—is stress on the spacetime medium at a four-dimensional point. I.e., it is a sort of pressure on the medium, but pressure on all four dimensions, not just on the three spatial dimensions.

Thus in the new standard theory, THETA-4 may have complex values.

In addition, a PHI-wave is to be interpreted as a scalar longitudinal wave in complex spacetime—directly in THETA-0, the normal average 4-space stress itself. And charge and charged mass must be recognized as two separate concepts.

This is the gist of what I finally recognized about Nikola Tesla's work and fundamental discovery.

This is exciting, for it means that Tesla stress waves can affect either space or time individually, or both space and time simultaneously, or even oscillate back and forth between primarily affecting time and primarily affecting space.

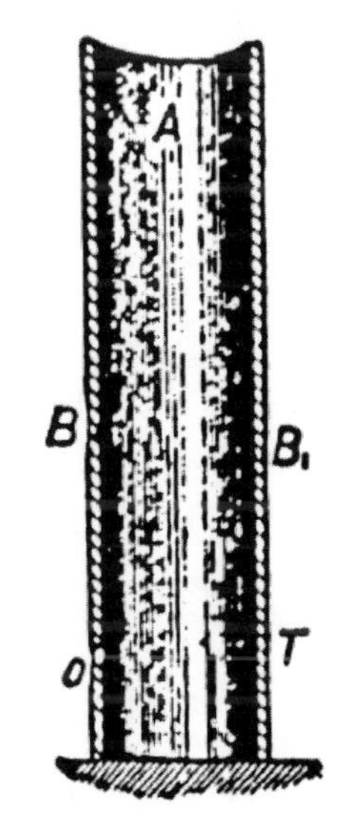

Tesla's waves were actually these THETA-field scalar waves. As such, they were fundamentally different from ordinary electromagnetic waves, and had entirely different characteristics, just as Tesla often stated. E.g., a Tesla wave can either move spatially, with time flowing linearly; move temporally only (sitting at a point and waxing and waning in magnitude—but changing the rate of flow of time itself in doing so, and affecting gravitational field, fundamental constants of nature, etc.), or move in a combination of the two modes.

In the latter case, the Tesla wave moves in space with a very strange motion—it oscillates between

(1) spatially standing still and flexing time, and

(2) moving smoothly in space while time flows smoothly and evenly.

I.e., it stands at one point (or at one columnar region), flexing for a moment; then slowly picks up spatial velocity until it is moving smoothly through space; then slows down again to a "standing column," etc.

This is Tesla's fabulous "standing columnar wave.

Another wild characteristic of the Tesla wave is that it can affect the rate of flow of time itself; hence it can affect or change every other field—including the gravitational field—that exists in time flow.

It can also affect all universal constants, the mass of an object, the inertia of a body, and the mind and thoughts as well!

All of these exist in the flow of time, and they are affected if the time stream in which they exist is affected.

This was the awful secret that Tesla partially discovered by 1900, and which he came more and more to fully realize as he pursued it nature and its ramifications into the 1920's and 1930's.

Tesla also found he could set up standing THETA-field waves through the earth. He in fact intended to do so, for he had also discovered that all charges in the highly stressed earth regions in which such a standing wave existed produced THETA-fields which would feed (kindle) energy into the standing THETA-field wave by pair coupling.

I.e., normal vector field energy would "assemble" onto the scalar matrix wave by means of pair-coupling. Thus by transmitting a scalar wave into the earth, he could easily tap the fiery scalar fields produced in the molten core of the planet itself, turning them into ordinary electromagnetic energy.

In such case, a single generator would enable anyone to put up a simple antenna and

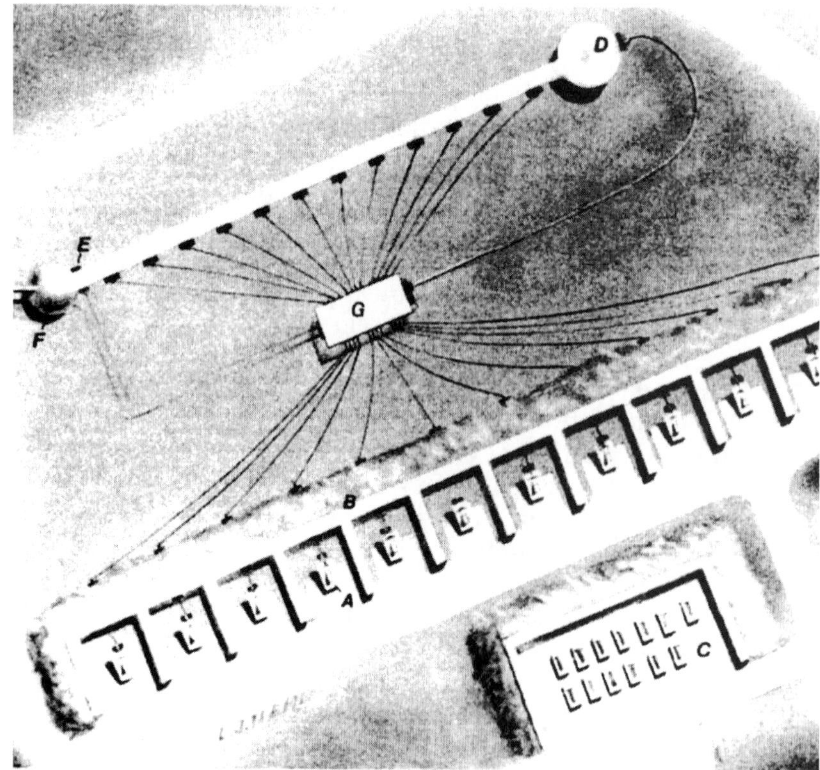

A U.S. satellite reconnaissance photo (enhanced) of the suspected Soviet beam weapon installation based on Tesla technology near Semipalatinsk. Published July 28, 1980 in *Aviation Week & Space Technology.*

extract all the free energy they desired.

When Tesla's alarmed financial backers discovered this was his real intent, they considered him a dangerous madman and found it necessary to ruthlessly stop his at all costs. And so his financial support was withdrawn, he was harassed in his more subtle patent efforts (and the patents themselves were adulterated), and his name gradually was removed from all the electrical textbooks.

By 1914 Tesla, who had been the greatest inventor and scientist in the world, had become essentially a nonperson.

A few other persons in the early 1900's also were aware that potential and voltage are different. And some of them even learned to utilize Tesla's PHI-field, even though they only vaguely understood they were utilizing a fundamentally different kind of electromagnetic wave.

For example, James Harris Rogers patented an undersea and underground communications system which Tesla later confirmed utilized Tesla waves.

The U.S. secretly used the Rogers communications system in World War I to communicate with U. S. submarines underwater, and to communicate through the earth to the American Expeditionary Force Headquarters in Europe.

The Rogers system was declassified after the War—and very shortly after that, it had mysteriously been scrubbed off the face of the earth.

Again, potential stress waves—Tesla waves—were eliminated and "buried." Probably the most brilliant inventor and researcher into Tesla's electromagnetics was T. Henry Moray of Salt Lake City, Utah.

Dr. Moray actually succeeded in tapping the limitless zero-point energy of vacuum (spacetime) itself. By 1939, Dr. Moray's amplifier contained 29 stages and its output stage produced 50 kilowatts of power from vacuum.

Interestingly, another 50 kilowatts could be tapped off any other stage in the device—which consequently could have produced almost 1.5 megawatts of electrical power! Dr. Moray's epoch-making work was suppressed also.

His device—which represented over 20 years of heartbreaking accumulation of 29 working tubes from thousands made—was destroyed by a Soviet agent in 1939, but not before the agent had obtained the drawing for building the tubes and the device itself.

Today the Moray amplifier is a standard component of many of the Soviet secret superweapons and Tesla weapons.

In the 20's and 30's, Tesla announced the final perfection of his wireless transmission of energy without losses—even to interplanetary distances.

In several articles (e.g., H. Winfield Secor, "Tesla Maps Our Electrical Future," *Science and Invention,* Vol. XVII, No. 12, pp. 1077, 1124-1126), Tesla even revealed he used longitudinal stress waves in his wireless power transmission.

Quoting from the article:

"Tesla upholds the startling theory formulated by him long ago, that the radio transmitters as now used, do not emit Hertz waves, as commonly believed, but waves of sound. ...He says that a Hertz wave would only be possible in a solid ether, but he has demonstrated already in 1897 that the ether is a gas, which can only transmit waves of sound; that is such as are propagated by alternate compressions and rarefactions of the medium in which transverse waves are absolutely impossible."

The wily Tesla did not reveal, of course, that such scalar waves nearly always immediately pair-coupled into vector waves when produced by normal means. Tesla himself was working with longitudinal scalar waves.

In the 1930's Tesla announced other bizarre and terrible weapons: a death ray, a weapon to destroy hundreds or even thousands of aircraft at hundreds of miles range, and his ultimate weapon to end all war—the Tesla shield, which nothing could penetrate.

However, by this time no one any longer paid any real attention to the forgotten great genius. Tesla died in 1943 without ever revealing the secret of these great weapons.

Unfortunately, today in 1981 the Soviet Union has long since discovered and weaponized the Tesla scalar wave effects.

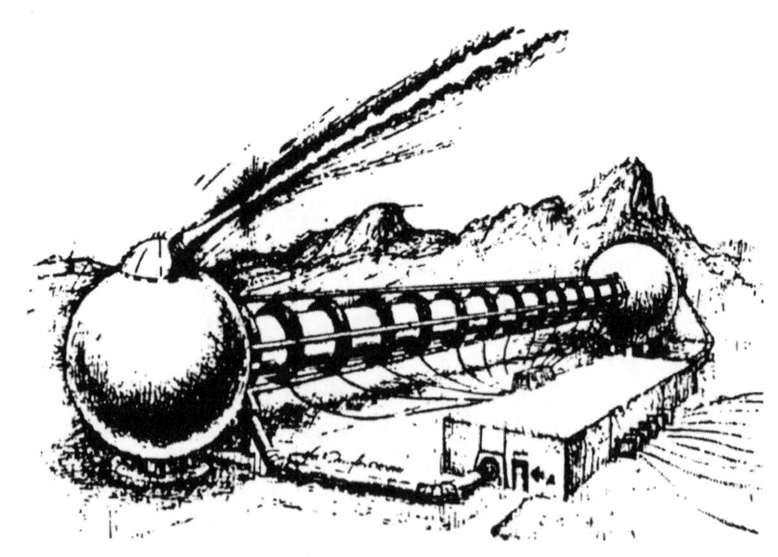

Artist Hal Crawford's drawing of the suspected Soviet beam weapon installation near Semipalatinsk.

Here we only have time to detail the most powerful of these frightening Tesla weapons—which Brezhnev undoubtedly was referring to in 1975 when the Soviet side at the SALT talks suddenly suggested limiting the development of new weapons "more frightening than the mind of man had imagined."

One of these weapons is the Tesla Howitzer recently completed at the Saryshagan missile range and presently considered to be either a high energy laser or a particle beam weapon. (See *Aviation Week & Space Technology,* July 28, 1980, p. 48 for an artist's conception.)

The Saryshagan howitzer actually is a huge Tesla scalar interferometer with four modes of operation.

One continuous mode is the Tesla shield, which places a thin, impenetrable hemispherical shell of energy over a large defended area. The 3-dimensional shell is created by interfering two Fourier-expansion, 3-dimensional scalar hemispherical patterns in space so they pair-couple into a dome-like shell of intense, ordinary electromagnetic energy.

The air molecules and atoms in the shell are totally ionized and thus highly excited, giving off intense, glowing light. Anything physical which hits the shell receives an enormous discharge of electrical energy and is instantly vaporized—it goes pfft! like a bug hitting one of the electrical bug killers now so much in vogue.

If several of these hemispherical shells are concentrically stacked, even the gamma radiation and EMP from a high altitude nuclear explosion above the stack cannot penetrate all the shells due to repetitive absorption and re-radiation, and scattering in the layered plasmas.

In the continuous shield mode, the Tesla interferometer is fed by a bank of Moray free energy generators, so that enormous energy is available in the shield. A diagram of the Saryshagan-type Tesla howitzer can be seen in the drawing.

In the pulse mode, a single intense 3-dimensional scalar Theta-field pulse form is fired, using two truncated Fourier transforms, each involving several frequencies, to provide the proper 3-dimensional shape. This is why two scalar antennas separated by a baseline are required.

After a time delay calculated for the particular target, a second and faster pulse form of the same shape is fired from the interferometer antennas.

The second pulse overtakes the first, catching it over the target zone and pair-coupling with it to instantly form a violent EMP of ordinary vector (Hertzian) electromagnetic energy.

There is thus no vector transmission loss between the howitzer and the burst.

Further, the coupling time is extremely short, and the energy will appear sharply in an "electromagnetic pulse (EMP)" striking similar to the 2-pulsed EMP of a nuclear weapon.

This type is what actually caused the mysterious flashes off the southwest coast of Africa, picked up in 1979 and 1980 by Vela satellites.

The second flash, e.g., was in the infrared only, with no visible spectrum. Nuclear flashes do not do that, and neither does super-lightening, meteorite strikes, meteors, etc.

In addition, one of the scientists at the Arecibo Ionospheric Observatory observed a gravitational wave disturbance—signature of the truncated Fourier pattern and the time-squeezing effect of the Tesla potential wave—traveling toward the vicinity of the explosion.The pulse mode may be fed from either Moray generators or—if the Moray generators have suffered their anomalous "all fail" malfunction—ordinary explosive generators. Thus the Tesla howitzer can always function in the pulse mode, but it will be limited in power if the Moray generators fail.

In the continuous mode, two continuous scalar waves are emitted—one faster than the other—and they pair-couple into vector energy at the region where they approach an in-phase condition.

In this mode, the energy in the distant "ball" or geometric region would appear continuously and be sustained—and this is Tesla's secret of wireless transmission of energy at a distance without losses.

It is also the secret of a "continuous fireball" weapon capable of destroying hundreds of aircraft or missiles at a distance.

The volume of the Tesla fireball can be vastly expanded to yield a globe which will not vaporize physical vehicles but will deliver and EMP to them to dud their electronics.

A test of this mode has already been witnessed, See Gwyne Roberts, "Witness to a Super Weapon?", the *London Sunday Times,* 17 August 1980 for several tests of this mode at Saryshagan, seen from Afghanistan by British TV cameraman and former War Correspondent Nick Downie.

If the Moray generators fail anomalously, then a continuous mode limited in power and range could conceivably be sustained by powering the interferometer from more conventional power sources such as advanced magneto-hydrodynamic generators.

In addition, of course, smaller Tesla howitzer systems for anti-tactical ballistic missile defense of tactical troops and installations could be constituted of more conventional field missile systems using paired or triplet radars, of conventional external appearance, in a scalar interferometer mode.

With Moray generators as power sources and multiply deployed reentry vehicles with scalar antennas and transmitters, ICBM reentry systems now can become long range "blasters" of the target areas, from thousands of kilometers distance.

Literally, "Star Wars" is liberated by the Tesla technology. And in air attack, jammers and ECM aircraft now become "Tesla blasters."

With the Tesla technology, emitters become primary fighting components of stunning power.

The potential peaceful implications of Tesla waves are also enormous. By utilizing the "time squeeze" effect, one can get antigravity, materialization and dematerialization, transmutation, and mind boggling medical benefits.

One can also get subluminal and superluminal communication, see through the earth and through the ocean, etc. The new view of Theta-field also provides a unified field theory, higher orders of reality, and a new super-relativity, but detailing these possibilities must wait for another book.

With two cerebral brain halves, the human being also has a Tesla scalar interferometer between his ears. And since the brain and nervous system processes avalanche discharges, it can produce (and detect) scalar Tesla waves to at least a limited degree.

Thus a human can sometimes produce anomalous spatio-temporal effects at a distance and through time. This provides an exact mechanism for psychokinesis, levitation, psychic healing, telepathy, precognition, postcognition, remote viewing, etc.

It also provides a reason why an individual can detect a "stick" on a radionics or Hieronymus machine (which processes scalar waves), when ordinary detectors detect nothing.

Most ordinary journals will not even accept material on such matters. Nonetheless, the area is of overwhelming importance—and I truly believe Tesla's lost secret will shortly affect the lives of every human being on earth.

Perhaps with the free and open release of Tesla's secret, the scientific and governmental bureaucracies will be shocked awake from their slumber, and we can develop defense before Armageddon occurs.

Perhaps there is hope after all—for even Brezhnev, in his strange July, 1975 proposal to the SALT talks, seemed to reveal a perception that a turning point in wear and weaponry may have been reached, and that human imagination is incapable of dealing with the ability to totally engineer reality itself.

Having tested the weapons, the Soviets must be aware that the ill-provoked oscillation of time flow affects the minds and thoughts—and the very life streams and even the collective species unconsciousness—of all life forms on earth.

They must know that these weapons are two-edged swords, and that the backlash from their use can be far more terrible to the user than was the original effect to his victim.

If we can avoid the Apocalypse, the fantastic secret of Nikola Tesla can be employed to

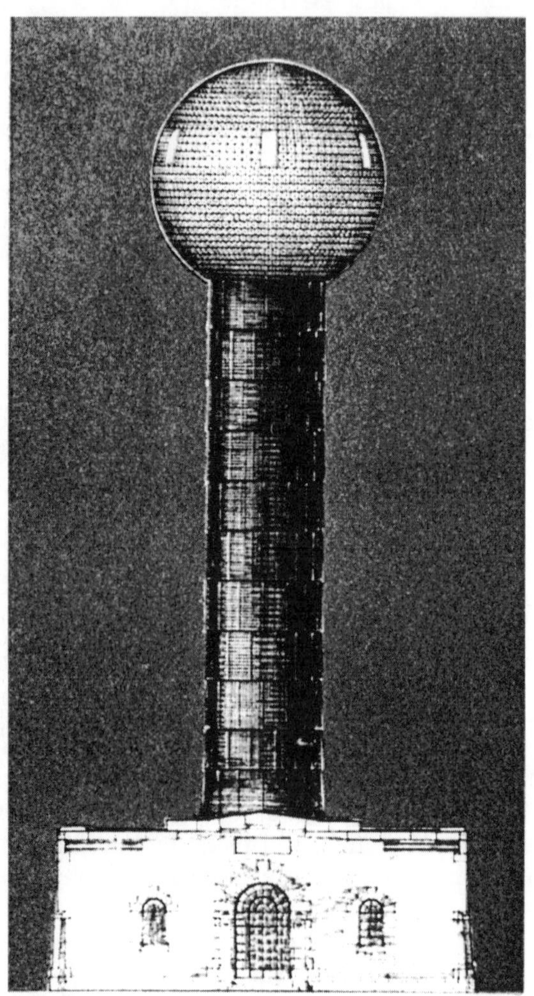

A 1934 architectural drawing showing Tesla's high potential terminal and powerhouse. This illustration was included in Tesla's beam weapon proposal to the U.S. government.

cure and elevate man, not kill him.

Tesla's discovery can eventually remove every conceivable external human limitation. If we humans ourselves can elevate our consciousness to properly utilize the Tesla electromagnetics, then Nikola Tesla—who gave us the electrical twentieth century in the first place—may yet give us a fantastic new future more shining and glorious than all the great scientists and sages have imagined.

Reference Articles for Solutions to Tesla's Secrets

The Electrical Engineer—London Dec. 24, 1909, p. 893:

NIKOLA TESLA'S NEW WIRELESS

Mr. Nikola Tesla has announced that as the result of experiments conducted at Shoreham, Long Island, he has perfected a new system of wireless telegraphy and telephony in which the principles of transmission are the direct opposite of Hertzian wave transmission.

In the latter, he says, the transmission is effected by rays akin to light, which pass through the air and cannot be transmitted through the ground, while in the former the Hertz waves are practically suppressed and the entire energy of the current is transmitted through the ground exactly as though a big wire.

Mr. Tesla adds that in his experiments in Colorado it was shown that a very powerful current developed by the transmitter traversed the entire globe and returned to its origin in an interval of 84 one-thousandths of a second, this journey of 24,000 miles being effected almost without loss of energy.

New York Times, Dec. 8, 1915, p. 8, col. 3:

TESLA'S NEW DEVICE LIKE BOLTS OF THOR

He Seeks to Patent Wireless Engine for Destroying Navies by Pulling a Lever.

To Shatter Armies Also.

"Impractical," He says of Westerner's Plan to Circle Country with Electric Fire.

Nikola Tesla, the inventor, winner of the 1915 Nobel Physics Prize, has filed patent applications on the essential parts of a machine the possibilities of which test a layman's imagination and promise a parallel of Thor's shouting thunderbolts from the sky to punish those who angered the gods.

Dr. Tesla insists there is nothing sensational about it, that it is but the fruition of many years of work and study. He is not yet ready to give the details of the engine which he says will render fruitless any military expedition against a country which possesses it.

Suffice to say that the destructive invention will go through space with a speed of 300 miles a second, and manless airship without propelling engine or wings, sent by electricity to any desired point on the globe on its errand of destruction, if destruction its manipulator wishes to effect.

Ten miles or a thousand miles, it will be all the same to the machine, the inventor says. Straight to the point, on land or on sea, it will be able to go with precision, delivering a blow that will paralyze of kill, as is desired.

A man in a tower on Long Island could shield New York against ships or army by working a lever, if the inventor's anticipations become realizations.

"It is not the time," said Dr. Tesla yesterday, "to go into the details of this thing. It is founded on a principle that means great things in peace, it can be used for great things in war. But I repeat, this is no time to talk of such things.

"It is perfectly practicable to transmit electrical energy without wires and produce destructive effects at a distance. I have already constructed a wireless transmitter which makes this possible, and have described it in my technical publications, among which I may refer to my patent 1,119,732 recently granted.

With transmitters of this kind we are enabled to project electrical energy in any amount to any distance and apply it for innumerable purposes, both in peace and war.

Through the universal adoption of this system, ideal conditions for the maintenance of law and order will be realized, for then the energy necessary to the enforcement of right and justice will be normally productive, yet potential, and in any moment available, for attack and defense.

The power transmitted need not be necessarily destructive, for, if existence is made to depend upon it, its withdrawal or supply will bring about the same results as those now accomplished by force of arms.

Dr. Tesla then said that it would be possible with his wireless mechanism to direct an ordinary aeroplane, manless, to any point over a ship or an army, and to discharge explosives of great strength from the base of operations.

Asked to express an opinion upon the announcement last Sunday of Charles H. Harris, and electrical engineer of Los Angeles, that he would be able to surround this country with an electrical wall of fire in time of war, Dr. Tesla gave it as his opinion that Mr. Harris was not practical.

"It is hard to stamp as impossible such results as those described in the press dispatches to which you refer. Granted, however, that the project is feasible, it would take more than all the motive power obtainable in the United States to throw a wall of fire around the country.

In fact, even the passage of small currents at considerable distances through air consumes a great deal of energy on account of the immense pressure required.

So, for instance, in lightening discharges, energy may be delivered at the rated of billions of horsepower, though the currents are of smaller volume than those developed by electrical generators in our power houses."

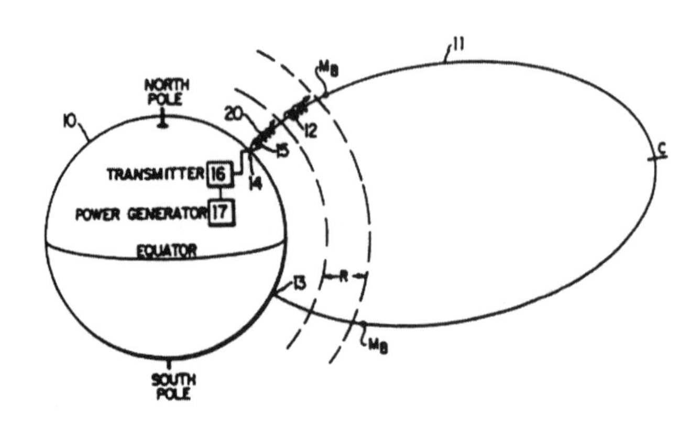

Bernard Eastlund's 1987 patent (number 4,686,605) depicting a method for altering a selected region of the earth's surface—a modified Tesla weapon?

NIKOLA TESLA: THE HISTORY OF LASERS AND PARTICLE BEAM WEAPONS

by Marc J. Seifer, Ph.D.
Box 32, Kingson, RI 02881

Figure 1. A simulated engraving of the inventor sending 250,000 volts through his body. From The World, 1894. ***CREDIT: Metascience Foundation.***

Electric Sorcerer of the Gilded Age

Tesla was thrust into prominence almost over night after a series of spectacular lectures which he presented around the world in the early 1890s. Not only was Tesla a leader in his specialized profession, writing down a bible of information for the generation of electricians to follow, he was also a showman, whose futuristic presentations, if performed today, almost a century later, would still astonish an audience.

The Tesla lectures which were given in America, England, France and Germany, were motivated to a great extent by the wish to counteract Tom Edison's attack on Tesla's invention of the alternating current polyphase system recently purchased by the Westinghouse Corporation of Pittsburgh. At the same time, Tesla displayed lights without filaments (i.e., fluorescent and neon lights), the principles to wireless communication, the means by which to create tuned circuits (patents no. 462,418 filed 2/4/1891), and the idea behind the radio tube. Where Edison had electrocuted dogs and cows with the Tesla AC currents in a negative propaganda campaign to demonstrate the viability of using the current in the electric chair, Tesla conceived of a way to send electricity through his own body to show that AC, when utilized correctly, was perfectly safe, and further, that electricity could even promote healing. This last invention quickly influenced the French medical doctor D'Arson-val, the father of electrotherapy [30, p. 286].

Referring to the Tesla lectures, Harpers Weekly, concluded, "*At one bound [Tesla] placed himself abreast ... such men as Edison, Brush, Elihu Thomson and Alexander Graham Bell*" [54, p. 524]. After describing the work of his predecessors and colleagues in the field of luminescence of vacuum tubes and Leyden jars by electromagnetic induction (i.e. William Crookes, Oliver Lodge, JJ Thomson, Elihu Thomson), Tesla proceeded to explain the real reason why commercial light bulbs gave off light. In essence, he stated that the air or lack of air had more to do with the phenomena than the filament. In a veiled attack against Edison's invention of the light bulb, Tesla showed that the action of the air was more important than the seemingly essential filament.

> *Two incandescent lamps exactly alike, one exhausted, the other not... were attached... [to] a current vibrating about one million times a second.... [These] passed through the filament.... The lamp which was exhausted glowed brightly, whereas the other one in which the filament was surrounded by air, at ordinary pressure, did not glow. This showed the great importance of the rarefied gas in the heating of a conductor, and it was pointed out that in incandescent lighting a high resistance filament [Edison's invention] does not at all constitute the really essential element of illumination.* [43, p. 249]

Tesla showed that the filament was superfluous for electric lights and thus he out 'Edisoned' Edison. However, the world (and also the lighting industry) would forget this fairly quickly while resurrecting time and again the "Edison sending men to the Amazon to look for the best filament" story.

In response, Edison said

> *[Tesla] has made no new discovery, but has shown considerable ingenuity in increasing vibrations. He gets his results from the induction coil and the Geissler tube. One great trouble is in the quality of the light if it is produced by that means. It is of a ghastly color... You cannot get a pleasant mellow light.* [12]

This last comment is as true today as then in terms of the inferiority of the quality of fluorescent lighting as compared to incandescent lamps. However, Tesla appears to have invented additional techniques to offset this common complaint, his laboratory having been described as "*flow[ing] with warmth of color.... Dazzling, pulsating clots of purple-violet light... sent out wave after wave of a strange, unearthly rich... hue that is not listed in the spectrum*" [20].

As a finale to the Tesla lectures, and as the coup de grace to Edison and the electric chair enthusiasts, Tesla sent hundreds of thousands of volts through his body to light various colored lamps, melt metal strips, and explode small globes. He also issued Svengali-like streams of lighting from his fingertips! Describing the experience, of subjecting his body to "the rapidly alternating pressure of an electrical oscillator of two and one half million volts," Tesla wrote:

[This] sight [is] marvellous and unforgettable. One sees the experimenter standing on a big sheet of fierce, blinding flame, the whole body enveloped in a mass of phosphorescent streamers, like the tentacles of an octopus. Bundles of light stick out from his spine. As he stretches out the arms, thus forcing the electrical fluid outwardly, roaring tongues of fire leap from his fingertips. Objects in his vicinity bristle with rays.... At each throb of the electric force, myriads of minute projectiles are shot off from him with such velocities as to pass through the adjoining walls. He is in turn being violently bombarded by the surrounding air and dust [31, pp. 94-95].

Tesla appeared as a real life sorcerer depicting once and for all the ability to tame mysterious natural phenomena to the will of the human species.

Before the general public he stands as a phenomenal inventor from the Eastern World, from whom is expected little less than if he carried Aladdin's lamp in his hand, which of course, is wrong, and an injustice to both the public and to Mr. Tesla. [27, p. 499]

Tesla's work spawned a new field of inexpensive lighting. Most office buildings with illuminated panels of fluorescent lights can trace their inception to Tesla.

Figure 2. From "The Diabolical Ray," by Hugo Gernsback, Practical Electronics, August, 1924. ***CREDIT: Metascience Foundation.***

Dematerialization Devices and Laser Beams

The word laser stands for Light Amplification by Stimulated Emission of Radiation. It was first developed in 1945 by Charles Townes. About ten years later, Townes, then at MIT and A. Prokhorov and N. Basov of the Soviet Union, further refined the device and all three received the Nobel Prize for the work in 1964. Hunt and Draper [22] have suggested that Tesla actually was the first inventor of the laser beam, which he called in the 1930s the "death ray." In the following passages a description of how the laser is constructed will be compared to various Tesla devices such as his button lamp which was developed in 1891, other lamps and oscillators, his X-ray tube and wireless apparatus.

Kock [23, pp. 28-35] describes two types of lasers:

1. A ruby laser which reflects energy back to its source, which in turn stimulates more atoms into emitting special radiation.
2. A gas laser, which consists of a tube filled with helium and neon. High voltage is applied across two electrodes near the ends of the tube causing a discharge to take place. In both instances the excited atoms are contained in an enclosure and then reflected into one specific direction. They differ from lights in two other ways:
   a. The ruby or the gas mixture must provide for a metastable (pausing) state in which the excited atoms can temporarily reside.
   b. Light must be emitted by exactly the same wavelength.

Tesla worked with lamps constructed in exactly these ways. The first he called a button lamp, and the second, an exhausted or phosphorescent tube. Their prime function was as efficient illumination devices. Their secondary function was as laboratory apparatus for a variety of experiments such as beam transmission and dematerializing matter. The following description was of a type of fluorescent light:

> *A small tube about 1 and 1/2 inches in diameter and 12 inches long has one of its ends drawn out into a fine fiber nearly 3 feet long.... The discharge passing through the tube first illuminates the bottom of same... but gradually the rarefied gas inside becomes warm and more conducting and the discharge spreads into the glass fibre. This spreading is... slow.... However, once the glass fibre is heated, the discharge breaks through its entire length instantaneously.* [43, p. 603]

The following passage [43, pp. 252-265] describes a device which could "vaporize" (dematerialize) any material including zirconia and diamonds. O'Neill [24] stated that the carbon-button lamp had a variety of uses, one being simply "just a lamp" (p. 146). Where Edison invented an "incandescent filament lamp... Tesla invented an absolutely original type of lamp... which gives twenty times as much light for the same amount of current consumed" (p. 147). Tesla began his discussion with a description of a "button lamp," which, in essence, was a globe coated with a reflective material (like the Leyden jar) and a "button" of any substance, most often carbon, attached to a source of power. Once electrified, the button would radiate energy which would bounce off of the interior of the globe and back onto itself thereby intensifying a "bombardment" effect. In this way the button would be "vaporized."

According to the author's belief, Tesla utilized this same basic principle to create ruby laser effects in 1894, although it is doubtful that Tesla completely understood the importance of the beam at the time, or its wide range of practical applications. The description of this invention is quite clear:

> *In an exhausted bulb we can concentrate any amount of energy upon a minute button... [of] zirconia... [which] glowed with a most intense light, and the stream of particles projected out... was of a vivid white....*
>
> *Magnificent light effects were noted, of which it would be difficult to give an adequate idea.... To illustrate the effect with a ruby drop..., at first one may see a narrow funnel of white light projected against the top of the globe where it produces an irregularly outlined phosphorescent patch.... In this manner an intensely phosphorescent, SHARPLY DEFINED LINE [emphasis added], corresponding to the outline of the drop [fused ruby] is produced, which spreads slowly over the globe as the drop gets larger....* [43, pp. 250-251]

Joan Bromberg [6], director of The Laser History Project, a book in progress celebrating 25 years since the discovery of the laser and supported by an advisory committee which includes Charles Townes, inventor of the laser, stated that she did not think that Tesla invented "a true laser." "It is not a matter of 'reflecting light back to its source,' but of creating an 'inverted population,' where there are a greater number of paramagnetic ions in higher than lower states.... There [seems to be] some confusion between massy particles and electromagnetic beams," she said. As an historian and not an engineer, Bromberg suggested I refer to a technical expert.

Terrence Feeley [14], owner of Laser Fare, an industrial laser company, stated that he thought Tesla did achieve a laser effect, but that it was unlikely that he, in 1894, understood the mechanism of population inversion. This state, he said, is highly unnatural and occurs when new energy is forced into the medium of the ruby. When this occurs, the laser effect takes place. To the direct question, do you think Tesla invented a laser, Feeley replied, "I think he achieved it."

Tesla also discussed the concept of "phase," or the obtaining of a single wavelength oscillating in such a way as to concentrate and amplify the effect (i.e. beam--although he did not refer to a beam for a few decades--still well before the 1940s).

> *The problem of producing light has been likened to that of maintaining a certain high pitch note by means of a bell.... We may deliver powerful blows at long intervals, waste a good deal of energy and still not get what we want; or we may keep up the note by delivering frequent taps, and get nearer to the object sought by the expenditure of much less energy.* [43, p. 252]

A number of years later, Tesla explains in a clearer fashion the extent of this research along these lines. It appears that he also conceived of a process quite similar to what today has evolved into digitalization transmissions via fiber optics, although Tesla is here referring to electromagnetic wavelengths:

> *The electromechanical process of producing isosynchronous oscillations is one of my earliest inventions. I have applied it in many ways with great success.... In my apparatus the isochrorism [sic] is so perfect and attunement sharp to such a degree that in the transmission of speech, pictures and similar operations, the frequency or wavelength is varied only a minute range which need not be more than 100th of a percent.... Any reasonable number of simultaneous and non-interfering messages is practicable and a speed of many thousands of words per minute can be attained in telegraphic messages.* [45, p.84]

THE RAY IN OPERATION EXPLODING GUNPOWDER.

Figure 3. Harry Grindell-Mathews. Drawing from "Diabolical Rays," <u>Popular Radio</u>, November, 1915. ***CREDIT: Metascience Foundation.***

The Diabolical Ray

In the mid 1890s, Tesla invented a peculiar single electrode X-ray tube. Using a principle which he also adapted in Colorado Springs whereby the outer primary never directly connects to the inner secondary, the Tesla X-ray tube was able to create hi-field emissions whenever the circuit was closed and the stored energy discharged. Developed from single terminal light bulbs, by 1897, Tesla was remotely discharging isolated capacitors at considerable distances and directing them to targets [9].

In articles beginning twenty years later in the New York Times and continuing for a period which spanned both world wars [30, 46, 47, 48], Tesla referred to various kinds of gross electrical manifestations which could be used in a variety of ways to protect a city or country from incoming invasions. His weaponry included defensive radar-like protective grids of light as well as offensive laser-like artillery.

TESLA'S NEW DEVICE LIKE BOLTS OF THOR
He Seeks to Patent Wireless Engine
For Destroying Navies by Pulling a Lever
-- To Shatter Armies Also --

"Impractical," he says of Westerner's plan to circle country with electrical fire ....

Nikola Tesla, the inventor, winner of the 1915 Nobel Prize, has filed patent applications on the essential parts of a machine the possibilities of which test a layman's imagination and promise a parallel of Thor's shooting thunderbolts from the sky to punish those who have angered the gods. Dr. Tesla insists there is nothing sensational about it but the fruition of many years of work and study.

Suffice it to say that the destructive invention will go through space with a speed of 300 miles/second, a manless airship without propelling engines or wings, sent by electricity to any desired point of the globe on its errand of destruction, if destruction its manipulator wishes to effect. Ten miles or a thousand miles, it will be all the same to the machine, the inventor says....

"It is perfectly practical to transmit electrical energy without wires and produce destructive effects at a distance... with precision." A man in a tower on Long Island could shield New York against ships or army by working a lever, if the inventor's anticipations become realized. [44]

At this same time, during World War I, while Tesla's partner John Hays Hammond Jr. was displaying remote controlled torpedo inventions to the U.S. Navy, at his coastal laboratory in Gloucester, Massachusetts, [18], in Great Britain, another Teslarian, Harry Grindell-Mathews [16], was provided with £25,000 by the British government for the creation of a search-light like beam which he said could control air craft. Grindell-Mathews, who had been wounded in the South African Wars while serving for the British Army a number of years earlier, eventually refined this invention and changed it into a "diabolical ray" which he said could destroy not only Zeppelins and aeroplanes, but immobilize marching armies and nautical fleets. Although he would not divulge the specifics of his creation, he made no secret of his admiration for Tesla, whose technologies had "inspired" the groundwork for this weapon. In July of 1924 he travelled to America to see an eye specialist. He probably met with electronics and science fiction publisher Hugo Gernsback [15] at this time, and thus may have also visited Tesla [17]. Staying at the Hotel Vanderbilt in New York, the British inventor was interviewed by a number of the local dailies:

Let me recall to you the air attacks on London during the [world] war. Searchlights picked up the German raiders and illuminated them while guns fired, hitting some but more often missing them. But suppose instead of a searchlight you direct my ray? So soon as it touches the plane this bursts into flame and crashes to the earth. [16]

Grindell-Mathews was also convinced that the Germans had such a ray. They were using a high frequency current of 200 kilowatts which as of yet they were "unable to control."

Working with the French government in Lyons, and performing successful tests before members of the British War Office, Grindell-Mathews instituted destructive effects at distances of over 60 feet, but was hoping to extend the force to a radius of about six or seven miles. Pried for specifics, he said that his device utilized two beams, one as a carrier ray, and the other as the "destructive current." The first beam would constitute a low frequency and would be projected through a lens; the second, of a higher frequency, would increase conductivity so that destructive power would be more easily transmitted. The motor of an aeroplane, for instance, could be the "contact point" where the paralyzing ray would do its handiwork. He admitted, however, that if the object were grounded, it would be protected against such a force [18].

Gernsback, who had published one of Grindell-Mathews articles, tried unsuccessfully to duplicate the effects using heat beams, X-rays and UV-rays. Along with Dr. W. Severinghouse, a physicist from Columbia University, he doubted the claims of the British engineer [15]. Nevertheless, leaders from other countries were concerned and many proclaimed that their scientists also had such diabolical rays. Herr Wulle, a member of the German Reichstag announced that "three German scientists have perfected apparatus that can bring down airplanes, halt tanks and spread a curtain of death like the gas clouds of the recent war" [17]. Not to be outdone, Leon Trotsky stated that the Soviets had also invented such a device. Warning all nations, Trotsky declared, "I know the potency of Grammachikoff's ray, so let Russia alone" (paraphrased).

This theme of all powerful efficient weaponry reappeared during the 1930s as the seeds of World War II began to be sown. The following passage is typical of a number of articles which graced the pages of many newspapers during this time.

TESLA, AT 78, BARES NEW 'DEATH-BEAM'

Invention Powerful Enough to Destroy 10,000 Planes 250 Miles Away, He Asserts.

DEFENSIVE WEAPON ONLY

Scientist, in Interview, Tells of Apparatus That He Says Will Kill Without Trace.

Nikola Tesla, father of modern methods of generation and distribution of electrical energy, who was 78 years old yesterday, announced a new invention, or inventions, which he said, he considered the most important of the 700 made by him so far.

He has perfected a method and apparatus, Dr. Tesla said yesterday in an interview at the Hotel New Yorker, which will send concentrated beams of particles through the free air, of such tremendous energy that they will bring down a fleet of 10,000 enemy airplanes at a distance of 250 miles from a defending nation's border and will cause armies of millions to drop dead in their tracks.

Times Wide World Photo.
NOTED INVENTOR 78.
Nikola Tesla.

Figure 4. From New York Times, July 11, 1934, 18:1,2. *CREDIT: Metascience Foundation.*

TESLA AT 78 BARES NEW "DEATH BEAM"
Invention Powerful Enough to Destroy
10,000 Planes at 250 Miles Away
He Asserts Defensive Weapon Only

> *He has perfected a method and apparatus, Dr. Tesla said yesterday in an interview at the Hotel New Yorker, which will send concentrated beams of particles through the free air, of such tremendous energy that they will bring down a fleet of 10,000 enemy airplanes from a defending nation's border and will cause armies to drop in their tracks.* [46]

In a 1937 paper which was only recently published, and which appears to be one of the so-called secret weaponry papers, Tesla [47] explained in vivid detail the schematics and mathematics of a particle beam weapon. It comprised four basic features:

1. An apparatus for manifesting electrical energy.
2. An apparatus for generating tremendous electrical potentials.
3. A method of amplifying the force developed.
4. A method for producing a tremendous repelling force. This would be the projector of gun of the invention.

In essence, it was a combination of a high powered electrostatic generator and an open-ended vacuum tube capable of producing a stream of highly charged "particles." These electrified bodies appear to be microscopic pellets of tungsten or mercury which were to be "shot" (actually repelled) at velocities ranging from 400 to 160,000 meters per second, depending on their size and the amount of electrical power generated. In order to eliminate the problem of dispersion, which had been caused when only rays of energy were transmitted, Tesla conceived of the ingenious idea of creating a "single row of minute bodies" (pp. 144-150).

> *Such a particle, notwithstanding its minute volume of 1/250,000 cubic centimeter, would be very destructive... produc[ing] intense heating effects.... It would pierce the usual protecting covering of aeroplanes, put machinery out of commission and ignite fuel and explosives.* (p. 147)

Tesla attempted to contact world leaders including Neville Chamberlain of England [26], Franklin Delano Roosevelt of the United States and perhaps even Joseph Stalin of the Soviet Union [39]. Although information concerning Tesla's connection to FDR is scanty, it is known that the president was familiar with the inventor from his tenure as Assistant Secretary to the Navy, during World War I. Further, Tesla had conversed with the president's wife, Eleanore on a number of occasions.

The FBI and the Tesla Papers

There were three major reasons why the federal government was interested in protecting the Tesla papers: (1) Tesla's announcements of his creation of secret weaponry, e.g. the "death ray;" (2) the advent of World War II and the quite understandable suspicion associated with the event; (3) the inventor's heritage, and thereby his connection to a state (Yugoslavia) which, although occupied by the Nazis at the time of Tesla's death, was in the process of adopting a communistic form of government.

Tesla was raised to the level of a national hero within the Slavic countries. Therefore, anyone in a blood relationship with the grand wizard could be considered practically of royal stock. For this and other reasons, Tesla's nephew, Sava Kosanovich, rose to become representative of the newly forming nation to the "Eastern European Planning Board" which met in Czechoslovakia [21]. Kosanovich, as Tesla, wanted a unified country, but their orientations were different.

In 1941, the Nazi Germans continued their policy of intimidation and deception by trying to force a treaty with King Peter of Serbia. Backed by popular support, Peter refused to agree to an alliance and thereby suffered a fatal blow as Germany masterminded a brutal invasion involving troops from Bulgaria, Italy and Hungary and also 300 Luftwaffe bombers [7, p. 258].

Professor Michael Markovitch of Long Island University stated that as a Serb living in Croatia during World War II, 90,000 Serbs were killed by the fascists, many of whom were of Croatian extraction [24]. As a young man, Markovitch watched the bodies "float down the river," his own survival, he attributing to sheer "luck." Concerning the Tesla mythology, Markovitch said that as a boy growing up in what became Yugoslavia, his fellow Serb was considered a great hero. Years later, as Hitler's invasion became imminent, Markovitch and his countrymen had expected that the wizard would return to the land and shield it from the Nazis by harnessing his death ray! Unfortunately, Tesla never came.

Kosanovich was not as romantic a figure, and although a Serb, he abandoned the exiled Serbian king in order to back the rising Croat leader, Joseph Tito (Josip Broz), and his Communist doctrine. Tito was a solid choice; although he was an ally of the Soviets, he was able to maintain autonomy. Further, he sought to unify the warring factions, his marriage to a Serbian woman being a powerful symbol for this goal.

The U.S.S.R. had become an ally of the United States in order to destroy the Axis nations. Thus, Kasonovich, as a Yugoslavian ambassador, was freely able to shuttle between America and Europe and discuss various diplomatic tactics with the new leadership. During the course of World War II, he was also able to attend to his ailing uncle in New York as he tried to finalize plans to set up a museum in Belgrade in order to honor the great inventor.

In 1942, Tesla became very ill and suffered from palpitations and fainting spells. On January 8, 1943, Tesla passed away. Upon his death, Sava Kasonavich, along with Kenneth Swezey, a writer and friend for many years, and George Clark, director of a museum and laboratory at RCA, entered the apartment. While Hugo Gernsback rushed to make a death mask, a locksmith was called in to enter the safe. With the hotel management present, various documents were removed [8]. Although the FBI alleged that "valuable papers, electrical formulas, designs, etc., were taken," other officials who were present (i.e. the hotel management) confirmed that Kosanovich removed only three pictures, and Swezey took the 1931 testimonial autograph book created from the occasion that he himself had organized to commemorate Tesla's 75th birthday.

These events were monitored by J. Edgar Hoover [21], a hardline anticommunist who considered himself to be a protector of American interests. Hoover wrote in a lengthy (and partially censored) memorandum under the heading "Espionage," that he feared that Kasonovich, as heir to the Tesla estate, "might make certain materials available to the enemy" (especially the Axis Powers). Hoover had identified Kosanovich as a member of the Eastern European Planning Board. However, due to the complicated condition in the Balkan states, there was really no way for Hoover to ascertain where Kosanovich's alliances rested. He could have been associated with King Peter of Serbia, the Communist Tito, fascist factions associated with Mussolini, Hitler, the Soviet Union or none of the above.

On January 8th, a Mr. Abraham N. Spanel, president of National Latex Corporation of Dover, Delaware, who was residing in New York City at the time, had called FBI agent Fredrich Cornels to discuss Tesla's secret experiments, especially the death ray, as the inventor had just died and Spanel feared that Kosanovich would obtain the papers and pass them to the enemy [8]. Spanel also contacted Dr. D. Lozado, an advisor to Vice President Henry Wallace, and a Mr. Bopkin of the Department of Justice who said he would personally call J. Edgar Hoover regarding the affair.

Spanel also contacted one Bloyce Fitzgerald, "an electrical engineer who was a protegé of Tesla's" who had also called Cornels. Having met Fitzgerald at an engineering meeting a few years earlier, Spanel became highly interested in the Tesla weapon and probably hoped to become involved in a profitable business developing and selling it to the military.

Fitzgerald, a young man still in his twenties, had been in postal communication with Tesla since the mid 1930s. But it had only been within the last year or two that he had been able to meet the inventor and borrow the various papers in which he was interested. Fitzgerald, who had worked at the Massachusetts Institute of Technology and also for the Ordinance Department for the U.S. Army, told FBI agent Cornels that he:

> *knows that the complete plans, specifications and explanations of the basic theories of these things are some place in the personal effects of Tesla... [and] that there is a working model [of the death ray]... which cost more than $10,000 to build in a safety deposit box of Tesla's at the Governor Clinton Hotel.* [8]

Fitzgerald also reported that Tesla had claimed that he had 80 trunks in different locations in the city containing inventions, manuscripts and plans of his various work.

To corroborate this story, another acquaintance of Tesla's, Charles Hausler, a hired hand who took care of the inventor's pigeons, wrote to Leland Anderson that Tesla

> *had a large box or container in his room near the pigeon cages, and he told me to be very careful not to disturb the box as it contained something that could destroy an airplane in the sky and he had hopes for presenting it to the world.* [7, p. 264]

Hausler also added that the device was later stored in the basement of a hotel.

Due, in part to problems of jurisdiction, and the connection with the foreign ambassador Kosanovich, the FBI transferred authority to the Office of Alien Property (OAP) on January 14, 1943. Now the situation could be "handled as an enemy custodian matter," Agent Donegan told Hoover. The FBI would take "no further action" [11].

Although this appears to be exactly what happened, due to the initial involvement by the FBI and their high visibility, numerous people continued to contact them through the years in attempts to gain access to the Tesla papers. The OAP would question the legality of their own jurisdiction, as Tesla was a naturalized citizen; however, as Kosanovich was probably legally entitled to his uncle's estate, and his American cousin, Nicholas Trbojevic had staked no claim, the OAP had some justification in considering the material as alien property. That probably was their rationale, as one way or another, they maintained authority over the papers for a full decade.

Walter Gorsuch, Alien Property Custodian, thereupon ordered all of Tesla's trunks in the basement of the Hotel New Yorker, the papers in his apartment and the device at the Hotel Governor Clinton to be sent to the Manhattan Warehouse (on 52nd St.) where other containers of Tesla's work were stored. It appears that the FBI also confiscated papers held by O'Neill and transferred them to the warehouse as well. Kosanovich took care of the $15 per month rent for the storage of the Tesla estate throughout this entire period.

Prompted by the highest levels of government, the National Defense Research Committee (NDRC) and the OAP commissioned John O. Trump, an MIT engineer, to conduct an investigation of the contents of the 80 trunks. He was accompanied by military personnel from the U.S. Marine Reserves and the Office of Naval Intelligence. [49]

The Trump papers, which included a synopsis of about one dozen articles by or about Tesla, began with an opening letter. Trump acknowledged that he and his colleagues investigated the Tesla trunks at Manhattan Storage on January 26 and 27, summarizing first that (1) "No investigation of the Tesla trunks held for 10 years in the basement of the Hotel New Yorker was conducted;" (2) "No scientific notes, descriptions of hitherto unrevealed methods of devices or actual apparatus... of scientific value to this country or which would constitute a hazard in unfriendly hands [was found].... I can therefore see no technical or military reason why further custody of the property should be retained." Nevertheless, Trump "removed... a file of various written materials which covers typically and fairly completely the ideas which he [Tesla] was concerned [with] during his later years" and forwarded copies to the OAP.

Trump concluded that the last 15 years of Tesla's life were "primarily of a speculative, philosophical and somewhat promotional character." With the permission of the OAP, Trump mailed off his report, including complete copies of Tesla's secret weaponry papers to Col. Holliday of Wright-Patterson Air Force Base, where they were studied by his engineers including Bloyce Fitzgerald, who was now stationed there.

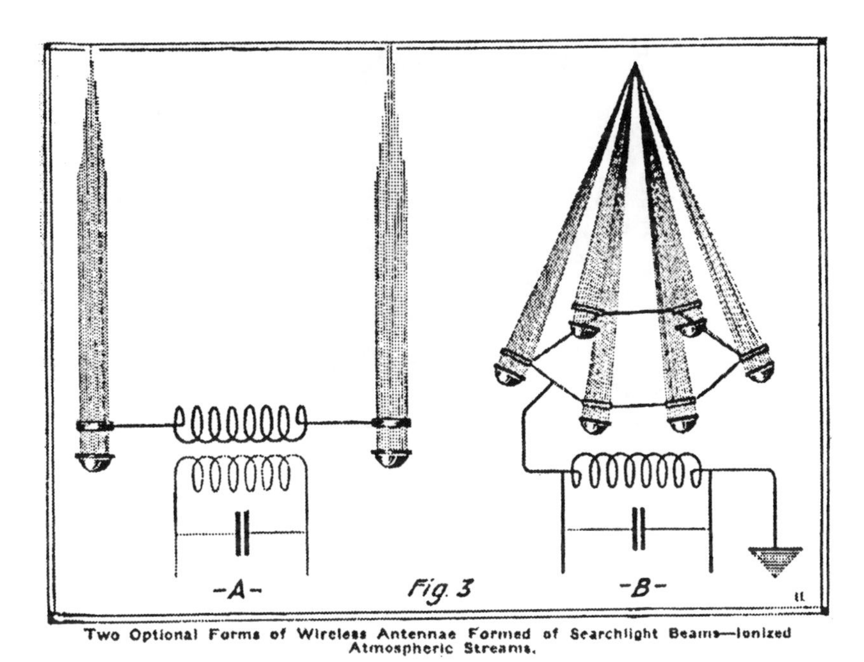

Figure 5. Directed ionized beam transmissions from "Wireless Transmission of Power Now Possible," <u>Electrical Experimenter</u>, March, 1920. *CREDIT: Metascience Foundation.*

Did Tesla Actually Build a Prototype of a Death Ray?

In lieu of money owed on rent, in the early 1930s, Tesla gave the management of the Governor Clinton Hotel a supposed invention of his to be used as collateral. He said that the device was very dangerous and worth $10,000. In 1943, Trump went to the hotel to retrieve it. He was told that the invention could "detonate if opened by an unauthorized person" [7, p. 276]. Trump stated that before he opened the container, he reflected upon his life, as he summoned his courage.

> *Inside was handsome wooden chest bound with brass... [containing] a multidecade resistance box of the type used for a Wheatstone bridge resistance measurements--a common standard item found in every electric laboratory before the turn of the century!* [7, p. 276]

Nikola Tesla, electrical wizard, foresees the day when airplanes will be operated by radio-transmitted power supplied by ground stations, as shown in the drawing above.

Figure 6. An updated design of the magnifying transmitter which may also have had defensive capabilities. From "Radio Power Will Revolutionize the World," by N. Tesla, Modern Mechanix & Invention, July 1934. *CREDIT: Metascience Foundation.*

An FBI report dated January 12, 1943, described the hotel managers' assessment of the inventor somewhat differently, and it does not appear that they took him as seriously as Trump alleges:

> *The Hotel managers report that he [Tesla] was very eccentric if not mentally deranged during the past ten years and it is doubtful if he has created anything of value during that time, although prior to that he probably was a very brilliant inventor.*

Tesla appears to have told both Hausler, his pigeon caretaker and Fitzgerald, the Army engineer, that he had built a working model. From an interview I had with Mrs. Czito [10], whose husband's father and grandfather were both trusted Tesla employees, she recalls that her father-in-law used to recount stories, circa 1918, of Tesla bouncing electronic beams off of the moon. This is not a death ray, but it certainly supports the hypothesis that the inventor created working models along these lines.

Joseph Alsop, who interviewed Tesla at the Hotel New Yorker, and who was one of the first to fully report Tesla's work with particle beam weapons, described an incident Tesla had when experimenting with cathode ray tubes:

Sometimes a particle larger than an electron, but still very tiny, would break off from the cathode, pass out of the tube and hit him. He said he could feel a sharp, stinging pain where it entered his body and again at the place where it passed out. The particles in the beam of force, [i.e. the] ammunition... will travel far faster. [46, p. 15]

As far back as the early 1890s, Tesla had replicated the work of J.J. Thomson who discovered the electron by shooting streams of electricity at a fluorescent screen and measuring sub-atomic particle deflections. Additional evidence has been reported by Corum and Corum [9]. They suggest that Tesla's particle beam work is an outcropping of his experiments with X-ray machines from the same period, as well as from his independent idea of illuminating the skies with his magnifying transmitter by beaming up electrical energy to the stratosphere. The Corums refer to concrete plans and "artist conceptions" drafted in the mid-1930s that were "made of a building with a tower in the form of a cylinder 16.5 feet in diameter, 115 feet tall. The structure was capped at the top by a 10 meter diameter sphere (covered with hemispheric shells as in the 1914 patent)" (their source, the Tesla Museum). The inventor had also contacted people at Alcoa Aluminum throughout 1935 who were "ready to start as soon as Tesla advanced the funds" (p. 47).

Two years later, at the age of 81, the inventor [31] stated at a luncheon attended by ministers of Yugoslavia and Czechoslovakia that he had constructed a number of beam transmission devices including the death ray for protecting a country from incoming invasions and a laser-like machine that could send impulses to the moon and other planets. He also said that he was going to take the death ray to a Geneva conference for world peace. When pressed by the columnists to "give a full description..., Dr. Tesla said..., "But it is not an experiment.... I have built, demonstrated and used it. Only a little time will pass before I can give it to the world" (p. 278). The recent publication of Tesla's 1937 article establishes that the inventor wrote out the entire specifications to this device.

Star Wars

Tesla's numerous inventions could be applied in a variety of ways for military purposes, e.g. particle beam weapons, world radar, earthquake contrivances, brain wave manipulation. One or more magnifying transmitters could theoretically send destructive impulses through the earth to any location. For instance, a well placed jolt of many millions of volts could theoretically destroy the communications network of any major city. Recent discourses on potential future warfare technologies stemming mainly from parapsychologists such as Tom Bearden [2, 3, 4] and Andrija Puharich [28, 29] suggest that the Soviets have already harnessed various Tesla weapons including apparatus for weather and mind control [41, 53].

In the fall of 1976, an ominous 10 Hz transmission picked up in Canada and in the northwestern part of the United States was attributed to Soviet-Tesla mind-control weaponry experiments [5]. These extremely low frequencies (ELF waves) labeled "woodpecker transmissions" are hypothesized to be Soviet attempts to "entrain" the brain waves of Americans. Through modulating this resonant neuronal frequency, detrimental effects in thought processes and even health problems could supposedly be augmented [3, 5]. Puharich [29] has invented a device to protect a person from such dangerous transmissions.

This research is highly controversial and speculative. Nevertheless, Bearden's work led this researcher to a 7,000 word article in Aviation Week on Soviet particle beam weapons [34]. The expose, which "shook Washington" when it appeared, was also abstracted in Science. [52] It contains a schematic drawing of a particle beam weapon which bears a remarkable resemblance to Tesla's drawing made four decades earlier, but published for the first time eight years after this article appeared.

Coupled with the realization that the Soviets are well advanced in this area, this is strong evidence in support of the claim that Tesla did, in fact provide schematic drawings of such devices to major countries like the Soviet Union, England and the United States in the mid-1930s.

The articles also mentioned the use of "young geniuses under [the age of] 29... located at the Wright-Patterson Air Force Base in Dayton Ohio" to try and conceive of a breakthrough in the technology; and also, surprisingly, that "the President [Jimmy Carter] was screened from vital technical developments by the bureaucracy of the CIA and Defense Intelligence Agency" [52]. The source was General George Keegan, former head of Air Force Intelligence.

Three intriguing points emerge: (1) the concept of high secrecy on particle beam weapons; (2) the mention of Wright-Patterson Air Force Base; and (3) the policy of utilizing bright young geniuses. All of these variables are evident in the FBI files on Tesla reviewed earlier. Great support is lent to the hypothesis that Tesla's work and papers were systematically hidden from public view in order to protect the trail of this top secret work, which today is known as Star Wars.

Particle beam weaponry at this date, 1988, is still more a dream than a reality. Therefore, if indeed the inner sanctum of one or both superpowers has access to Tesla's plans, why is it that no death ray has ever been constructed? Perhaps there are prototypes that I do not know about, but it seems to me that they should have been tried out in the field at such times as during the Vietnam War or the recent conflict in Afghanistan. This part of the story must remain a mystery.

References

[1] Anderson, Leland. Personal correspondence. Denver, CO, 1988.

[2] Bearden, Thomas E.. Solutions to Tesla's Secrets and the Soviet Tesla Weapon. Milbrae, CA: Tesla Book Company, 1981. (pp. 1-45)

[3] Bearden, Thomas E.. "Tesla's Secret." Planetary Association For Clean Energy, No. 3; December, 1981. (pp. 12-24)

[4] Bearden, Tom. "Tesla's Electromagnetics and Soviet Weaponization." In E. Rauscher & T. Grotz (Eds.), 1984. (pp. 119-138)

[5] Bise, W. "The Soviet 10 Hz Pulses." Specula, vol. 1, no. 3; July-September 1978. (pp. 24-28)

[6] Bromberg, Joan. Private correspondence. Laser History Project, MA, 1984.

[7] Cheney, M. Tesla: Man Out of Time. Englewood Cliffs, NJ: Prentice Hall, 1981.

[8] Cornels, Frederich. FBI memorandum. Washington, DC: FOIA, January 9, 1943.

[9] Corum, Jim & Corum, Ken. "A Physical Interpretation of the Colorado Springs Data." In E. Raucher, & T. Grotz (Eds). (pp. 50-58)

[10] Czito, Nancy. Personal interview. Washington, DC, 1984.

[11] Donegan, T.J. FBI memorandum. Washington, DC: FOIA, November 14, 1943.

[12] Edison, Thomas. "Light Without Heat," The Morning Journal, July 26, 1891.

[13] Federal Bureau of Investigation. Archives, Washington, DC.

[14] Feeley, Terrence. Personal interview. Smithsfield, RI: Laser Fare Inc, 1984.

[15] Gernsback, Hugo. "The Diabolic Ray." Practical Electrics, August, 1924. (pp. 554-555, 601)

[16] Grindell-Mathews, Harry. "The Death Power of Diabolical Rays." New York Times. May 21, 1934. (1:2, 3:4,5)

[17] Grindell-Mathews, Harry. "Three National Seek Diabolical Ray." New York Times. May 28, 1924, (25:1,2)

[18] Grindell-Mathews, Harry. "Diabolical Rays." Popular Radio. August, 1924. (pp. 149-154)

[19] Hammond, John Hays, Jr. "We Need 2,000 Aeroplanes." New York Times. November 15, 1915. (3:1)

[20] Hawthorne, Julian. "In Tesla's Lab." Philadelphia North American. 1901.

[21] Hoover, J. Edgar. FBI memorandum. Washington, DC: FOIA, January 21, 1943.

[22] Hunt, W. and Draper, I. Lightning In His Hands: The Life Story of Nikola Tesla. Hawthorne, CA: Omni Publ, 1977. (Originally published, 1964).

[23] Kock, W. Engineering Applications Of Lasers and Holography. New York: Plenum Press, 1975.

[24] Markovitch, Michael. Personal interview. Belgrade, Yugoslavia, 1985.

[25] O'Neill, John. Prodigal Genius. New York: Ives Washburn, 1944.

[26] O'Neill, John. "Tesla Tries to Prevent World War II." Tesla Coil Builders Association, Vol 7, #3, 1988. (pp. 13-15)

[27] Patten, F. "Nikola Tesla and His Work." Electrical World. April 14, 1894. (pp. 496-499)

[28] Puharich, Andrija. "The Tesla Weaponry Papers." Presentation before the Tesla Centennial Symposium. Colorado Springs, CO. 1984.

[29] Puharich, Andrija. "How to Protect Yourself From Omnipresent Electromagnetic Pollution." In Tesla Magnifying Transmitter. Dobson, NC, 1985

[30] Ratzlaff, John, (Ed). Solutions To Tesla's Secrets. Milbrae, CA: Tesla Book Co, 1981.

[31] Ratzlaff, John, (Ed). Tesla Said. Milbrae, CA: Tesla Book Co, 1984.

[32] Ratzlaff, John, & Anderson, Leland. Dr. Nikola Tesla Bibliography: 1884-1978. Palo Alto, CA: Ragusan Press, 1979. (First edition published by Anderson, 1956.)

[33] Rauscher, Elizabeth & Grotz, Toby (Eds). Proceedings of the Tesla Centennial Symposium. Colorado Springs, CO: International Tesla Society, 1985.

[34] Robinson, C. "Soviet Push for a Beam Weapon." Aviation Week. May 2, 1977. (pp. 16-27)

[35] Seifer, Marc. "Nikola Tesla: The Man Who Fell to Earth." Ancient Astronauts. May, 1977. (pp. 23-26) (Harry Imber, pseudonym.)

[36] Seifer, Marc, "The Battle of the Currents." Metascience Quarterly. Vol 1, no. 1. Spring, 1979. (pp. 101-102)

[37] Seifer, Marc. "The Belief in Life on Mars, A Turn of the Century Group-fantasy." In J. Dorinson, & J. Atlas (Eds) Psychohistory: Persons & Communities. Long Island University and the International Psychohistory Society Press, 1983. (pp. 209-231)

[38] Seifer, Marc. Tesla: Grand Wizard of the Gilded Age (Video). New York: Windsor Total Video, 1984.

[39] Seifer, Marc. Nikola Tesla: Psychohistory of a Forgotten Inventor. (Doctoral Dissertation.) San Francisco, CA: Saybrook Institute, 1986.

[40] Seifer, Marc. "Nikola Tesla: Interplanetary Communicator?" Popular Electronics. December, 1988. (pp. 62-66, 102)

[41] Sunday Times. "Russians Secretly Controlling World Climate?" Scranton, PA: November 6, 1977. (p. 14)

[42] Tesla, Nikola. "On Light and Other High Frequency Phenomena." Electrical Review, 1893. (pp. 407-417; 579-582, 626-627)

[43] Tesla, Nikola. The Inventions, Researches and Writings of Nikola Tesla. T.C. Martin, (Ed). New York: Electrical Review Publishing Co, 1894.

[44] Tesla, Nikola. "Tesla's Device Like Bolts of Thor." New York Times. 1915. (8:3)

[45] Tesla, Nikola. "World System of Wireless Transmission of Energy." 1927. In Ratzlaf (Ed.), 1981. (pp. 83-86)

[46] Tesla, Nikola. "Tesla at 78 Bears New Death-beam." New York Times. July 11, 1934. (18:1)

[47] Tesla, Nikola. "The New Art of Projecting Concentrated Non-dispersive Energy Through the Natural Media." 1937. (pp. 280-281) (In E. Raucher & T. Grotz (Eds), 1985.)

[48] Tesla, Nikola. "Death Ray for Planes." New York Times. September 22, 1940. (2:7)

[48] Tesla, Nikola. Lectures, Patents & Articles. Belgrade: Tesla Museum, 1956.

[49] Trump, John. "The Trump Report." Washington, DC: FBI Archives - FOIA, January 30, 1943.

[50] Trump, John. Letter to F.A. Jenkins. January 2, 1976. (In Cheney, 1981)

[51] Wade, N. Charge Debate Over Russian Beam Weapons." Science. May 1977. (pp. 957-959)

[52] Washington Star. "Are Soviets Testing Wireless Electric Power?" January 1, 1977. (1:5)

[53] Wetzler, M.. "Nikola Tesla Lecture." Harpers Weekly, 1891. (p. 524)

About The Author:

Dr. Marc J. Seifer, editor of MetaScience, teaches psychology at Bristol Community College, Falls River, Massachusetts and psychohistory at Providence College School of Continuing Education in Providence, Rhode Island. He has a Bachellor's degree from the University of Rhode Island, post graduate work in graphology at the New School for Social Research, New York City, a Masters degree from the University of Chicago and a Doctorate from Saybrook Institute, San Francisco. Dr. Seifer has produced and directed a short video entitled *Tesla: Grand Wizard of the Gilded Age*. He has presented slide lectures on Tesla before numerous groups in America, Canada, Israel and Yugoslavia. Additionally, Dr. Seifer has had articles published in a variety of publications.

Marc is also the author of the definitive biography *Wizard: The Life & Times of Nikola Tesla, Biography of a Genius,* published by Citadel Press/Kensington.

The Secret History of the Wireless

Marc J. Selfer, Ph.D.
Box 32
Kingston, RI 02881

A Silent Tribute...[1]

Marconi's talk, before 1100 members of the New York Electrical Society, was held on the very day the Titanic sank. Frank Sprague, who gave the eulogy, "visibly affected Mr. Marconi when he credited him with saving the lives of from 700 to 800 persons."[2] Unfortunately, Marconi was unable to save 1500 other individuals including Colonel John Jacob Astor, who went down with the ship after helping his new bride board one of the remaining lifeboats.

> *April 18, 1912*
>
> *My dear Mr. Tesla:*
>
> *I attended the Marconi meeting last night, in company with illustrious society.... Mr. Marconi gave the history, as he sees it, of wireless up to this date.... [He] does not speak any more of Herzian Wave Telegraphy, but accentuates that messages he sends out are conducted along the earth....*
>
> *Pupin had the floor next, showing that not only was wireless a great achievement, but that it was due entirely to one single man who was too modest to have spoken in the first person...*
>
> *The only speaker of the evening who understood Mr. Marconi's merit and who did not hesitate to vent his opinion... was Steinmetz. In a brief historical sketch... he maintained that while all elements necessary for the transmission of wireless energy were available, yet it was due to Marconi that intelligence was actually transmitted...*
>
> *That evening was, without any question, the highest tribute that I ever have heard paid, to you in the language of absolute silence as to your name.*
>
> *Sincerely yours,*
> *Fritz Lowenstein*

Litigation

If ever an event epitomized the loss of innocence, the myopic condition of humanity, it was the sinking of the Titanic. Remindful of Tesla's own voyage, this watershed, recapitulated the story of Icarus, the prideful aeronaut who crumbled, because of lack of respect for his limitations. With Tesla's ponderous wish to transmit unlimited energy to the far reaches of the world and bring rain to the deserts, to become, as it were, a master of the universe, it was inevitable that he too would succumb. On the positive side, the tragedy prompted Congress to pass into law an act requiring the use of wireless equipment on all ships carrying 50 or more passengers. On the negative side, the event helped spurn a new wave of litigation causing company after company to elbow themselves into debt and melt away.

A cartoon depicting Marconi saving people on the Titanic.
Credit: MetaScience Foundation

Tesla was not the only casualty in the wireless war. Reginald Fessenden's concern "all but ceased functioning in 1912," due to his erratic nature, internal quarrelling with his backers, and "prolonged litigation." And Lee DeForest, who by now, had nearly 40 patents in wireless, also went under, when he was convicted with officers of his company in a stock fraud.[3]

Concerning Lowenstein, Tesla's former employee from his days at Colorado Springs, Tesla, for a modest royalty, was backing his protegè's attempts to install equipment on U.S. Navy ships. "He is much abler than the rest of the wireless men," he wrote his secretary, George Scherff, "[so] this has given me great pleasure."[4] The one edge Lowenstein had over Marconi, was the Italian's insistence of an all or nothing deal with the military. Either all ships were hooked into his system, or none were. The U.S. Government, however, was loathe to relinquish their upper hand, so Marconi had great difficulty integrating his system into the American military arena.

Nevertheless, Marconi was still the only major competitor, so Tesla set out to reestablish his legal right. Conferring with his lawyers, the pioneer began a strategy to sue the pirate in every country he could.

In England, Tesla had allowed an important patent to lapse, so progress there was halted. Oliver Lodge, on the other hand, was able to prevail, and received 1,000 £/year for seven years from the Marconi concern there.[5] In the United States, where Tesla was applying for a renewal of his most fundamental patent, he had yet to formalize the suit, but in "the highest court in France" the inventor achieved a resounding success. Sending his written testimony to Judge M. Bonjean in Paris, Tesla explained his work in 1895, where he "erected a large wireless terminal above the building... [and] employed damped and undamped oscillations." He also enclosed two patents from 1897, and specifications for his telautomaton which confirmed that he had displayed it before G.D. Seely, the Examiner in Chief of the U.S. Patent Office in Washington, DC in 1898. Concerning Marconi's June 2, 1896 patent, Tesla testified:

> *[This patent is] but a mass of imperfection and error. It describes apparatus shown and experimented with long before and a known method of signaling Hertzian rays, the energy of which is absorbed within a very short distance from the transmitter, and which are wholly unsuited for practical use. This patent has absolutely nothing to do with the successful application of the wireless art and, if anything, it has been the means of misleading many experts and retarding progress in the right direction.... It gives no hint as to the length of the transmitting and receiving conductors and the arrangements illustrated preclude the possibility of accurate tuning.... [Marconi replaced] the old-fashioned Rhumkorff coil [with] ... the Tesla coil.*[6]

Speaking in support of Tesla's cause for Popoff, Ducretet & Rochefort, the French company who had initiated the action, was electrical engineer M.E. Girardeau who stated:

> *Indeed, one finds in the American patent extraordinary clearness and precision, surprising even to physicists of today.... Tesla as the inventor of the four circuits in resonance... even gives, by way of illustration, the numerical value of the condenser (4/100 of a microfarad) and says that the discharges of the condenser could be effected by a mechanical break.... Tesla is also the inventor of the contact detector which is everywhere employed today... and all of this [can be found] in his 1897 patents and articles of 1898 and 1899.*
>
> *What a cruel injustice would it be now to try to stifle the pure glory of Tesla in opposing him scornfully.*[7]

Judge Bonjean struck down Marconi's patents, and reestablished Tesla's as superseding. Most likely, compensation was also paid to him from the French company who won the case.

The German Connection

Marconi's greatest rival in the legal arena may have been Nikola Tesla, but in the battleground of the marketplace, it was Telefunken, the German wireless concern. Although Marconi had patents in Germany, the Telefunken syndicate, had too many important connections on its home front, and easily maintained a monopoly there. Formed through a forced merger under orders of the Kaiser, of the Braun-Siemens-Halske and Arco-Slaby systems, Telefunken had fought the Marconi conglomerate vigorously on every front. It was, without doubt, the number two competitor in the world. Although Marconi had achieved a recent coup in Spain, in America, Telefunken gained an edge when it constructed two enormous transatlantic systems in Tuckertown, New Jersey, and Sayville, New York.

For nationalistic reasons, Tesla had been prohibited from obtaining his rightful royalties in Germany, but Professor Slaby, never hid the fact that he considered Tesla the patriarch of the field. Thus, when Telefunken came to America, Slaby sought out his mentor, not only on moral grounds, but also for gaining a legal foothold against Marconi, and to obtain the inventor's technical expertise.

In 1913, a meeting was held between Tesla and the principles of Telefunken's American holding company, the innocuous sounding Atlantic Cable Company at 111 Broadway, the location of their offices. Present was the director, Dr. Karl George Frank, "one of the best known German [American] electrical experts," and his two managers, Richard Pfund, a frequent visitor to Tesla's lab, who was head of the Sayville plant, and "the monocle," Lieutenant Emil Meyers, head of operations at Tuckertown.

Tesla asked for an advance of $25,000 and royalties of $2,500/month, but settled for $1,500/month with a one month advance.[8] Edwin Armstrong also struck a similar deal for his feedback circuit.

Tesla met with Pfund to discuss the turbine deal with the Kaiser, and also fix a transmitter that the German was working on at the Manhattan office. Shortly thereafter, he travelled out to the two transmitting stations with his plan to institute his latest refinements in order to boost their capabilities.[9]

J.P. Morgan Jr.

Shortly after J. Pierpont Morgan's death in 1913, Tesla began to court his son Jack, trying not to fall into the trap he had laid for himself the first time around. He sent J.P. Morgan Jr. an articulate proposal outlining his plans in the field of "fluid propulsion," i.e., his bladeless turbines, and also his continuing work in wireless.

To reassert his dominance as the preeminent inventor in wireless, the inventor forwarded the entire French litigation proceedings where Marconi's work was overturned in favor of his. If Jack could help fight the legal battle in the U.S., the wireless enterprise, which they contractually shared, could be revived. Jack, however, was not smitten by the same vision of destiny, and graciously declined to become involved in Wardenclyffe in any way. He had not ruled out the turbines, however, and advanced Tesla $20,000 for the endeavor over the next few months.

The wizard returned to the Edison Waterside Station where his turbines were located with a new transfusion of 23 Wall Street blood. To reflect the reactivation of the resurrected alliance, the inventor set out to search for more fashionable chambers. Within a few months he took up residence in the loftiest skyscraper in the world, the Woolworth Building, which soared above the city to the dizzying height of 800 feet.

The Woolworth Building where Tesla moved his offices in 1913. Credit: MetaScience Foundation

Throughout the latter half of 1913, the inventor prepared a careful marketing plan to exploit his turbines. His plan was to use the profits from this endeavor to finance his triumphant return to Wardenclyffe and prevail at last in the field of wireless. Tesla's best lead in marketing the turbines came from former Edison associate, Sigmund Bergmann who was now one of the leading manufacturers for the Kaiser in Germany.

By the end of the year, however, the inventor was forced to admit that his turbines were not yet perfected, and requested from Morgan an additional $5,000 to help secure the German deal. Morgan agreed to defer interest payments, but decided not to increase the loan. Tesla, however, simply required the funds, and followed up the letter with a testimonial from Excellenz von Tirpitz, Minister of Marine, "who has been requested to the German Emperor relative to the Tesla Turbine who is greatly interested in this invention." Von Tirpitz had "promised his Excellency that the machine will certainly be here on exhibition about the middle of January, so you know what that means." Tesla also informed Morgan that if the deal was consummated, Bergmann would come through with royalties on the turbine of $100,000/year.

Considering Jack's keen antipathy for the Germans, it would seem unlikely that he would reverse his decision. However, unlike his father, the son was able to compromise. He graciously changed his mind and forwarded the additional funds.[10]

The War to End All Wars

Within two weeks after World War I started, Germany's transatlantic cable was severed by the British. The only reasonable alternative for communicating with the outside world was through Telefunken's wireless system. Suddenly, the Tuckertown and Sayville plants became of paramount concern. The Germans obviously wanted to maintain the stations to keep the Kaiser abreast of President Woodrow Wilson's intentions, but the British wanted them shut down.

In July of 1914, Marconi was decorated by the King of England at Buckingham Palace. Now, the fight against Telefunken would be fought on military as well as commercial grounds as it became clear that the Germans were using their plants to help coordinate submarine and battleship movements. The wireless lines also marked the burgeoning alliance forming between Italy and the English empire.[11]

The German Kaiser during WW I. Credit: MetaScience Foundation

As a pacifist, Wilson maintained a strict policy of neutrality although the sentiments of the majority of the American population was with England, particularly after Germany stormed through the peaceful kingdom of Belgium. Nevertheless, fully one-tenth of the country was of German stock, and their sentiments were with the other side. To avoid the problem of having the Germans use its American based transmitters for war purposes Wilson ordered the U.S. Navy to appropriate the Tuckertown plant so that the United States could send its own "radio coded messages abroad." Wilson's presidential decree was published in the papers:

> *Whereas an order has been issued by me this day, August 5, 1914, declaring that all radio stations within the jurisdiction of the United States of America were prohibited from transmitting or receiving... messages of an unneutral nature and from in any way rendering to any one of the belligerants any unneutral services.... It is ordered, by virtue of authority vested in me by the Radio Act... that one or more of the high powered radio stations within the jurisdiction of the United States... shall be taken over by the Government.*[12]

The Legal Front

Throughout the beginning of the war, Tesla stepped up his legal campaign against Marconi, as he continued to advise and receive compensation from Telefunken. As the country was still officially neutral, (America would not enter the war for another three years), the arrangement was en-

Tesla must have gotten a kick out of this drawing , as it depicts him using Marconi's primitive wireless equipment to blow up ships by remote control. From a Sunday newspaper, 1905. **Credit: MetaScience Foundation**

tirely above board. Nevertheless, few people knew about the German/Tesla link, although the inventor made no secret of it to Jack Morgan.

February 19, 1915

Dear Mr. Morgan,

I am expecting to embody in their plant at Sayville some features of my own which will make it practicable to communicate with Berlin by wireless telephone and then royalties will be very considerable. We have already drawn papers.[13]

Camouflaged by the smokescreen of the American sounding Atlantic Cable Company, Telefunken swiftly moved to increase the power of its remaining station at Sayville. Located near the town of Patchogue out on the flats of Long Island, just a few miles from Wardenclyffe, the Sayville complex encompassed 100 acres, and employed many German workers. With its main offices in Manhattan, and its German director, Dr. Karl George Frank, an American citizen, Telefunken was legally covered, as no foreigner could own a license to operate a wireless station in the country. (Thus, Marconi also had an American affiliate.) It was an easy matter for Tesla to confer with Atlantic in the city and also go out to the site of the plant.

Within two months of Tesla's letter to Morgan, the plant at Sayville tripled its output by erecting two more 500-foot-tall pyramid-shaped transmission towers. Resonating accoutrements spread out over the land for thousands of more feet. Telefunken's output was thus boosted from 35 Kilowatts to over 100. Germany was now perceived as jumping into the number one spot in the wireless race. The **NEW YORK TIMES** reported on their front page,

Few persons outside radio officials knew that Sayville was becoming one of the most powerful transatlantic communicating stations in this part of the world.[14]

Tesla Sues Marconi on Wireless Patent

Alleges That Important Apparatus Infringes Prior Rights

*A suit brought by Nikola Tesla against the Marconi Wireless Telegraph Company of America, seeking the annulment of one of the chief Marconi patents, will have its first public argument in the United States District Court tomorrow. ...**New York Times,** August 4, 1915*[15]

The years preceding America's entrance into World War I contained an overwhelming quagmire of litigation involving most countries, and virtually every major inventor in the wireless field. In August of 1914, Lowenstein was sued for violating Marconi's patent 763,772. However, this was only one of a number of legal battles the Marconi Wireless Telegraph Company would have with the Tesla legacy, as Tesla himself sued his arch rival that year, as did the U.S. Navy and also Atlantic Cable Company.

Michael Pupin... fancied himself as the inventor of the wireless!
Credit: MetaScience Foundation

During the following spring, Marconi was subpoenaed by Telefunken. Due to the importance of the case, he sailed off for America on the *Lusitania* arriving in April of 1915 to testify. "We sighted a German submarine periscope," he told astonished reporters and his friends at dockside.[16] As three merchant ships had already been torpedoed without warning by the German U-Boats the month before, Marconi's inflammatory assertion was not taken lightly.

The **BROOKLYN EAGLE** reported that this suit brought "some of the world's greatest inventors on hand to testify."[17] Marconi had been declared a victor in a Brooklyn district court by Judge Van Vechter Veeder, during some of the Lowenstein proceedings. Although he certainly had the aura of the press behind him, Signor Marconi was beaten by the Navy in one patent dispute that year. [18] This German case, with all the heavyweights in town, promised to be portentous for establishing, once and for all, the true legal rights.

Aside from Marconi, there was, for the defense, Columbia Professor Michael Pupin, whose testimony was even quoted in papers in California. With braggadocio, Pupin declared, "I invented wireless before Marconi or Tesla, and it was I who gave it unreservedly to those who followed!"[19]

"Nevertheless," Pupin continued, "it was Marconi's genius who gave the idea to the world, and he taught the world how to build a telegraphic practice upon the basis of this idea. [As I did not take out patents on my experiments], in my opinion, the first claim for wireless telegraphy belongs to Mr. Marconi absolutely, and to nobody else."[20]

Watching his fellow Serb upon the stand, Tesla's jaw dropped so hard, it almost cracked upon the floor!

When Tesla took the stand for Atlantic, he came with his attorney, Drury W. Cooper of Kerr, Page & Cooper. Unlike Pupin, who could only state abstractly that he was the original inventor, Tesla proceeded to explain in clear fashion all of his work from the years 1891-1899. He documented his assertions with transcripts from published articles, from the Martin text, and from public lectures such as his well known wireless demonstration which he had presented to the public in St. Louis in 1893. The inventor also brought along copies of his various requisite patents which he had created while working at his Houston Street Lab during the years 1896-1899.

> **COURT:** *What were the [greatest] distances between the transmitting and receiving stations?*
>
> **TESLA:** .... *From the Houston laboratory to West Point, that it, I think, a distance of about thirty miles.*
>
> **COURT:** *Was that prior to 1901?*
>
> **TESLA:** ... *Yes, it was prior to 1897 ....*
>
> **COURT:** *Was there anything hidden about [the equipment), or were they open so that anyone could use them?*
>
> **TESLA:** *There were thousands of people, distinguished men of all kinds, from kings and greatest artists and scientists in the world down to old chums of mine, mechanics, to whom my laboratory was always open. I showed it to everybody; I talked freely about it.*[21]

John Stone Stone

I think we all misunderstood Tesla. We thought he was a... visionary.... He did have visions but they were of a real future, not an imaginary one.... It has been difficult to make any but unimportant improvements in the art of radiotelegraphy without traveling part of the way, at least, along a trail blazed by this pioneer.... The apparatus he devised and constructed... was so far ahead of [its] time that the best of us then mistook him for a dreamer. [22]

Another jolt to Marconi came from John Stone Stone, (his mother's maiden name, by coincidence, was also Stone). Having traveled with his father, a general in the Union Army, throughout Egypt and the Mediterranean as a boy, Stone was educated as a physicist at Columbia University and Johns

Hopkins University, where he graduated in 1890. A research scientist for Bell Labs in Boston for many years, Stone had set up his own wireless concern in 1899.

The following year, he filed for a fundamental patent on tuning, which was allowed by the U.S. Patent Office over a year before Marconi's. [23] Its essential features included his pioneer claim for the idea of "adjustable tuning by means of a variable inductance of the closed circuits of both transmitter and receiver."[24]

Stone, who never considered himself the original inventor of the radio, as President of the Institute of Radio Engineers, and owner of a wireless enterprise, put together a dossier of inventor priorities in "continuous-wave radio frequency apparatus." He wanted to determine for himself the etiology of the invention. Adorned in a formal suit, silk ascot, high starched collar and pince-nez attached by a ribbon around his neck, the worldly aristocrat took the stand:

> *Marconi, receiving his inspiration from the experiments of Hertz and Righi... [was] impressed with the electric radiation aspect of the subject... and it was a long time before he seemed to appreciate the real role of the earth in the operation of his system, though he early recognized that the connection of his oscillator to the earth was very material value.... Tesla's electric earth waves explanation was the more serviceable in that it explained the important and useful function of [potential waves in] the earth, whereby the waves were enabled to travel over and around hills and were not obstructed by the sphericity of the earth's surface, while Marconi's view led many to place an altogether too limited scope to the possible range of transmission.... With the removal of the spark gap from the antenna, the development of earthed antenna, and the gradual enlargement of the size of stations as it was realized... greater range could be obtained with larger power used at lower frequencies, [and] the art returned to the state to which Tesla developed it.*

Attributing the opposition, and alas, even himself, to having been afflicted with "intellectual myopia," Stone concluded that although he had been designing wireless equipment and running wireless companies since the turn of the century, it wasn't until he "commenced with this study" that he really understood Tesla's contribution to the development of the field.[25]

The FDR Connection

Another case which did not receive much publicity, but which became vital to the Supreme Court's ruling in Tesla's favor over Marconi in 1943, was Marconi vs. the United States Navy which Marconi brought on July 29, 1916. Marconi was suing for infringement of a fundamental wireless patent (#763,772), which had been allowed in June of 1904. The Italian was seeking $43,000 in damages. Acting Secretary of the Navy, E.F. Sweet, and also Assistant Secretary Franklin D. Roosevelt began a correspondence in September of 1916 to review Tesla's 17 year old file to the Light House Board.

John Stone Stone, Jonathan Zenneck, Lee DeForest, Nikola Tesla, and Fritz Lowenstein at a dinner in Stone's honor. David Sarnoff is outside of this cropped photo on the left.
Credit: MetaScience Foundation

Roosevelt wrote: *This [Light House Board Tesla File] may be made use of in forthcoming litigation in which the Government is involved... [It is] the discovery of suitable proof of priority of certain wireless usages by other than Marconi [and thus] might prove of great aid to the Government.*[26]

Writing from his Experimental Laboratory in Colorado Springs, in 1899 to the Navy, Tesla described seven features of his wireless system which established unequivocally his stature as primary inventor. This discovery, Tesla pointed out, was announced "a few years ago [when] I laid down [these] certain novel principles: (1) an oscillator; (2) a ground and elevated circuit; (3) a transmitter; (4) a resonant receiver; (5) a transformer "that scientific men have honored me by identifying it with my name" [Tesla coil]; (6) a powerful conduction coil; and (7) a transformer in the receiving apparatus." [27]

The history of Marconi's patent applications to the U.S. Patent Office provided additional ammunition. In 1900, John Seymour, the Commissioner of Patents who had protected Tesla against the demands of Michael Pupin for an AC claim at this same time, disqualified Marconi's first attempts at achieving a patent because of prior claims of Lodge and Braun and particularly Tesla.

"Marconi's pretended ignorance of the nature of a 'Tesla oscillator' [is] little short of absurd," wrote the Commissioner. "Ever since Tesla's famous [1891-1893 lectures]... widely published in all languages, the term 'Tesla oscillator' has become a household word on both continents." The Patent Office also cited quotations from Marconi himself admitting use of a Tesla oscillator.

Two years later, in 1902, Stone was granted a patent on tuning which the government cited as anticipating Marconi, and two years after that, after Seymour retired, Marconi was granted his infamous 1904 patent.[28]

The Fifth Column

Due to the dangers that existed on the high seas, and the rumors that the Germans were out for Marconi's head, the "Senatore" did not sail back on the *Lusitania*, but rather returned on the *St. Paul* in a disguised identity and under an assumed name.

A week after Marconi's departure, in May of 1915, a German submarine torpedoed the great ocean liner, the *Lusitania*, killing 1,134 individuals; only 750 survived. The sinking, in lieu of the alternative procedure of boarding unarmed passenger ships by military vessels, was unheard of. Quite possibly, Marconi could have been a target, however, the Germans used as their reason the cargo of armaments on board headed for Great Britain.

In July of 1915, the Senate chambers in Washington was rocked by a terrorist bomb. The following day, the fanatic who planted it, Dr. Erich Muenter, alias Frank Holt, a teacher of German from Cornel

The sinking of the Lusitania.
Credit: MetaScience Foundation

University, and wife murderer, walked into Jack Morgan's Long Island home toting a six gun in each hand. He wanted Morgan to stop the flow of arms to Europe. With his wife and daughter leaping at the assailant, Morgan charged forward. Shot twice in the groin, Morgan and his wife were able to wrestle the guns from the man, and get him arrested. Tesla sent the overnight hero a get well letter,[29] as Holt committed suicide in a Long Island jail cell.

The Fifth Column had emerged. German spies were everywhere. Reports started filtering in that the Germans were creating a secret submarine base around islands off the coast of Maine. It was also alleged that the broadcasting station out at Sayville was not merely sending neutral dispatches to Berlin, but rather coded messages to battleships and submarines. The front pages of the papers were saturated with alarming headlines:

20 OR MORE AMERICANS LOST WHEN GERMANS SINK [PEACEFUL FREIGHTER]

NAVY MAY SEIZE SAYVILLE WIRELESS

Plant Under Suspicion Officers Think German Station May Send Messages to Submarines

Evidence Before Congress

Wilson Hears of New Disaster

New Submarine Attack, However a Surprise in View of Hope from Berlin

Base for German Submarines Here?

Von Tirpitz Said to be Launching a New Campaign to Sink Munitions Ships

With Tesla, just a few months earlier, boasting to Morgan that he was working for the Germans, and "Grand Admiral von Tirpitz contemplat[ing] a more vigorous campaign against freight ships... [and planning] a secret base on this side of the Atlantic"[30] it is quite possible that the inventor's name became tainted in some inner circles.

A week later, on Tesla's 59th birthday, the **TIMES** reported that not only were the Germans dropping bombs over London from Zeppelins, they were also "controlling air torpedoes" by means of radio dynamics. Fired from Zeppelins, the supposed "German aerial torpedo[es] can theoretically remain in the air three hours, and can be controlled from a distance of two miles.... Undoubtedly, this is the

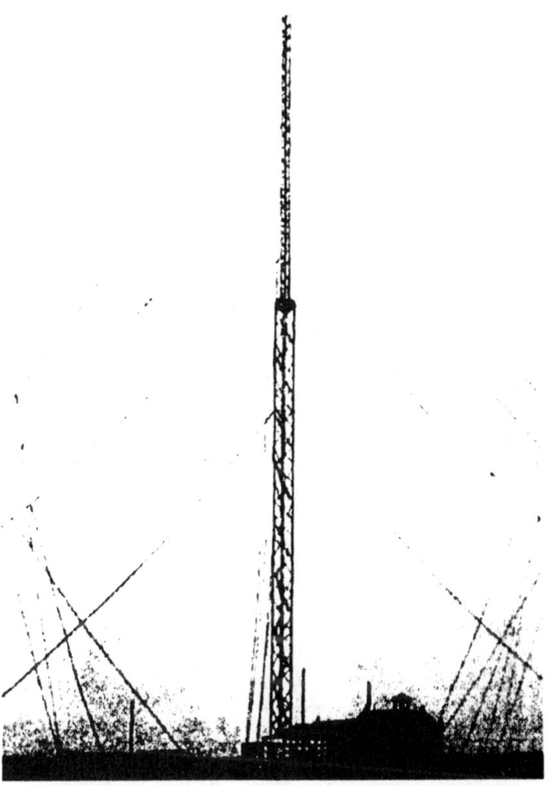

One of Marconi's wireless towers which resembled Telefunken's towers at Tuckertown, NJ and Sayville, NY.
Credit: MetaScience Foundation

secret invention of which we have heard so many whispers that the Germans have held in reserve for the British fleet."[31] Although it seemed as if Tesla's telautomatic nightmare prognostication of 1898 had come to be, Tesla himself announced to the press that "the news of these... magic bombs... cannot be accepted as true, [though] they reveal just so many startling possibilities."

"Aghast at the pernicious existing regime of the Germans," Tesla accused Germany of being an "unfeeling automaton, a diabolic contrivance for scientific, pitiless, wholesale destruction the like of which was not dreamed of before." No doubt, Tesla stopped doing business with von Tirpitz, although he probably continued his relationship with Professor Slaby, who may have been morally opposed to the war.

Tesla's solution to war was twofold: (1) a better defense, through an electronic star-wars type shield he was working on, and (2) "the eradication from our hearts of nationalism." If blind patriotism could be replaced with "love of nature and scientific ideal... permanent peace [could] be established."[32]

Spies had infiltrated the Navy Yard in Brooklyn to use the station to send secret coded messages to Berlin; through Richard Pfund, head of the Sayville plant, they had apparently also installed equipment on the roof at 111 Broadway, the building that housed Telefunken's offices.[33] Lieutenant Meyer, "who ran the Tuckertown operation... [was placed] in a Detention Camp in Georgia," suspected of spying.[34] As Tesla had been working with Pfund, Meyer and von Tirpitz, and as he had continued to receive royalty payments from a division of Telefunken right through the first month of 1917,[35] it is possible that he was placed on a list of potential subversives in the mad quest to ferret out all German agents, even though his lengthy condemnation of Germany was published as a major treatise in **THE SUN**.

The job of taking over all wireless stations fell to Josephus Daniels, Secretary of the Navy; his assistant was Franklin Delano Roosevelt. In the summer of 1915, Daniels placed Thomas Edison head of a civilian think-tank called the "Naval Consultant Board." Working with Franklin Roosevelt, Edison appointed numerous inventors to various positions such as Frank Sprague, Elihu Thomson and Michael Pupin, with Tesla's name conspicuously missing from the list.

HARRISBURG TELEGRAPH

RADIO SECTION

HARRISBURG, PA., SATURDAY EVENING, MARCH 22, 1924.

Radioed Light, Heat and Power Perfected by Tesla

INVENTOR ANNOUNCES FINAL SUCCESS OF EXPERIMENTS BEGUN THIRTY YEARS AGO

NEW HOOK-UPS FOR RADIO FANS

Tesla envisioned a world of prosperity — energized by the unleashed wireless power provided by his towers.
Credit: MetaScience Foundation

Wizard Swamped by Debts

Inventor Testifies He Owes the Waldorf
Lives Mostly on Credit Hasn't a Cent in Bank and "Hocked" Stock in His Company [36]

As 1915 was drawing to a close, Tesla began to find himself in deeper and deeper financial troubles. Although an efficient water fountain which he designed that year, was received favorably, [37] his overhead was still too high. Expenses included outlays for the turbine work at the Edison Station, his office space at the Woolworth Building, salaries to his assistants and secretary, past debts to investors, maintenance costs for Wardenclyffe, legal expenses on the wireless litigation, and his accommodations at the Waldorf-Astoria.

With the publication of his wretched state in the public forum, came also a deep sense of anger and corresponding shame; for now the world had officially branded him a dud. If success is measured in a material way, it became clear that Tesla was the ultimate failure. On the exterior, the inventor kept up appearances, but this event would mark the turning point in his life. He now began a slow, but steady turning away from society. One way or another he rejected the Nobel Prize which was supposedly offered to him at this time, (although it is apparent now from the recently released Swedish Academy files that he was never nominated during that period); and Tesla also dismissed the coveted Edison Medal, although one of his friends, B.A. Behrend, finally convinced him to accept the honor.

The Waldorf-Astoria

Tesla was aghast that Mr. Boldt, manager of the Waldorf-Astoria, had not protected Wardenclyffe adequately as it was valued at a minimum of at least $150,000. Even though he had signed it over to the Hotel, he had done so, according to his understanding, so as to honor his debt "until [his] plans matured." As the property, when completed, would yield $20,000 or $30,000 a day, Tesla was simply flabbergasted that Boldt would move to destroy the place. Boldt, and/or "the Hotel Management" saw Wardenclyffe now as theirs, free and clear, even though Tesla offered as proof "a chattel mortgage" on the machinery that the inventor had placed at his own expense. The Hotel's insurance was only $5,000, whereas Tesla's covered the machinery valued at $68,000. Why would Tesla independently seek to protect the property if he didn't still have an interest in it? Tesla saw the contract as "a security pledge,"[22] but the paper he signed did not specify any such contingency. According to the Hotel's lawyer, Frank Hutchins of Baldwin & Hutchins, "it was bill of sale with the deed duly recorded two years ago. We fail to see what interest you have," Hutchins callously concluded. Tesla found out that the Smiley Steel Company would be in charge of salvage operations. [38]

The United States Navy

The wizard decided that the only way to save Wardenclyffe was to extol its virtues as a potential defensive weapon for the protection of the country. Capitalizing upon the excellent Nobel Prize publicity, as it had been announced on the front page of the November 6, 1915 edition of the **NEW YORK TIMES** that he was to share it with Tom Edison, the inventor, once again strained the reader's creditability with another startling vision.

TESLA'S NEW DEVICE
LIKE BOLTS OF THOR
He Seeks to patent Wireless Engine for Destroying Navies by Pulling a Lever.
TO SHATTER ARMIES ALSO

Nikola Tesla, the inventor, winner of the 1915 Nobel Physics Prize, has filed patent applications on the essential parts of a machine the possibilities of which test a layman's imagination and promise a parallel to Thor's shooting thunderbolts from the sky to punish those who had angered the gods. Dr. Tesla insists there is nothing sensational about it....

"Ten miles or a thousand miles, it will be all the same to the machine," the inventor says. Straight to the point, on land or on sea, it will be able to go with precision, delivering a blow that will paralyze or kill, as it is desired. A man in a tower on Long Island could shield New York against ships or army by working a lever, if the inventor's anticipations become realizations.[39]

In "a serious plight," with nowhere else to turn, the inventor contacted Morgan once again to ask for assistance. He pointed out to Morgan that the military was using $10-million of wireless equipment based upon his patents, and that he hoped, some day, to gain compensation for their use. But he had exhausted all other avenues. Morgan provided his last chance to protect their commonly held wireless patents and save the tower. "Words cannot express how much I have deplored the cruel necessity which compelled me to appeal to you again," the inventor explained, but it was to no avail.[40] He still owed Jack $25,000 plus interest on the turbines; the financier ignored the request, and quietly placed Tesla's account in a bad debt file.[41]

In February of 1917, the United States broke off all relations with Germany, and seized the wireless plant at Sayville. "Thirty German employees... were suddenly forced to leave, and enlisted men of the American Navy have filled their places." Guards were placed around the plant as the high command decided what to do with the remaining broadcasting stations lying along the coast.[42] Articles began springing up like early crocus to announce the potential "existence of [yet another] concealed wireless station [able] to supply information to German submarines regarding the movements of ships."[43]

19 MORE TAKEN AS GERMAN SPIES
Dr. Karl George Frank, Former Head of Sayville Wireless Among Those Detained [44]

Wardenclyffe was to be the first of Tesla's towers that would bring wireless power and communication to the world.
Credit: MetaScience Foundation

On April 6th, 1917, President Wilson issued a proclamation "seiz[ing] all radio stations. Enforcement of the order was delegated to Secretary Daniels.... It is understood that all plants for which no place can be found in the navy's wireless system, including amateur apparatus, for which close search will be made, are to be put out of commission immediately."[45] Clearly, an overt decision had to be made about the fate of Wardenclyffe.

Tesla's expertise was well known to Secretary Daniels and Assistant Secretary Franklin Roosevelt, as they were actively using the inventor's scientific legacy as ammunition against Marconi in the patent suit. Coupled with the inventor's astonishing proclamation that his tower could provide an electronic aegis against potential invasions, Wardenclyffe must have been placed in a special category.

However, there were two glaring strikes against it. The first was that Tesla had already turned over the property to Mr. Boldt to cover his debt at the Waldorf; and the second was the transmitter's record of accomplishment: *nonexistent*. What better indication of the folly of Tesla's dream could there be than the tower's own perpetual state of repose. To many, Wardenclyffe was merely a torpid monument to the bombastic prognostications of a not very original mind gone astray. From the Navy's point of view, Tesla may have been the original inventor of the radio, but he was clearly not the one who made the apparatus work.

A History of Navy Involvement

In 1899, the U.S. Navy, via Rear-Admiral Francis J. Higginson, requested Tesla to place "a system of wireless telegraphy upon Light-Vessel No. 66 [on] Nantucket Shoals, Massachusetts, which lies 60 miles south of Nantucket Island."[46] Tesla was on his way to Colorado and was unable to comply. Further, the Navy did not want to pay for the equipment, but rather wanted Tesla to outlay the funds himself. Considering the great wealth of the country, Tesla feigned astonishment at the penurious position of John D. Long, Secretary of the Navy, via Captain Perry, who brazenly forwarded the financial disclaimer on U.S. Treasury Department stationary.

Be that as it may, upon Tesla' s return to New York in 1900, he wrote again of his interest in placing the equipment aboard their ships. Rear Admiral Higginson, chairman of the Light House Board, wrote back that his committee would meet in October to discuss with the Congress "the estimates of cost."[47] Higginson, who had visited Tesla in his lab in the late 1890's, wanted to help, but he had been placed in the embarrassing position of withdrawing his offer of financial remuneration because of various levels of bureaucratic inanity. Tesla spent the time to go down to Washington to confer face-to-face with the high command, but he was given the proverbial "runaround" and he returned to New York empty handed and disgusted with the way he was treated.

From the point of view of the Navy, this was an entirely new field, and they were unsure what to do. Further, they may have been turned off by Tesla's haughty manner, particularly when it came to being "compared" to Marconi, who Tesla refused to be compared with.

In 1902, the Office of Naval Intelligence called upon Commodore F.M. Barber, who had been in retirement in France, back to the States to be put in charge of the acquisition of wireless apparatus for testing. Although still taking a frugal position, the Navy came up with approximately $12,000 for the purchase of wireless sets from different European companies for testing. Orders were placed with Slaby-Arco and Braun-Siemans-Halske of Germany and Popoff, Ducretet and Rochefort of France. Bids were also requested from DeForest, Fessenden, and Tesla in America and also Lodge-Muirhead in England. Marconi was excluded, because he arrogantly coveted an all or nothing deal.[48]

Fessenden was angry with the Navy for obtaining equipment outside the United States, and so did not submit a bid. Tesla was probably too upset with his treatment from the past, and too involved with Wardenclyffe, which was under active construction at that time, to get involved, and so the Navy purchased additional sets from DeForest and Lodge-Muirhead.

In 1903, a mock battle with the North Atlantic Fleet was held 500 miles off the coast of Cape Cod. The "White Squadron" was commanded by Rear Admiral J.H Sands, and the "Blue Squadron" by Tesla's ally, Rear Admiral Higginson. The use of wireless played a key role in determining the victor. Commander Higginson, who won the maneuver, commented, "To me, the great lesson of the search we ended today is the absolute need of wireless in the ships of the Navy. Do you know we are three years behind the times in the adoption of wireless?"[49]

Based upon comparison testing, it was determined that the Slaby-Arco system out performed all others, and the Navy ordered 20 more sets. Simultaneously, they purchased an 11 year lease on the Marconi patents.[50]

With the onset of World War I, the use of wireless became a necessity for organizing troop movements, surveillance, and intercontinental communication. While the country was still neutral, the Navy was able to continue their use of the German equipment up until sentiments began to shift irreversibly to the British side. Via the British Navy, Marconi had his transmitters positioned in Canada, Bermuda, Jamaica, Columbia, the Falkland Islands, North and South Africa, Ceylon, Australia, Singapore and Hong Kong. His was a mighty operation. In the United States, the American Marconi division, under the directorship of the politically powerful John Griggs, former governor of New Jersey and Attorney General under President McKinley, had transmitters located in New York, Massachusetts, and Illinois.[51] One key problem, however, was that the Marconi equipment was still using the outmoded spark-gap method.

In April of 1917, when the U.S. Navy took over all wireless stations, this included their allies, the British plants as well. At the same time, Marconi was in the process of purchasing the Alexanderson alternator, which was, in essence, a refinement of the

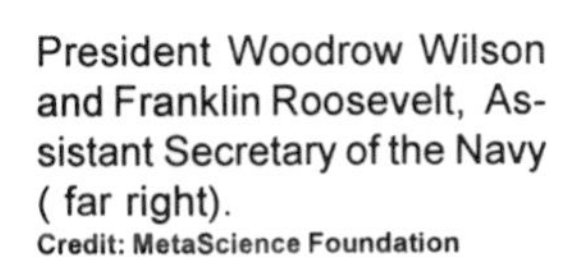

President Woodrow Wilson and Franklin Roosevelt, Assistant Secretary of the Navy (far right).
Credit: MetaScience Foundation

Tesla oscillator. Simultaneously, the Armstrong feedback circuit was becoming an obvious necessity for any wireless instrumentation. However, the Armstrong invention, created a judicial nightmare not only because it used as its core the DeForest audion, but also because DeForest's invention was overturned in the courts in favor of an electronic tube developed by Fessenden — never mind that Tesla, as far back as 1902, had beaten Fessenden in the courts for this development. With the Fessenden patent now under the control of Marconi, the courts would come to rule that no one could use the Armstrong feedback circuit without the permission of the other players. *The most important ruling, concerning the true identity of the inventor of the radio, became neatly sidestepped by the war powers act of President Wilson calling for the suspension of all patent litigation during the time of the war.* France had already recognized Tesla's priority by their high court, and Germany recognized him by Slaby's affirmations and Telefunken's decision to pay royalties; but in America, the land of Tesla's home, the government backed off and literally prevented the courts from sustaining a decision. The Marconi syndicate, in touch with kings from two countries, with equipment instituted on six continents, was simply too powerful.

The Farragut Letter

With the suspension of all patent litigation and the country in the midst of a world war, Franklin Roosevelt, Assistant Secretary of the Navy, penned the famous "Farragut Letter." This document allowed such major companies as AT&T; General Electric, Westinghouse and American Marconi the right to pool together to produce each other's equipment without concern for compensating the rightful inventors. Further it "assured contractors that the Government would assume liability in infringement suits."[52]

On July 1, 1918, Congress passed a law making the United States financially responsible for any use of "an invention described in and covered by a patent of the United States." By 1921, the United States Government had spent $40 million dollars on wireless equipment, a far cry from Secretary Long's policy of refusing to pay a few thousand dollars for Tesla's equipment 18 years before. Thus, the Interdepartmental Radio Board met to decide various claims against it. Nearly $3 million in claims were paid out. The big winners were Marconi Wireless, who received $1.2 million for equipment and installations taken over (but not for their patents); International Radio Telegraph received $700,000; AT&T, $600,000, and Edwin Armstrong, $89,000. Tesla received a minuscule compensation through Lowenstein who was awarded $23,000.[53]

In 1921, the Navy published a list of all the inventors in wireless who received compensation from them. The list contained only patents granted after 1902. Inventors included: Blockmen, Braun, Blondel, DeForest, Fuller, Hahnemann, Logwood, Meissner, Randahl, Poulsen, Schiessler, von Arco and Watkins. Note that both Tesla's and Marconi's name is missing.[54] Marconi's could be missing either because his patents had lapsed, or more likely, because they were viewed as invalid from the point of view of the government. In the case of Tesla, all of his 12 key radio patents had "expired and [were] now common property."[55] However, Tesla had renewed one fundamental patent in 1914,* and this should have been on the list, as should have Armstrong's feedback patent.

Radio Corporation of America

The United States government, through Franklin Roosevelt, knew that Marconi had infringed upon Tesla's fundamental patents. They knew the details of Tesla's rightful claims, through their own files and through the record at the patent office. In point of fact, it was Tesla's proven declaration which was the basis and central argument that the government had against Marconi when Marconi

* *Patent 1,119,732,* ***Apparatus for transmitting electrical energy,*** *was applied for January 16, 1902. The* ***application*** *was renewed May 4, 1907, and granted December 1, 1914. This patent, in essence, contains all of Tesla's key ideas behind the construction of Wardenclyffe. Armstrong's invention, although a necessity, was still tied up in court because of DeForest.*

sued in the first place, and it was this same claim, and the same Navy Lighthouse Board files that would eventually be used by the U.S. Supreme Court to vindicate Tesla three months after he died, nearly 25 years later in 1943.

Rather than deal with the truth, and with a difficult genius whose present work appeared to be in a realm above and beyond the operation of simple radio telephones and wireless transmitters, Roosevelt, Daniels, President Wilson and the U.S. Navy took no interest in protecting the Tesla tower.

It was during the height of the world conflagration, when the Smiley Steel Company's explosives expert approached the gargantuan transmitter soaring above as he circled it to place a charge around each major strut. With the Associated Press recording the event, and military personnel apparently present, the magnifying transmitter was leveled, the explosion alarming many of the Shoreham residents. Tesla became essentially a non-person the day his magnifying transmitter was leveled.

September, 1917 293

THE ELECTRICAL EXPERIMENTER

H. GERNSBACK EDITOR
H. W. SECOR ASSOCIATE EDITOR

Vol. V. Whole No. 53 September, 1917 Number 5

U. S. Blows Up Tesla Radio Tower

The destruction of Wardenclyffe's tower!
Credit: MetaScience Foundation

With the death of the Tesla World Telegraphy Center came the birth of the Radio Broadcasting Corporation, a unique conglomerate of private concerns under the auspices of the U.S. government. Meetings were held behind closed doors in Washington between President Wilson, who wanted America to gain "radio supremacy,"[56] Navy Secretary Daniels, his assistant Franklin Roosevelt, and representatives from General Electric, American Marconi, AT&T and the Westinghouse Corporation. With J.P. Morgan & Company on the Board of Directors, and the Marconi patents as the backbone of the organization, RCA was formed. It would combine resources from the mega-corporations of General Electric, American Marconi, AT&T, and Westinghouse, all who had cross-licensing agreements with each other,* and all who co-owned the company.[57]

Here was another entente cordial reminiscent of the AC polyphase days, which was not so for the originator of the invention. It was a second major time Tesla would be carved from his creation.** A secret deal was probably concocted which absolved the government for paying any licensing fee to Marconi in lieu of them burying their Tesla archives. David Sarnoff, as managing director, would soon take over the reigns of the entire operation.

The **NEW YORK SUN** somewhat inaccurately reported:

U.S. BLOWS UP TESLA RADIO TOWER

Suspecting that German spies were using the big wireless tower erected at Shoreham, L.I., about twenty years ago by Nikola Tesla, the Federal Government ordered the tower destroyed and it was recently demolished with dynamite. During the past month several strangers had been seen lurking about the place. [58]

* Cross-licensing agreements also existed with the government who also owned some wireless patents.

** Tesla would also be cut out of a secret agreement between GE and Westinghouse to hold back production of efficient fluorescent lighting equipment as they did not want to undermine the highly profitable sale of normal Edison light bulbs, nor "cut too drastically the demand for current" (Gilfillan, S.C. Invention & the Patent System. Washington, DC: U.S. Printing Office, 1964, p. 100)

> *The destruction of Nikola Tesla's famous tower... shows forcibly the great precautions being taken at this time to prevent any news of military importance of getting to the enemy.*[59]

"One of the first actions of the Board of Directors was to invite President Wilson to nominate a naval officer of the rank of captain or above...to present the Government's views and interests concerning matters pertaining to radio communication." Simultaneous with the end of the war, President Wilson also returned all confiscated radio stations to their rightful owners. American Marconi, now RCA, of course, was the big beneficiary.[60]

In 1920, the Westinghouse Corporation, was granted the right to "manufacture, use and sell apparatus covered by the [Marconi] patents."[61] Westinghouse also formed an independent radio station which became as prominent at RCA. At the end of the year, Tesla wrote a letter to E.M. Herr, president of the company, offering his wireless expertise and equipment. Westinghouse replied:

> *November 16, 1920*
>
> *Dear Mr. Tesla,*
>
> *I regret that under the present circumstances we cannot proceed further with any developments of your activities.*[62]

A few months later, Westinghouse requested that Tesla "speak to our 'invisible audience' some Thursday night in the near future [over our...] radio telephone broadcasting station."[63]

> *November 30, 1921*
>
> *Gentleman,*
>
> *Twenty-one years ago I promised a friend, the late J. Pierpont Morgan, that my world-system, then under construction... would enable the voice of a telephone subscriber to be transmitted to any point of the globe....*
>
> *I prefer to wait until my project is completed before addressing an invisible audience and beg you to excuse me.*
>
> *Very truly yours,*
> *N. Tesla*[64]

END NOTES

1. Fritz Lowenstein/Nikola Tesla 4/18/1912 [Swezey Papers, Smithsonian].
2. Marconi lecture before NY Electrical Soc. Electrical World, 4/20/1912, p. 835.
3. Sobel, Robert. RCA. New York, NY: Stein & Day, 1986, pp. 19-20; Harding, Robert. George H. Clark Radiona Collection. Smithsonian Institute, 1990.
4. Nikola Tesla/George Scherff 1/18/1913 [Library of Congress Microfilm].
5. Jolly, W. Marconi. New York, NY: Stein & Day, 1972, p. 190.
6. Nikola Tesla/JP Morgan Jr. 3/19/1914 [Archives, J. Pierpont Morgan Library].
7. Nikola Tesla/JP Morgan Jr. 7/23/1913 re: Girardeau, M.E. testimony [Archives, J.Pierpont Morgan Library].
8. 1917 royalty payment from Hochfrequenz Maschiemen Aktievgesell [no doubt, a division of Telefunken] for $1567 in 1917 [Swezey Collection, Smithsonian Institute].
9. Nikola Tesla/JP Morgan Jr. 2/19/1915 [Archives, JPM Library]; Nikola Tesla/Frank and Nikola Tesla/Pfund corresp., circa 1912-1922 [Nikola Tesla Museum]; 19 more taken as German spies, New York Times, I, 1:3; Find radio outfit in Manhattan tower, New York Times, 3/5/1918, 4:4.
10. Nikola Tesla/JP Morgan Jr. 1/6/1914 [Morgan Library].
11. Jolly, 1972.
12. Nation to take over Tuckertown plant. New York Times. 9/6/1914, II, p.14:1.
13. Nikola Tesla/JP Morgan Jr. 2/19/1915 [Morgan Library.].
14. Germans treble wireless plant. New York Times, 4/23/1915, 1:6.
15. Tesla sues Marconi. New York Times, 8/4/1915, 8:1.
16. Jolly, 1972, p. 225.
17. R & A, p. 100.
18. Marconi loses Navy suit, New York Sun, 10/3/1914 [Nikola Tesla/JP Morgan Jr. corresp.,Morgan Library.].
19. Prof. Pupin now claims wireless his invention. Los

Angeles Examiner, May 13, 1915; Ratzlaff, J. & Anderson, L. Dr. Nikola Tesla Bibliography 1884-1978. Palo Alto, CA: Ragusen Press, 1979, p. 100 [literary licence on quote].

20. When powerful high-frequency electrical generators replace the spark-gap. New York Times, 10/6/1912, VI, 4:1.

21. Marconi Wireless vs. Atlantic Communications Co., 1915 [Archives, L. Anderson].

22. Anderson, Leland (Ed.). John Stone Stone on Nikola Tesla's Priority in Radio and Continuous-Wave Radiofrequency Apparatus. The Antique Wireless Review, Vol 1, 1986.

23. Dunlap, Orin. Radio's 100 Men of Science. New York, NY: Harper & Brothers, 1944.

24. Marconi Wireless vs. United States. Cases Adjudged in the Supreme Court. 10/1942, v. 320, p. 17. This feature was obviously also part of Tesla's design, although the court eventually ruled Stone as the originator.

25. Anderson, L. John S. Stone on Nikola Tesla, 1986, pp. 37 to 40.

26. E.F. Sweet and FDR correspondence re: Tesla, 9/14/1916; 9/16/1916; 9/26/1916 [Nat. Arch.].

27. Nikola Tesla/Light House Board corresp., 8/11/1899 [Nat. Arch.].

28. Anderson, Leland. Priority in the invention of the radio: Tesla vs. Marconi. The Tesla Journal, vol. 2/3, 1982/83, pp. 17-20.

29. Nikola Tesla/JP Morgan Jr. 7/1915 [Library of Congress Microfilm].

30. Germany to sink the Armenian. Navy may Seize Sayville Wireless. New York Times, 7/1/1915, 1:4-7.

31. Wireless controls German air torpedo. New York Times, 7/10/1915, 3:6,7.

32. Nikola Tesla. Science and discovery are the great forces which will lead to the consummation of the war. New York Sun, 12/20/1914, in Nikola Tesla, 1956, pp. A-162-171.

33. Federal Agents raid offices once occupied by Telefunken. Former employee Richard Pfund charged; no arrests made. New York Times, 3/5/1918, 4:4.

34. Nikola Tesla/George Scherff 12/25/1917 [Library of Congress Microfilm].

35. Royalty check to Nikola Tesla for $1567 from Hochfrequenz Maschienen Aktievgesell Schaft for drachlose Telegraphic, 1917 [Swezey Collection, Smithsonian Institute]. Tuckertown was still owned by the Germans, although seized by the U.S. Navy, and Tuckertown, with full knowledge of the "Director of Naval Communicatons," had agreed to pay Tesla royalties, see Nikola Tesla/George Scherff 10/12/1717 [Library of Congess Microfilm].

36. Tesla no money; Wizard swamped by debts. New York World, 3/16?/1916.

37. Nikola Tesla's Fountain. Sci Am, 1915.

38. Lester S. Holmes was represented for the Hotel as owner of said Tesla property. Baldwin & Hutchins/Nikola Tesla corresp., 7/13/1917, from: Wardenclyffe property foreclosure proceedings, NY Supreme Ct., circa 1923 [L. Anderson files].

39. Tesla's new device like bolts of Thor. New York Times, 12/8/1915, 8:3.

40. Nikola Tesla/JP Morgan Jr. 4/8/1916 [L. Anderson files].

41. Nikola Tesla/JP Morgan Jr. 2/19/1915 [Morgan Library.].

42. Reason for seizing wireless. New York Times, 2/9/1917, 6:5.

43. Spies on ship movements. New York Times, 2/17/1917, 8:2.

44. 19 More taken as German spies. New York Times, 4/8/1917, 1:3.

45. Navy to take over all radio stations. New York Time, 4/7/1917, 2:2.

46. FJ. Higginson/Nikola Tesla 5/11/1899 [Nat. Arch.].

47. F. Higgenson/Nikola Tesla 8/8/1900 [Nat. Arch.]. For the full correspondence of this event, see Chapter 27, Tesla Trilogy by Marc Seifer.

48. Howeth, L.S. History of Communications-Electronics in U.S. Navy. Washington, DC: U.S. Government Printing Office, 1963, pp. 518-519; Hezlet, A. Electronics & Sea Power. New York,, NY: Stein & Day, 1975, p. 41.

49. Howeth, 1963, p. 64.

50. Hezlet, 1975, pp. 41-42.

51. Sobel, 1986, p. 43; Hezlet, 1975, p. 77.

52. Howeth, 1963, p. 256.

53. Howeth, 1963, pp. 375-376; George Scherff/Nikola Tesla corresp.[Library of Congress Microfilm].

54. Howeth, 1963, pp. 577-580.

55. Nikola Tesla. Electric drive for battleships. New York Herald, 2/25/1917; in Nikola Tesla, 1956, p. A-185.

56. Howeth, 1963, p. 354.

57. The breakdown was as follows: GE 30%, Westinghouse 20%, AT&T 10%, United Fruit 4%, others 34%. Sobel, 1986, pp. 32-35.

58. US. blows up Tesla radio tower. Electrical Experimenter, 9/1917, p. 293.

59. Destruction of Tesla's tower at Shoreham, LI hints of spies,'New York Sun, 8/5/1917.

60. Howeth, 1963, pp. 359-360.

61. Howeth, 1963, p. 361.

62. E.M. Herr/Nikola Tesla 11/16/1920 [Library of Congress Microfilm].

63. GW Corp./Nikola Tesla 11/28/1921 [Library of Congress Microfilm].

64. Nikola Tesla/GW Co. 11/30/1921 [Library of Congress Microfilm].

ABOUT THE AUTHOR:

DR. MARC J. SEIFER is one of the world's leading Tesla experts. A graduate of Saybrook Institute and the University of Chicago, he has lectured at the United Nations, Brandeis University, West Point, in Zagreb Yugoslavia at the University of Vancouver, in Jerusalem Israel, at Cambridge University, Oxford University and at numerous conferences throughout the United States. Featured in *The New York Times, New Scientist, Rhode Island Monthly, The Washington Post* and on the back cover of Uri Geller's book *Mind Medicine,* his publications include articles in *Civilization, Lawyer's Weekly* and *Wired.* His works include ***STARETZ ENCOUNTER*** (novel), ***THE BIG GUY*** (true crime), ***THE GURU WITHIN*** (metaphysics) and the definitive Tesla biography ***WIZARD : THE LIFE & TIMES OF NIKOLA TESLA*** (Citadel Press/Kensington). HIGHLY RECOMMENDED by the American Academy for the Advancement of Science, ***WIZARD*** has also been performed as a screenplay reading at Producer's Club Theater in New York, screenplay co-written with Tim Eaton, visual effects editor at Industrial Light & Magic.

DR. MARC J. SEIFER

Tesla Coil Instrumentation

Dr. Gary L. Johnson
Electrical Engineering Department
Kansas State University
Manhattan, KS 66506

Abstract

A fiber optic system is described for measuring the voltage at the top of a Tesla coil secondary or extra coil. An APEX PA-19 op amp, rated at 40 V and 3 A, slew rate of 650 V/μs, is used to drive the coil as a quarter-wave antenna above a ground plane. A capacitor divider is used to get a small voltage proportional to the electric field at the top of the coil. A battery powered circuit with a Motorola MFOE76 LED converts the small voltage to an optical signal which is sent by plastic optical fiber to a receiver a safe distance away. A Motorola MFOD71 PIN diode converts the optical signal back to electrical for display on an oscilloscope. The system works well for examining waveforms and relative voltage magnitudes, but needs a separate calibration to measure absolute values of voltage.

Introduction

Tesla coils need to be tuned or optimized for maximum performance. Even after a century of development, this still remains an art rather than a science. A typical instruction is "Tune for most smoke" (a saying I heard from Harry Goldman, editor of the Tesla Coil Builders Association, who in turn heard it from Gary Legel and Bill Wysock). That is, a person building a Tesla coil will apply nearly full voltage to it and visually observe the length and color of the discharges. Some parameter will be adjusted and the process repeated. While testing is proceeding, components are being stressed because of the non-optimum tuning conditions. Eventually, an acceptable operating condition is reached. Sometimes, however, satisfactory coil performance is never obtained. When Robert Golka moved his large coil from Utah to Colorado, performance was less than expected, partly because of a lack of appropriate equipment to tune the coil.[1,2]

High powered radio transmitters will typically be tuned in a low power mode. Not all sections of a Tesla coil can be tuned at low power because the spark gap, for example, only functions at voltages near rated. However, some parts of the system can be optimized at low power, especially the secondary or extra coil.

A power amplifier suitable for such tuning will be described, followed by a detailed discussion of fiber optics since this material may not be well known by the readers. Specific circuits for a fiber optic transmitter and a fiber optic receiver will then be shown.

The Secondary or Extra Coil

Tesla coils typically have either two or three stages of voltage increase. Smaller units will have an iron-core transformer that steps up the 60 Hz supply voltage to perhaps 12 kV, then a loosely coupled air-core transformer to provide the final increase. Larger systems will have the iron-core transformer, a tightly coupled air-core transformer, and then an 'extra' coil for a third stage of voltage increase.

The last stage of voltage increase, either the air-cored transformer secondary or the extra coil, looks like a vertical helical coil above a ground plane. This is difficult to analyze theoretically. A somewhat related geometry which is much easier to analyze is that of a helical conductor in a coaxial cable. This is a standard procedure for electromagnetic theorists. The really interesting problems are unsolvable, so we find a related problem that we can solve, and then argue that the solution provides insight to the original problem. The geometry for this case is shown in Figure 1.

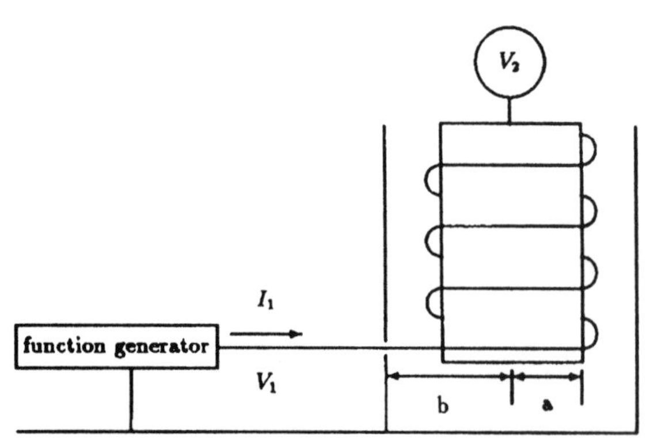

Figure 1. Short Helical Coil Transmission

Some arguments for the last step come from the Corums. They present[3] a number of sketches showing how the walls of a coaxial helical resonator can be folded down to become a ground plane. That is, we start with a coaxial helical resonator, which looks somewhat like a transmission line with a helical inner conductor. In turn, this becomes a helical antenna above a ground plane if we stretch and push on the cylindrical outer conductor until it lays down flat into a ground plane.

We will not go through this analysis here. Rather, we want to discuss driving the air-cored transformer or the extra coil with a function generator, at voltages below the breakdown of air, and experimentally examining the voltage waveforms.

The characteristic impedance of a transmission line with a helical inner conductor is quite high, in the range of 5000 to 10,000 ohms. However, the input impedance of the secondary or extra coil is the characteristic impedance divided by the standing wave ratio on the coil, which may be 100 or more. The input impedance is therefore on the order of 50 ohms or less.

At resonance, the input impedance Z_{in} is real, so the input voltage and current are in phase. The real power input to the coil is then easy to calculate.

$$P_{in} = V_1 I_1 = \frac{V_1^2}{Z_{in}} = I_1^2 Z_{in} \qquad (1)$$

For the above example, if the rms input voltage is 20 V and the input impedance is 40Ω, the input power is $20^2/40 = 10$ W. The output voltage will be the input voltage multiplied by the standing wave ratio, or perhaps 2000 V. This is far below the operating voltage of a Tesla coil, but still too high to measure with a conventional oscilloscope probe. The fiber optic measurement technique presented here is a way to avoid using an oscilloscope probe.

The APEX PA-19 Power Amplifier

Standard small laboratory function generators have trouble driving a high Q, low impedance circuit like the input to a Tesla air-cored transformer. As one tunes through resonance, the generator frequency tends to jump a few kHz. This makes it difficult to find the exact resonant point. It is also desirable to drive the air-cored transformer with a somewhat greater voltage than is typically available from the small function generators, so the difference between tuning voltage and operating voltage is not so great. One solution to this problem is to build a power amplifier to go between the readily available small function generator and the driven secondary or extra coil, as shown in Figure 2.

The power amplifier needs to be untuned or at least broad band, so it can operate over the range of perhaps 30 kHz for very large coils up to about 1 MHz for the small coils. Power op amps are now available which cover this frequency range. One example is the APEX PA—19. These are rather expensive devices, costing $220 each. The slew rate is 650 V/μs in the appropriate circuit configuration. The op amp will deliver ±30 V at 3 A up to a frequency of 3 MHz. To fabricate the circuit, a ±36 V, 4 A open frame power supply was obtained from PDS Technology for $525. The output capacitors of the power supply were moved over so there was room for the PA—19 and its heat sink on the power supply frame. Two BNC connectors were mounted on the frame, one for input from a small function generator and one for output to the air-cored transformer.

The circuit used was very conventional, as shown in Figure 3. The two resistors can be the quarter-

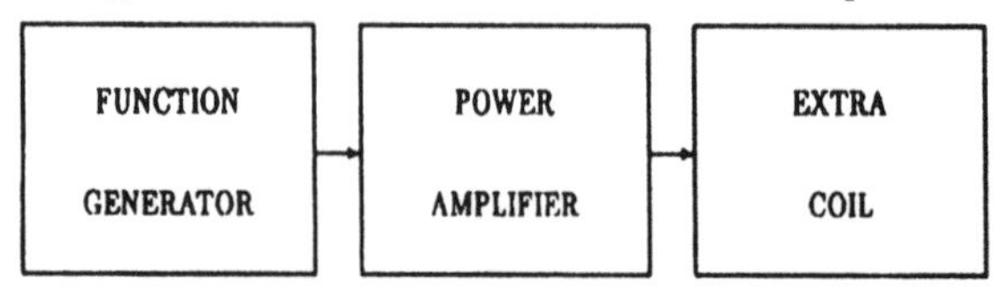

Figure 2. Extra Coil Driven by Power Amplifier

watt size. The function generator sees a constant resistive load of 220Ω, which it can drive easily.

One of the disadvantages of high performance op amps is that they tend to oscillate. Anyone who does circuit layout with the casualness appropriate to a 741 op amp will build an oscillator rather than an amplifier. When the application note says that circuit layout is critical, it means it. What finally worked for us was a piece of what is sometimes referred to as waffle board. This is a type of printed circuit board with a heavy ground plane on one side of a dielectric board and a pattern of small conducting squares on the other side. Holes are drilled through the board to accommodate the eight pins of the op amp socket. A small amount of ground plane is scraped off around each pin to prevent accidental shorts to ground. The ground plane is next to the socket so the waffle side is used for components. The waffles are bridged across with

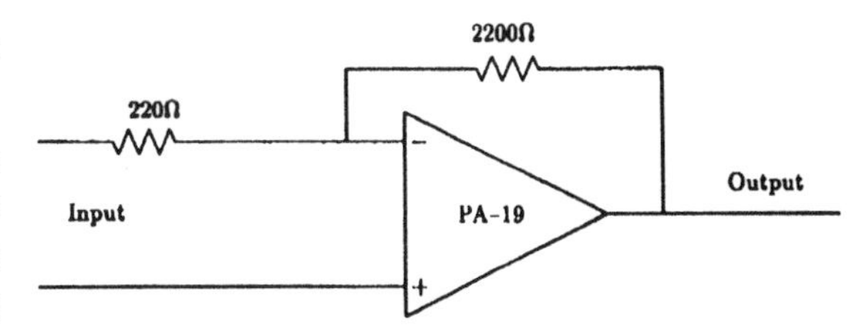

Figure 3. Power Amplifier Circuit

solder to form pads as necessary. Holes are drilled through the board for the ground lead of the bypass capacitors, which need to be high frequency 0.1 μF ceramics. All leads are kept as short as possible. Short pieces of coaxial cable are used to connect between the board and the BNC connectors.

The final result was an amplifier that was able to drive loads of as low as 10 Ω input impedance in a relatively robust manner.

Fiber Optic Technology

We are using plastic optical fibers to carry information about the voltage at the top of a Tesla coil to an oscilloscope several meters away. This application of fiber optics is not well known, and good reference books are difficult to find, so the broad field of fiber optics will be reviewed here. Glass fibers will be included also in order to make some important comparisons.

Fiber optic technology is a method of carrying information from one point to another. An optical fiber is a thin strand of glass or plastic that serves as the transmission medium over which the information passes. It serves the same function as copper wires carrying a signal. But the signal is flowing as photons of light rather than as electrons in a wire. In most cases, the signal starts as an electrical waveform in a wire and needs to end in the same way, so we must convert from electrons to photons at the transmitting end, and from photons to electrons at the receiving end, as indicated by the block diagram in Figure 4. The actual device that converts from an electrical waveform to light is called an *emitter* while the device that converts from light to an electrical signal is called a *detector*.

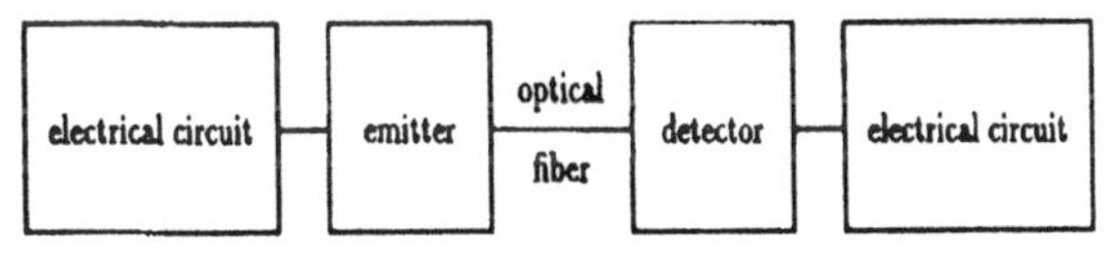

Figure 4. Fiber Optic Circuit

These extra steps must be justified by one or more of the following advantages of fiber optics[4] over wire communications:

1. ***Wide bandwidth.*** A high quality coaxial cable can carry up to perhaps 1000 telephone conversations. A similar quality fiber optic system can easily carry over 10,000 telephone channels.
2. ***Electromagnetic immunity.*** Optical fibers do not radiate electromagnetic waves to interfere with others, not can they receive electric, magnetic, or optical signals from outside the fiber. There is no counterpart to radio frequency interference (RFI) or electromagnetic interference (EMI) that is common with electrical circuits.
3. ***Low loss.*** The best transatlantic telephone coaxial cables require repeaters about every 10 km to boost the signal and overcome the effects of line losses. Similar telephone fiber optic systems need repeaters only every 30 km or even further apart.
4. ***Light weight.*** A coaxial cable weighs about ten times as much as an optical fiber with the same channel capacity. This can be significant in applications like aircraft or satellites.
5. **Small size.** Optical fibers have cross sectional ar-

eas well under ten percent of the area of comparable coaxial cables. More fibers can be placed in a given space, or fibers can go in smaller places than possible for coaxial cables. This can be very important in aircraft, submarines, underground telephone conduits, especially in large cities. This is also very helpful in making connections between computers.

6. ***Safety.*** A broken optical fiber does not produce sparks, so fibers can be run directly through fuel tanks and other hazardous locations where copper wires may not be advisable. Fibers can be used to monitor variables in high voltage locations since fibers do not conduct electricity.
7. ***Security.*** Optical fibers do not radiate energy that can be received by a nearby spy antenna. It is very difficult to tap a fiber, and even more difficult to make the tap undetected. A fiber optics system is considered to provide a secure communications path.
8. ***Ruggedness.*** Optical fibers are not chemically affected by most of the substances that react with copper or aluminum.

Different advantages are important to each application. For modern communication systems involving telephone, television, and computer networks, the wide bandwidth and low loss are essential. Another application might require making measurements in a hostile environment, such as near a Tesla coil, where immunity from interference is the most important advantage. The advantage of safety is certainly important in measuring voltage and current on high voltage power transmission lines.

Optical fibers are made of two different materials, glass and plastic. The glass used is ultrapure, ultratransparent silicon dioxide or fused quartz. Impurities are added to achieve the desired index of refraction. Germanium or phosphorous increases the index of refraction while boron or fluorine decreases the index. These impurities and others also tend to absorb or scatter the light traveling down the fiber, thus increasing the attenuation of the signal.

Glass fibers are usually made of a central glass core surrounded by a glass cladding. There will also be a plastic buffer coating around the cladding for mechanical protection. The outer diameter of the cladding is typically 125 μm, which is thin enough that the glass can be easily bent without breaking.

Plastic fibers have both a plastic core and a plastic cladding. The cladding serves as mechanical protection also, so no buffer layer is needed. The standard core diameter is 1000 μm, much thicker than a glass core. This makes plastic fibers easier to handle and to make connections. Plastic fibers and the associated emitters and detectors are also cheaper than their glass counterparts. Unfortunately, plastic fibers have much higher losses, as shown in Table 1. This basically restricts their use to applications involving electromagnetic immunity and safety where data rates and bandwidths do not have to be high.

Plastic fibers are also used in motor vehicles rather than glass because of their resistance to effects of temperature and humidity variations. Glass fibers tend to develop microcracks during vibration, which then attract moisture during high humidity conditions. This moisture increases the losses dramatically for the glass fibers, possibly to where glass is no better than plastic.

Single mode refers to the case of a circular waveguide where only a single solution to Maxwell's equations is allowed by the combination of frequency and waveguide size. A light ray can take only a single path in traveling down the fiber. Single mode fibers have bandwidths of several GHz and allow transmission of signals for tens of kilometers between repeater stations.

As the core diameter gets larger, there are many more allowable solutions to Maxwell's equations. These are usually covered in a second course in

TABLE 1

Optical Fiber Specifications

From *Technician's Guide to AMP Fiber Optics*

Type	Core Dia(μm)	Cladding Dia(μm)	Attenuation (dB/km)
Single mode	8	125	0.5 @ 1300 nm
Single mode	5	125	0.4 @ 1300 nm
Graded index	50	125	4 @ 850 nm
Graded index	63	125	7 @ 850 nm
Graded index	85	125	6 @ 850 nm
Graded index	100	140	5 @ 850 nm
Step index	200	380	6 @ 850 nm
Step index	300	440	6 @ 850 nm
PCS	200	350	10 @ 790 nm
PCS	400	550	10 @ 790 nm
PCS	600	900	6 @ 790 nm
Plastic	—	1000	400 @ 650 nm

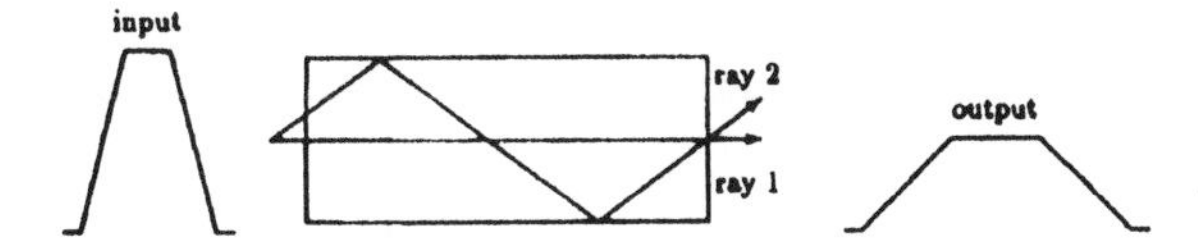

Figure 5. Dispersion in a Multimode Fiber

electromagnetic theory and are beyond the scope of this paper. Multimode fibers allow light rays to take different paths along the fiber, as illustrated in Fig. 5. The ray that goes straight down the center of the core arrives at the receiving end first. Other rays reflect off the sides of the waveguide (or refract in regions where the index of refraction is different) and therefore travel a longer distance than the ray going down the center. If the index of refraction is constant, as is the case for step index, plastic-clad silica (PCS), and plastic fibers, then these other rays take longer to get to the other end. A tall and thin pulse at the sending end becomes a short and wide pulse at the receiving end because of this *dispersion*.

For example, ray 2 takes a long path that requires 15 ns longer to travel one km than ray 1 taking the shortest path. Consider a series of 5 ns pulses with 5 ns spacing. At the 1 km point, the original 5 ns pulse has spread out to be 5 + 15 or 20 ns long. It now overlaps the following pulse, making it difficult to extract information from the pulse train. The dispersion actually limits the product of frequency and fiber length. The 1 km fiber would handle 100 ns pulses with 100 ns spacing nicely since the pulses are much longer. Likewise the same fiber would handle the original pulse train in lengths up to 100 or 200 m where the cumulative effect of dispersion is much smaller.

The *graded index* fiber has an index of refraction that is highest at the center and gradually decreases until it matches that of the cladding. There is no sharp break between the core and the cladding. The core actually has many concentric layers of glass, somewhat like the annular rings of a tree. Each successive layer outward from the central axis of the core has a lower index of refraction. The reason for this grading is that light travels faster in a lower index of refraction. Therefore, the further the light is from the center axis, the greater is its speed. The light experiences a series of refractions rather than a single sharp reflection as in the step index fiber. The light traveling the longer path has just the right amount of increased speed so it arrives at the other end at the same time as a ray traveling directly down the center of the fiber. It is possible to reduce dispersion to 1 ns/km or less with this technique. The effective bandwidth, after safety factors are considered, is on the order of 200 MHz-km.

The right hand column of Table 1 refers to the attenuation of light traveling down the fiber in dB/km. Attenuation is defined as

$$\text{Loss} = 10\log\frac{P_{out}}{P_{in}} \tag{2}$$

where P_{in} is the optical power entering the fiber and P_{out} is the optical power leaving the fiber. The logarithm is to base 10. The calculation results in a negative number, but the minus sign is often omitted when there is no possibility of confusion. The purist would be perfectly correct in saying that the loss is -6 dB/km, but most engineers call it 6 db/km since everyone knows that the light output is smaller than the light input.

Table 2 shows the amount of power remaining for different decibel values. A 3-dB loss represents a loss of 50% of the power. A 10-db loss represents a loss of 90% of the original power; only 10% remains. Another 10-dB loss means that 90% of the remaining power is lost, or 99% of the original power. A total loss of 30-dB means that 99.9% of the original power is lost. The emitter produces 1000 μW of power but only 1μW reaches the detector. Many fiber optic links will perform quite acceptably with 30-dB of loss.

TABLE 2
Power Loss in dB

Loss(dB)	Power Remaining (%)
0.1	97.7
0.5	89.1
1	79.4
2	63.1
3	50.1
6	25.1
10	10.0
20	1.0
30	0.1
40	0.01

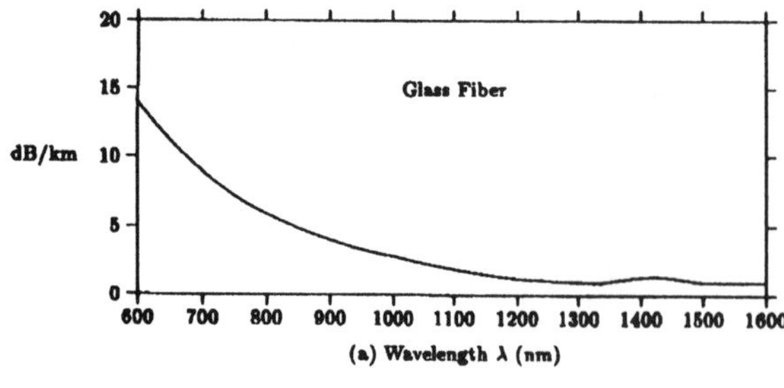

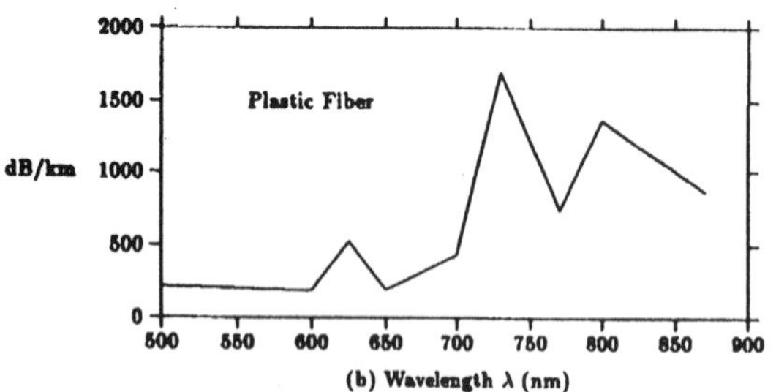

Figure 6. Fiber Attenuation Versus Wavelength

Light is attenuated in a fiber by impurities, surface roughness, moisture, and by the dielectric material itself. The attenuation in glass increases rapidly with frequency in the near infrared and visible region, which means that we need to operate glass fibers at low frequencies or long wavelengths. Plastic, on the other hand has its lowest attenuation in the visible range. The attenuation in both glass and plastic as a function of wavelength[4] is shown in Figure 6. The visible spectrum extends from approximately 380 to 720 nm. Glass has an attenuation greater than 5 dB/km in the visible range, but which drops to less than 1 dB/km by around 1300 nm. The highest performance fiber optic systems therefore operate at wavelengths near 1300 nm, as indicated in Table 1.

Plastic has much higher attenuation than glass for all wavelengths, ranging from 200 to 400 dB/km in the visible region to nearly 2000 dB/km at some wavelengths in the infrared region. The dip around 650 nm, in the visible red range, is therefore the primary wavelength for plastic fibers.

The glass fibers can be used for communications at rates from near dc to over 1 GHz over distances from a few meters to many kilometers. The plastic fibers can be used for frequencies up to perhaps 20 MHz and distances up to perhaps 100 m. If we need, or think we might need, high data rates or long distances, then we must use glass fibers. For this reason, most of the written material on fiber optics does not even mention plastic fibers. The assumption is that high data rates might be required, in which case there is no point in discussing plastic fibers.

There are many applications in high voltage systems, however, where neither high data rates nor long distances are required. The choice can be made between glass and plastic based on cost, convenience, power consumption, and other factors. Much of the time plastic fibers will be an appropriate choice, but there will be cases, especially in certain types of sensors, where glass must be used. If very fast rise times must be present, such as in power MOSFET gate drivers, then it may be necessary to use glass fibers to take advantage of the high speed light sources, even though the switching frequency is relatively low. We will therefore continue to discuss the technology of both glass and plastic fibers in enough detail that informed decisions can be made between them.

Sources

The source of light in a fiber optic system is either a light emitting diode (LED) or a laser diode. Both are small semiconductor chips that emit light when current flows through them. They serve as electrooptic transducers by converting an electrical signal to an optical signal.

We recall from our first electronics course that a forward biased diode has holes and electrons sweeping across the pn junction and recombining. Energy is emitted when a hole and electron recombine. This energy may or may not be in the form of light, depending on the atomic structure. The recombination energy for silicon, for example, is in the form of heat or lattice vibrations.

Lasers and LEDs use elements from Groups III and V of the periodic table. Gallium arsenide (GaAs) and gallium aluminum arsenide (GaAlAs) are commonly used. Gallium atoms have three valence electrons while arsenic atoms have five valence electrons. The wavelength can be varied by adding other materials (such as aluminum, indium, or phosphate). When more than two elements are used, the wavelength depends on the ratios of the elements. Motorola, for example, has two different emitters fabricated from GaAlAs, one operating at 660 nanometers (nm), in the visible red region, and the other at 850 nm in the near infrared region.

They also have a GaAs emitter operating at 940 nm. Sources made of gallium indium arsenide phosphate (GaInAsP) operate at the even longer wavelength of 1300 nm, and would be the proper source for the very low loss single mode glass fibers. The 940 nm emitters are the cheapest, but this wavelength is not optimum for commonly used detectors. Performance is sufficient, however, to cause these emitters to be used in nearly all optoisolators. The 850 nm emitters are more efficient, faster, and more expensive than the 940 nm emitters. They are used where efficiency and speed are more important than cost.

The high attenuation of plastic fiber is more important than the detector efficiency, so most plastic fibers operate at wavelengths around 650 nm for this reason.

LEDs

A typical rectifier diode is not very effective as a LED, even when built of the proper materials, since it radiates light in all directions. Only a small fraction of the light produced can be coupled into an optical fiber. A different type of structure, called a heterojunction, is needed. Such a structure is shown in Figure 7. Instead of one p-type layer and one n-type layer, it has two p-type layers and three n-type layers. The layers above and below the pn junction itself serve to refract the light back toward the junction. Most of the light is emitted from the front and back edges of the LED. If a reflective material is deposited on the back surface, then the light is forced to leave from the front edge.

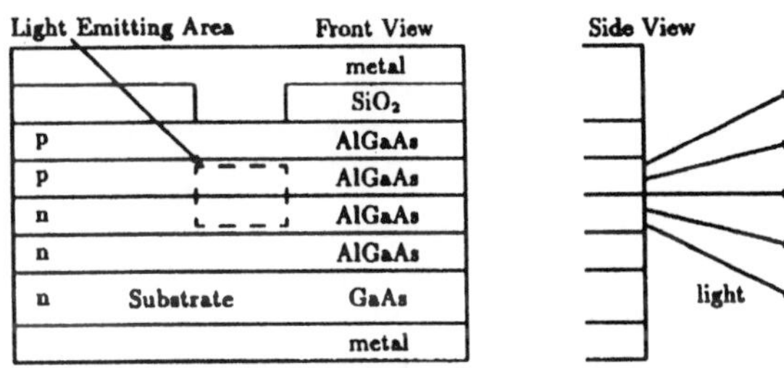

Figure 7. Edge Emitting LED

A slot is etched in the silicon dioxide insulation along the top of the junction and metal is deposited in the slot to form a narrow electrode. The electrode width determines the width of the emitting area. Emission is therefore from a rectangular area, although practical construction details cause the emission pattern to actually be oval or elliptical in nature.

The layers of material with different refractive indices form what is called an optical waveguide. The waveguide effect is similar to that observed in a cylindrical optical fiber. (The mode structure is slightly different, of course).

Lasers

The word *laser* is an acronym for Light Amplification by the Stimulated Emission of Radiation. A semiconductor laser is made in basically the same manner as a LED except that the two ends of the optical waveguide are very highly reflective. This forms an optical cavity that is necessary for laser operation. The semiconductor chip is made a little larger than necessary and the two ends are split off by a cleaving operation. If this is done properly, the chip splits along a plane of the crystal lattice so that the two surfaces are exactly parallel to each other and also have a highly reflective mirrorlike finish.

At low drive currents, the laser functions like a LED. Light is emitted spontaneously as holes and electrons recombine. A threshold level is reached as the current increases, above which laser action begins. Some of the photons emitted by the recombinations are trapped in the optical cavity, and reflect back and forth between the two end mirrors. These photons can initiate recombination of holes and electrons. A photon hits an excited atom (electron in the conduction band) and *stimulates* the electron to emit a photon while moving to the valence band and recombining with a hole. This stimulated photon is a duplicate of the first photon, with the same wavelength, phase, and direction of travel. Stimulated photons release other stimulated photons in an amplification process. Some photons escape from the end faces in an intense

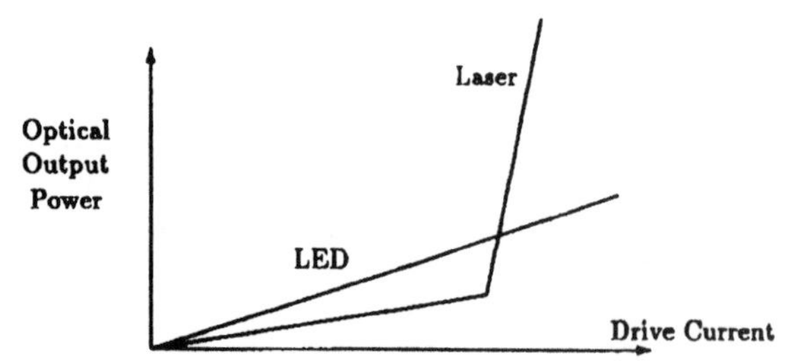

Figure 8. Optical Power versus Drive Current

beam of light. Since emission of light is usually desired to be in only one direction, the rear face is usually coated with additional reflective material to reduce the losses at that surface and increase the light emission from the front face.

Laser action only occurs when a high percentage of the atoms near the pn junction are in the excited state, which only occurs with a high drive current. This situation is called *population inversion.* When most of the atoms are in the ground state, an arriving photon merely moves a valence electron into the conduction band. Thus photons are absorbed by ground level atoms. The extra photons released by excited atoms have to overcome the photon losses to ground level atoms, hence the need for a high drive current to produce many excited atoms. The variation in light produced as a function of drive current is shown in Figure 8, for both the laser and LED. For low drive levels, the LED emits more light because of the lack of highly reflective surfaces. Once laser action begins, however, the laser produces several times as much light as the LED.

In addition to the greater output optical power, there are several other important differences between LEDs and lasers. In particular, the laser is:

1. ***Monochromatic.*** The emitted light is in a narrow band of wavelengths, as compared to a much wider band for LEDs.
2. ***Coherent.*** All the light is in phase, as compared to random photon phases for the LED.
3. ***Directional.*** Light is emitted in a highly directional pattern with little divergence. That is, the beam of light does not spread out very much as it leaves the laser.

Semiconductor lasers typically operate in the infrared region, where the light is invisible to the human eye. Even though it cannot be seen, it is still powerful enough to permanently damage the retina of the eye. One should never look directly into a laser or even a high-radiance LED unless it is certain that the device is off. The same holds true for a glass fiber which has low enough attenuation to transmit possibly hazardous light levels to the output end.

Detectors

The detector is an optoelectronic transducer which converts optical energy to electrical energy. The simplest detector is an ordinary pn junction made so light can enter the depletion region. This *photodiode* will be operated in a reverse bias mode as shown in Figure 9. The applied voltage will produce a wide depletion region, so only a small current will flow under dark conditions.

Photons incident on the depletion region will produce hole-electron pairs. These charge carriers will move under the influence of the applied electric field, and thus will cause the reverse bias current to increase. This current can then be amplified to produce the original signal in the light.

The standard PN junction is not suitable as a photodiode for most applications for two reasons. First, the depletion region is a relatively small portion of the total diode volume, so many photons are absorbed outside the depletion region. The resulting charge carriers do not experience a very large electric field so tend not to move very fast. Most of them recombine in the diode without contributing to the external current, which lowers the efficiency of the photodiode. Second, the remaining charge carriers that eventually migrate to the depletion region before recombining contribute to a delayed current. This slow response is called the *slow tail response* and limits operation of the standard PN diode to the kHz range.[4]

A much more effective device is the PIN photodiode shown in Figure 10. A relatively wide intrinsic

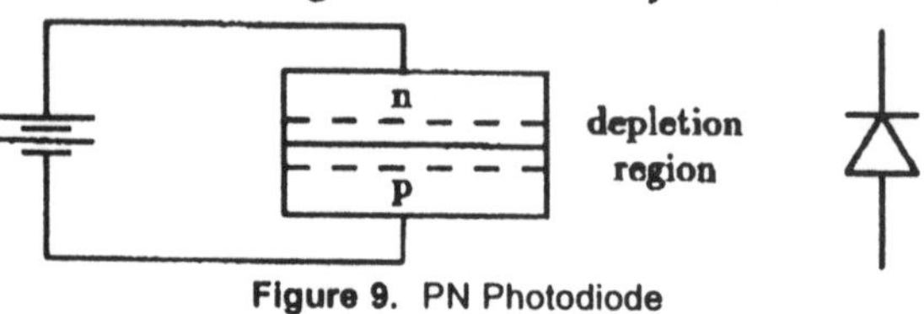

Figure 9. PN Photodiode

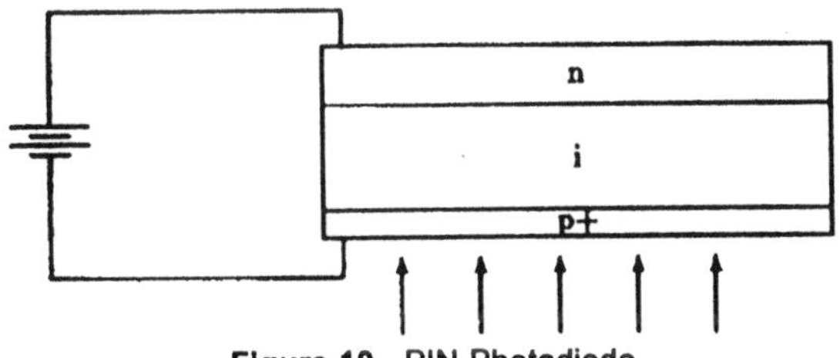

Figure 10. PIN Photodiode

layer is placed between the p and n layers. The letters describing the three layers give the device its name. The depletion region extends from the p region across the intrinsic region and into the n region, hence is quite large. The p layer is heavily doped and is relatively thin. Most of the incident photons get past the thin portion of the p layer that is not in the depletion region, but then find themselves in a thick depletion region in which the large majority will be absorbed.

The ratio of the diode current to the input optical power is called the *responsivity R*, with units of A/W (or μA/μW).

$$R = \frac{\text{output current}}{\text{input optical power}} \quad (3)$$

Typical values for R range from 0.2 to 0.6 μA/μW. A responsivity of 0.2 μA/μW means that an incident light of 100 μW will produce a current flow of 20 μA.

Responsivity varies with wavelength and with the type of material used in the photodiode. Plots of responsivity for silicon, germanium, and indium gallium arsenide are shown in Figure 11. Silicon has a wider bandgap than either germanium or InGaAs, so it takes more energy to create a hole electron pair. Photon energy is proportional to frequency or inversely proportional to wavelength, so a photon at 700 nm has twice the energy as a photon at 1400 nm. It therefore takes twice as many 1400 nm photons to make the same power level as a given number of 700 nm photons. But the number of electrons is proportional to the number of photons, so we get more current from two 1400 nm photons striking germanium than we do from one 700 nm photon on silicon.

Once the energy of each photon is adequate to move every hit electron from the valence band to the conduction band, any additional energy is wasted, so to speak. The optical power is increasing, but the number of electrons or the amount of current is remaining constant, so the responsivity is decreasing. This is shown as the slope of the responsivity curves to the left of the peaks.

An important parameter of a photodiode is the *dark current*. It is a thermally generated current or noise current that increases about 10% for every increase of 1°C of temperature. It is much lower in silicon photodiodes around 800 nm than in germanium or indium gallium arsenide photodiodes around 1400 nm.[4] This limits the minimum optical power that can be detected. The minimum detectable power, also called the *noise floor*, is the ratio of dark current to responsivity.

$$\text{Minimum detectable power} = \frac{\text{dark current}}{\text{responsivity}} \quad (4)$$

For example, the Motorola MFOD71, a silicon photodiode, has a typical dark current of 10 nA at 85°C and a typical responsivity of 0.2 μA/μW. The minimum detectable power would then be

$$\text{Minimum detectable power} = \frac{10 \times 10^{-9}}{0.2} = 0.05\mu\text{W}.$$

Voltage Measurement

Now that we have seen the broad picture of fiber optics, we return to the specific problem of measuring the voltage V_2 at the top of the Tesla coil. In particular we are interested in three aspects of the voltage.

1. Waveform
2. Relative Amplitude
3. Absolute Amplitude

A classical Tesla coil is rich in harmonics and beat frequencies. The waveform can be used to deter-

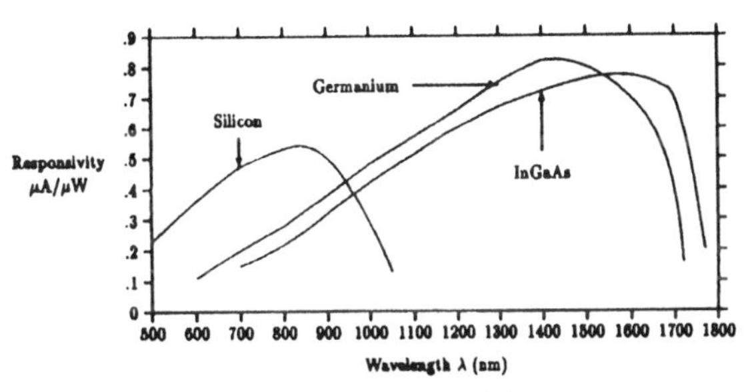

Figure 11. Responsivity

mine such things as the proper timing of the spark gap opening and closing. Proper operation would indicate a strong fundamental component and a barely detectable second harmonic. A double humped waveform indicates a strong second harmonic and improper operation. This can happen easily when using both a secondary and an extra coil. Both the secondary and the extra coil are tuned to their quarter wave resonant frequencies. When connected together, we have two quarter wave sections in series, and presumably a large voltage increase in each section. However, unless very careful attention is paid to relative coil sizes and impedances, what we actually have is a half wave structure at the driving frequency and therefore a quarter wave structure at half the driving frequency. There is much greater voltage increase in a quarter wave system, so relatively small subharmonics in the driving voltage will appear as a large voltage at the top. The waveform will be double humped. There are many such subtle problems in Tesla coils, and one is strictly groping in the dark without knowing the waveform at the top.

The relative magnitude is important in tuning the coil. We can change the primary inductance or capacitance, the height, size, and shape of the top loading element (toroid or sphere), the quality of the grounding system, and other parameters in order to get the highest possible voltage out of the coil. This can be done at tuning voltages rather than full operating voltages, and is more accurate than trying to estimate spark lengths.

The fiber optic system to be described does not measure the absolute voltage directly. A separate measurement must be made to determine the calibration factor. This is not easy to do, and hopefully will be the subject of another paper. Of course, knowing that a particular coil operates at 1.4 MV rather than 1.2 MV, for example, has its major application in boosting our egos, rather like bragging about the speed and memory size of our computer. There has not been any method of measuring this voltage in the past, so we have extrapolated from data collected at lower voltages. If voltage breakdown occurs at 30 kV between two clean brass spheres one cm apart in dry air, then the assumption is that a spark of 1 m length must correspond to a voltage of 3 MV. This figure has always been recognized as too high for Tesla coil sparks, but a value of 1 MV per meter has been used in the past. The range of possible values is illustrated in the following exchange in the **TCBA News**.[5]

> ***Question.*** *Harry [Goldman], it seems that most of the Tesla coil builders are confused by the voltage-versus-spark dilemma. Thanks to some previous comments, we no longer have a standard to go by. I have spent time-and-effort researching this subject and have been in contact with so-called experts. None were able to give me a pat answer. It seems that there are just too many variables. My vote would be for a 25 kilovolt-per-inch [1 MV/m] value as a good average.*
>
> ***Answer.*** *[by Harry Goldman]. I know from what direction you are coming. I can recall giving lectures using the 'million-volts-per-meter' statistic. That sure impressed my audiences. Now, we have been deflated. The new standard for high voltages at high frequencies is somewhere between 5,000 and 10,000 volts per inch [200 to 400 kV/m].*

We see that extrapolations by knowledgeable people can differ by a factor of five (200 kV/m to 1 MV/m). It is therefore probably more meaningful to compare Tesla coils by other parameters that can be readily measured, such as coil diameter and height, input power, and spark length.

Any voltmeter attached to the top of a Tesla coil, even for observing waveform and relative amplitude, must have several characteristics. These include a high input impedance, a minimal effect on coil operation, and an upper operating frequency several times the fundamental frequency of the Tesla coil secondary. We are all familiar with oscilloscope probes, and it is tempting to use one to measure the top voltage. To see why it is not appropriate, consider the case of a probe with 10 MΩ resistance paralleled by a 20 pF capacitor, and connected to the top of a Tesla coil operating at 100 kV and 200 kHz. The power dissipated in the resistor is

$$P = \frac{V^2}{R} = \frac{\left(10^5\right)^2}{10^7} = 10^3 \text{ W}$$

The probe would obviously vaporize at this power level. But even if high wattage resistors were used, this would still represent a load that is much too large for our Tesla coil.

One way of determining an appropriate resistance to ground would be to decide the maximum acceptable power dissipation for a given voltage level, and solve for the resistance. If we decided that 10 W would be acceptable at 100 kV, for example, the corresponding resistance would be 1000 MΩ.

The current flow in the 20 pF capacitor is

$$I_C = \frac{V}{X_C} = \omega CV = 2.51\text{ A}$$

We might then calculate the reactive power in the 20 pF capacitor as

$$VI_C = (2.51\text{A})(100{,}000\text{V}) = 251\text{ kVAR}.$$

The current of 2.51 A might seem acceptable, but 251 kVAR is certainly excessive. Most input 60 Hz power transformers used for Tesla coil work are rated less than 10 kVA, with a few very large Tesla coils powered by 25 or 50 kVA transformers, so supplying this much reactive power is just not possible. Not only that, a 20 pF capacitance would lower the resonant frequency substantially (perhaps a factor of two to five), making any measurements with such a probe of very little value.

For tuning purposes, and depending on the size of the Tesla coil, we might be able to get useful results with a 2 pF probe capacitance. At operating voltages of 100 kV and up, however, we must have a probe capacitance less than 0.02 pF. This would still result in a reactive power flow of 251 VARs at 100 kV. Such a probe has never been built, to the author's knowledge. Therefore we need a different method of measuring voltage, not like the physical probes with which we are familiar.

Voltage Divider/Fiber Optic Transmitter

We will be using a capacitive voltage divider to measure the voltage V_2 at the top of a Tesla coil. First we need to discuss the intrinsic coil capacitance. The top element (usually a toroid or sphere) of a secondary or extra coil has a capacitance C to ground as shown in Figure 12.

This capacitance increases as the top element is made larger. It decreases slightly as the element is moved further away from the ground plane, which is useful for tuning purposes. The capacitance C can be considered as a collection of smaller capacitors in parallel. If the top toroid was divided into ten equal sections, for example, then each section would have a capacitance of $0.1C$ to ground. This concept is illustrated in Figure 13.

Suppose now we cover a portion of the top element with a thin layer of dielectric material and place a conducting sheet on top the dielectric. We have added a capacitance in series with the original capacitance of that portion to ground, as shown in Figure 14.

I use 4.5 mil polyethylene, available from the local paint store as a heavy drop cloth, for the dielectric, and 16 oz. copper flashing, available from the local plumbing store, for the conducting plate. The advantage of copper is that one can file a brass screw head flat and solder directly to the copper. The screw can be inserted through a somewhat larger hole drilled in the toroid to bring V_x inside the toroid. This technique keeps the outside surface of the extra plate as smooth as possible, to prevent unwanted corona.

The voltage V_x across the capacitor C_x is given by the standard voltage division equation for capacitors in series.

$$V_x = \frac{V_2(.1C)}{C_x + .1C} \tag{5}$$

Figure 12. Tesla Coil Top Capacitance

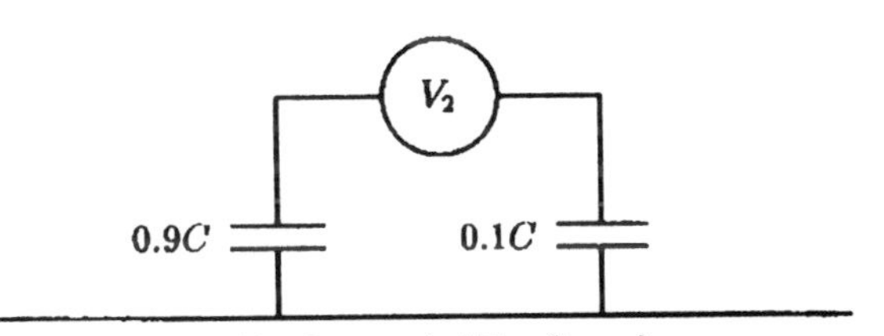

Figure 13. Segmented Top Capacitance

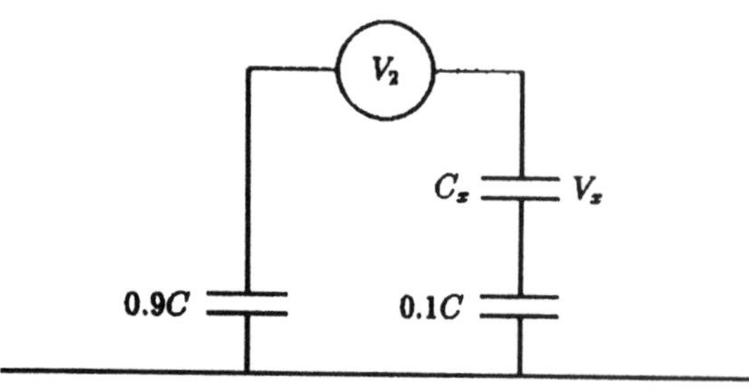

Figure 14. Capacitive Divider

If $C_x >> .1C$, as it always will be, V_x is given by

$$V_x = \frac{V_2(.1C)}{C_x} \quad (6)$$

We do not know C nor the fraction of C covered by the extra plate, so we should perhaps refer to the quantity $0.1C$ as C_a or some other unknown quantity. The important point is that V_x will be directly proportional to V_2. Measuring V_x immediately gives us the waveform shape and the relative amplitude.

A simple fiber optic transmitter circuit for converting V_x to an optical signal is shown in Figure 15. The input resistance of the amplifier, shown as 10 MΩ, needs to be as large as possible to avoid shorting out C_x. Since we have plenty of voltage available, we do not really need any voltage gain. A feedback resistor of 2 MΩ gives a gain of 0.2. This means that a V_x of 40V will produce an output voltage of 8V, well within the capabilities of the op amp.

The 1 kΩ resistor serves to limit the current to the MFOE76 emitter. Note that the other end of the emitter is tied to -9 V, so even with no input voltage there will be an output current and light in the fiber. The intensity of the light gets greater as the input voltage goes through its negative half cycle, due to the inverting op amp circuit. All the amplitude and frequency information is sent down the fiber in analog (rather than digital) form. This particular emitter operates in the visible red portion of the spectrum so it is easy to tell when the circuit is turned on.

The circuit is powered by two 9V transistor batteries, which will last several hours in continuous operation. It is important to have a switch on the circuit board so the transmitter can be turned off when not in use.

Range switching can be accomplished by the capacitor C_y. Adding additional capacitance in parallel with C_x lowers the voltage seen by the op amp input. If C_x is the proper value to get a useful signal out for a tuning voltage of, say, 1 kV, then the op amp will be in saturation for an operating voltage of, say, 100 kV. For this example, C_y would need to be about 100 times as large as C_x to get the same amplitude signal out of the op amp. Of course, several capacitors could be mounted on the pc board, in ratios of five or ten, and switched in for different operating voltages. One would look at the received signal for saturation or clipping, and switch in additional C_y as necessary to get an unsaturated signal. This switching operation would be done with the Tesla coil drive circuits turned off, of course.

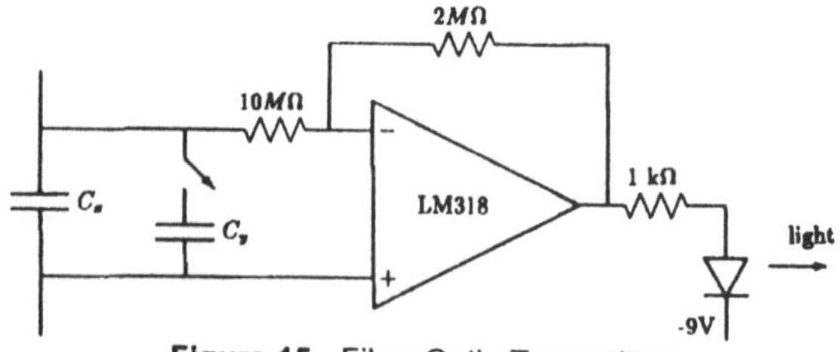

Figure 15. Fiber Optic Transmitter

Fiber Optic Receiver

A fiber optic receiver is shown in Figure 16. The Motorola MFOD71 is a PIN diode that converts photons of light into a flow of current, as discussed earlier. This current flows through the 56 kΩ resistor to produce a voltage proportional to current at the output of the LM318. This voltage is then amplified by the second op amp stage. The LM318 has a tendency to oscillate, which is damped out by the 3 pF capacitor. This capacitor reduces the frequency response of the circuit, hence should be as small as possible. The AD846 is a different type of op amp for which no feedback capacitor is needed to eliminate oscillation.

The second stage is made with variable gain to aid in calibration. A variable resistor in the feedback resistor position is not particularly good op amp design because of the tendency of a variable resistor to get dirty and noisy, but I have not had significant problems with this practice.

The circuit is located well away from the Tesla coil, hence can be operated with a conventional ± 12 or ± 15 V power supply.

The circuit will work with two op amps of the same type, but best performance is obtained by using both the LM318 and the AD846. Both op amps are relatively high performance devices. The LM318 has a slew rate of 70 V/µs, and a small signal bandwidth of 15 MHz, while the AD846 has a slew rate of 450 V/µs and a small signal bandwidth of 46 MHz. The performance is high enough that it is difficult to get proper operation with perf board, Protoboard, or wire wrap techniques. One usually has to prepare a compact printed circuit board and use short leads for all components.

Plastic fiber optic components are relatively inexpensive and easy to use. The Motorola MFOE76 was priced at $2.20 in mid-1992, and the MFOD71 at $1.36. They are available from Hamilton-Avnet

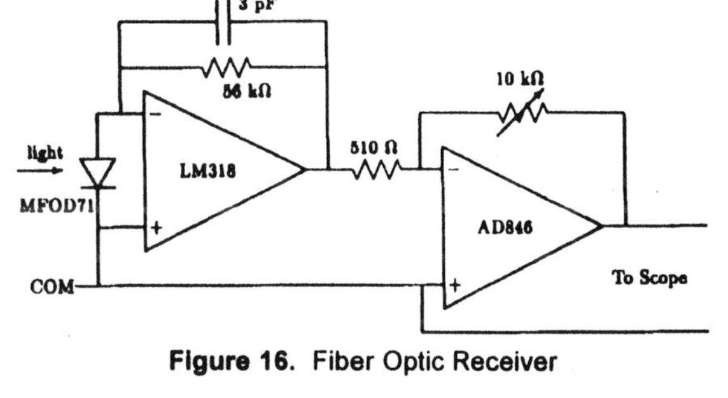

Figure 16. Fiber Optic Receiver

or any other Motorola supplier. The 1000 µm fiber costs about $2/m, depending on length purchased, and is available from Digi-Key. Usually cutting the fiber with a sharp knife is adequate, but the fiber end can be polished with #600 sandpaper if a greater signal strength is desired. Connections are quite easy. Anyone with the ability to insert an IC chip into a socket can make all necessary connections.

The circuits shown will operate over optical fibers up to about 50 m in length and at frequencies up to about 1 MHz. These values should be quite adequate for most Tesla coil work.

Conclusions

A function generator has been described which will drive a Tesla coil secondary or extra coil to produce top voltage on the order of 1 kV. A fiber optic transmitter and receiver have been presented which will measure a voltage proportional to the top electrode voltage, transmit the signal over an optical fiber, and detect it for examination by an oscilloscope. Both devices should be helpful in improved tuning of Tesla coils.__ ***GJ***

References

1. Elswick, Steven, "Inside Project Tesla," *Extraordinary Science*, Vol. 1, No. 1, January/February/March 1989, pp. 12-23.
2. Grotz, Toby, "Project Tesla—An Update," *TCBA News*, Vol. 9, No. 1, January/February/March 1990, pp. 16-18.
3. Corum, James F., Daniel J. Edwards, and Kenneth L. Corum, "TCTUTOR—A Personal Computer Analysis of Spark Gap Tesla Coils", published by Corum and Associates, Inc. First Edition, 1988.
4. Sterling, Donald J., Jr., *Technician's Guide to Fiber Optics*, Delmar Publishers, 2 Computer Drive West, Box 15-015, Albany, NY 12212-9985, 1987.
5. Goldman, Harry. *TCBA News*, Volume 10, No. 3, July/August/September 1991, p. 7.
6. Wu, Francis, Kevin Brown, and Michael Stern, *Fiber-Optic Fundamentals for Motion Systems*, PCIM (Power Conversion Intelligent Motion), Vol. 17, No. 12, Dec. 1991, pp. 21-25.
7. Johnson, Gary L., "A Fiber Optic System for Measuring High Voltages," *Extraordinary Science*, Vol. 1, No. 1, January/February/March 1989, pp. 5-9.

Tesla's Contributions To Electrotherapy

Patton H. McGinley, Ph.D.
Emory Clinic--Radiation Therapy
1365 Clifton Road
Atlanta, GA 30322

Abstract

The purpose of this paper is to review Nikola Tesla's research in the field of electrotherapy. It is clear from his lectures and publications beginning in 1891 that he was the first to discover that radio frequency currents could be employed to safely heat tissue and to point out that such heating may have therapeutic benefits. Tesla found that radio frequency currents could be produced in tissue by the use of induction coils as well as by the capacitive technique where the patient acts as the dielectric of a condenser. It is of interest that both modalities are now employed with considerable success for the treatment of cancer. Tesla also suggested that radio frequency currents could be used for other medical purposes such as the sterilization of wounds, as an anesthesia, for stimulation of the skin, and to produce surgical incisions.

The author of this paper will summarize his research related to Tesla's techniques for the production of hyperthermia as well as clinical results obtained when heat treatment of cancer patients was carried out.

Introduction

There has long been an interest in the use of electromagnetic energy to diagnosis and treat diseases. Luigi Galvani in the eighteenth century conducted the first experiments demonstrating the effects of electrical current on biological systems. Of major importance was his observation that the leg muscles of a frog could be stimulated at some distance from a spark produced by a static electric generator. The subject of remote stimulation of nerves by electrical means lay dormant for over a hundred years. Sometime before 1891, Tesla made the discovery that alternating currents of frequency of 10kHz or greater could pass through the body without the sensation of electrical shock being produced. In fact, levels of electrical energy that would prove fatal at a reduced frequency could be tolerated when the frequency was above 10kHz. Tesla predicted that medical use would be made of this phenomenon in his lecture before the American Institute of Electrical Engineers, in May, 1891. [14] At about the same time as Tesla's discovery, d'Arsonval [5] independently reported similar observations on the physiological effects of high frequency currents before the Society of Biology in Paris. Certainly the electrical oscillators produced by Tesla for electrotherapy were superior to those devised by d'Arsonval. [2]

History has not been kind to Tesla in the sense that the credit for all of the pioneering work in the field of electrotherapy has gone almost exclusively to d'Arsonval.

Early Development of the Medical Use of High Frequency Currents

Using high frequency currents greater than 10kHz Tesla soon found that living tissue could safely be heated and other effects such as blood pressure and psychological changes could be produced. Figure 1 was prepared from one of Tesla's drawings which was obtained by the author on a recent trip to Yugoslavia. In the diagram are depicted three circuits for medical use. The one on the left is to be used to sterilize wounds, the center circuit is that of an electrosurgical knife, and the device on the right is used to produce whole body heating by induction.

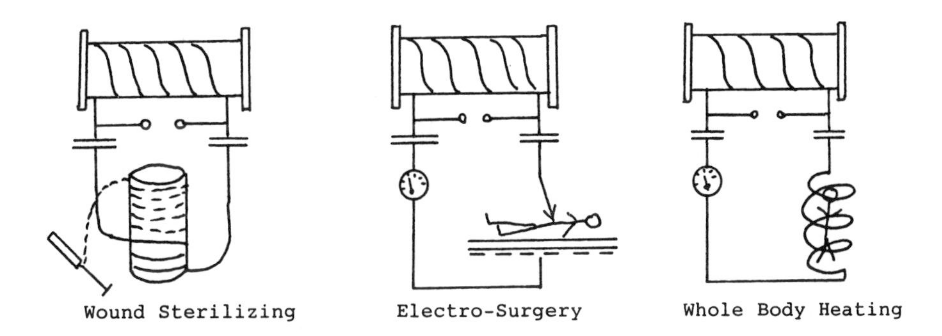

Figure 1. Electrotherapy devices

In early 1892, Tesla traveled to England and France to deliver lectures on his recent research. While in France he met d'Arsonval and they became good friends. Tesla was pleasantly surprised to find that d'Arsonval used his oscillators to investigate the physiological effects of high frequency currents.

Tesla's interest in the therapeutic uses of high frequency currents continued for a number of years and in 1898 he read a paper before The American Electro-Therapeutic Association reviewing various oscillator designs. [15] Several modes of application of the electromagnetic energy such as capacitive coupling, direct contact electrodes, and inductive coupling to the patient were described. All of theses techniques are used today for cancer therapy and diathermy. In Figure 2 is shown a simplified schematic of one of the devices Tesla suggested for patient treatment by use of contact electrodes.

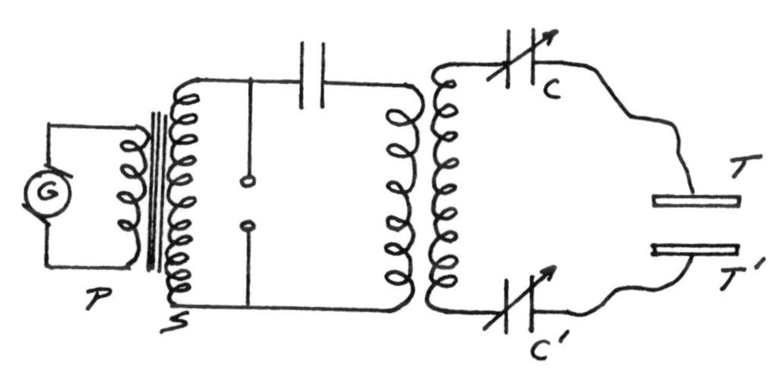

Figure 2. RF Heating with contact plates.

Radio frequency currents are produced by a Tesla coil; adjustable condensers C and C' are used to match the impedance of the power supply to that presented by the patient. Metal contact electrodes T and T' are placed on the patient and relatively large currents can be passed through the patient without discomfort providing the frequency is of the order of 10kHZ or greater. The tissue between the electrodes is elevated in temperature as a result of the joules heating produced by the current. The major components of some modern hyperthermia units perform similar functions (rf power supply, impedance matching network, and coupling electrodes).

An interesting device described by Tesla in his lecture is shown in Figure 3.

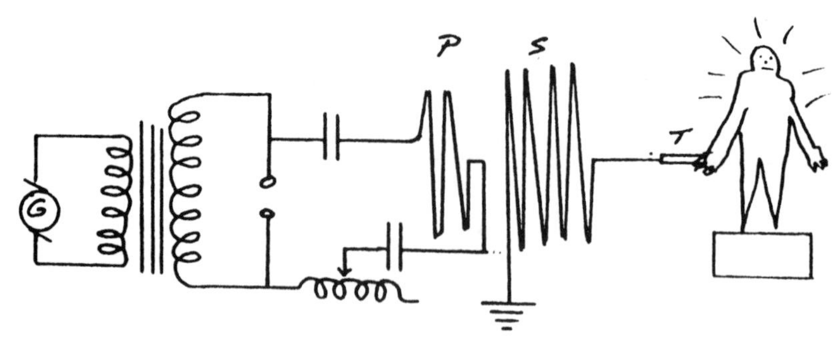

Figure 3. Single electrode electro-massager.

Voltages as large as 2MV can be applied to a person if he is insulated and touches the terminal T. Superficial or surface rf currents are produced as well as luminous streamers which issue from the body. It was reported that a feeling of warmth over the body surface was produced and soon after the person perspires freely. Prolonged use of the devise causes a feeling of fatigue and induces deep sleep. Since the device acts as a quarter wavelength coil, the position of maximum voltage can be shifted along the person's body at will by varying the frequency of the primary circuit. Tesla felt the device would find use as a mode of hygienic treatment to instantaneously clean the skin.

The suggestion for medical use of radio frequency currents caught on and within a few years attempts were made to treat many disorders by electrical means. Figure 4 outlines the early development of the applications of high frequency currents in the medical field. Early on most of the interest was centered on nonthermal effects which produced physiological and psychological changes. After the early 1920's this approach seems to have been abandoned. In fact, the present protection standards are based exclusively on thermal effects.

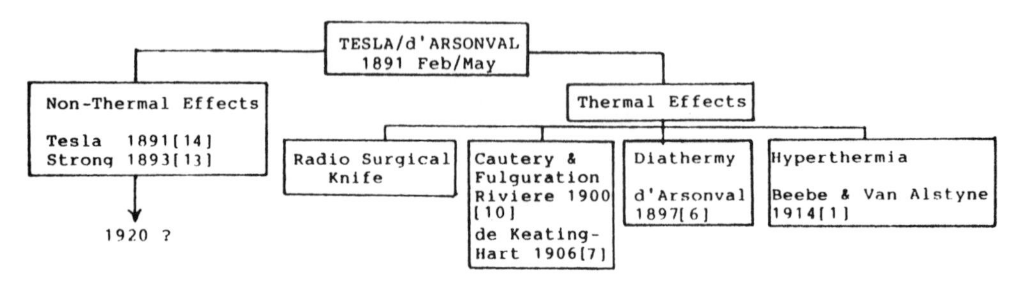

Figure 4. Early history of electrotherapy.

By 1900 the effects of heat production in tissue were under full investigation and attempts were made by Riviere [10] to treat cancer by use of rf currents. Other uses such as diathermy and electrosurgery were under development. Diathermy was in medical use as early as 1897 [6] and was initially referred to as d'Arsonval current or autocondensation.

In 1903 [3] after funding for the Wardencliffe Worldwide Communication System was cut off, Tesla used the facility to produce medical oscillators. Curtis [4] in his book on high frequency apparatus has an excellent review of the equipment used by the physician for electrotherapy and x-ray work. Figure 5 shows an electrotherapy unit that could be used to carry out all of the popular high frequency modalities in use by 1916.

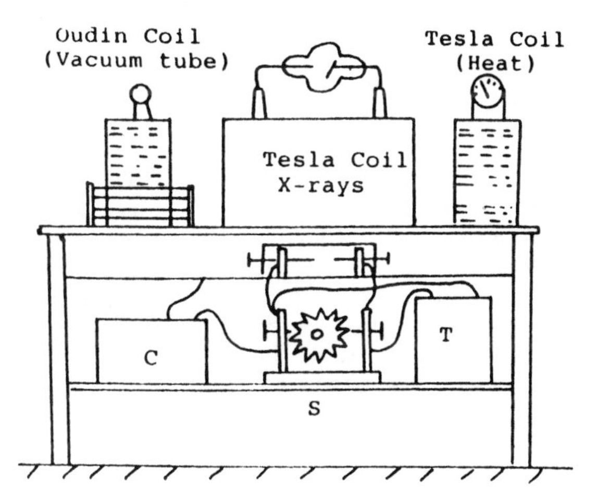

Figure 5. 1916 Electrotherapy unit from Curtiss[4], C=condenser, S=spark gaps, T=transformer

The unit operates with 110 or 220 v ac and 1 kw of input power. Three high frequency coils are used to carry out electrotherapy treatment and x-ray production. The coil on the left is of the Oudin type [11] and was capable of producing high frequency current of low or high strength with moderate to high voltage. Fulguration was carried out with large currents and spark spray therapy was conducted with the Oudin coil by operating at very high frequency and voltage. Vacuum tube or Tesla current treatment was done with medium potential and high frequency applied to the electrode shown in Figure 6.

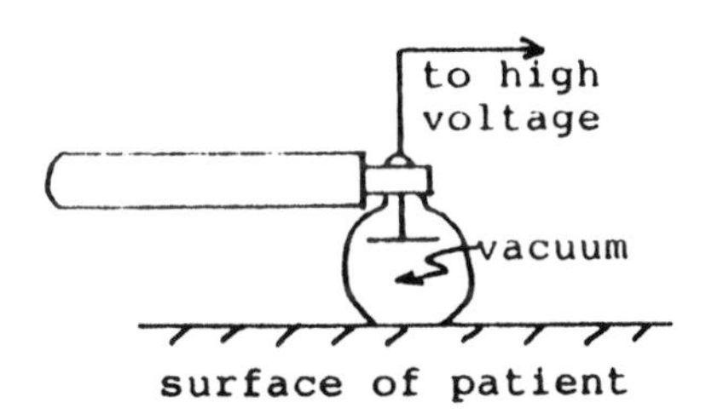

Figure 6. Vacuum tube electrode (Tesla current).

This interesting device which is no longer used consisted of a single electrode vacuum tube with an insulated handle. The vacuum tube was moved along the patient's skin. The device produced general warming to the region treated, increased blood flow, increased arterial tension, and a sedative effect when operated at high frequency and a stimulating effect when operated at low frequency. It was used to treat many forms of skin disorders as well as a tonic. A Tesla coil is shown in the center of Figure 5 and it was used for x-ray production. The coil on the right side of Figure 5 is also of the Tesla type and it was used for diathermy or d'Arsonval treatment. High frequency currents of up to 2 amp were applied to a pair of metal plates which were placed on the patient to produce tissue heating.

Later Developments

Tissue heating by electrical current before 1929 was carried out with oscillators producing frequencies on the order of one megahertz. Schliephake [12] introduced what was soon to be called shortwave diathermia in 1929 when he began patient therapy with a 400 watt oscillator which produced 3 meter radio frequency waves. Similar to long wave diathermia, plate type electrodes and induction coils were employed to couple the electrical energy into the patient. Microwave diathermia was initiated in 1936 by Denier [8] who used a magnetron oscillator. With the development of shortwave and microwave oscillators, the use of long wave diathermia faded. It is of interest that the present author can find discussion on the therapeutic use of Tesla type oscillator in medical text published as late as 1958.

After 1971 world wide interest developed in the use of heat for the treatment of malignancies. Electromagnetic techniques for the production of elevated temperature in tumors have been extensively used for the last 15 years. Patient treatment has been conducted with plate electrodes and inductive coils similar to those suggested by Tesla. Microwave antennas and wave guides operating at 915 and 2450 MHz have also been employed. In Figure 7 is shown the various methods of applying microwave and rf energy to the patient.

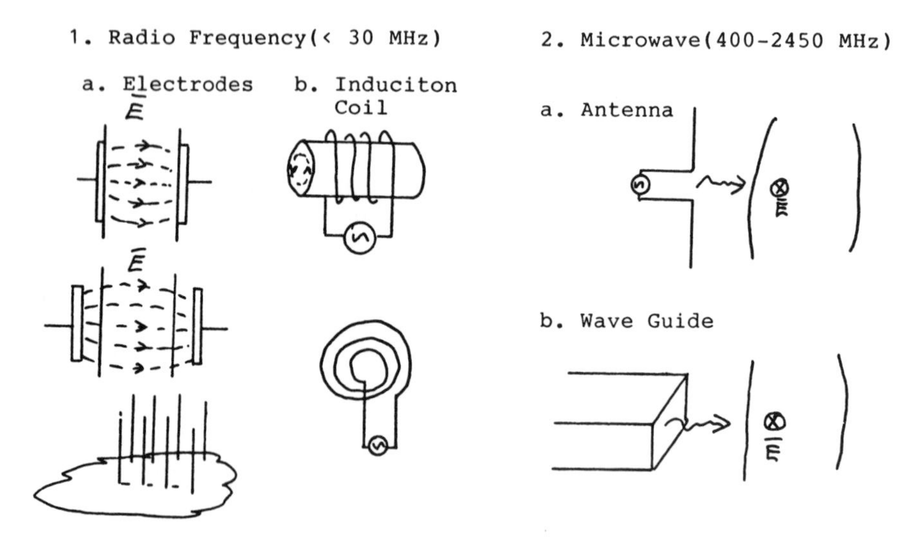

Figure 7. Modes of electromagnetic heating.

The heat generated in tissue can be predicted from the electrical field strength (E) at the point in question and the conductivity of the tissue (σ) by use of Equation 1 [9].

$$\text{Power deposited per unit volume of tissue} = \frac{\sigma E^2}{2} \qquad (1)$$

The evaluation of the electric field strength can be difficult for the case where the electromagnetic energy penetrates various layers of tissue with complex geometry. Tissues with high water content such as muscle have high conductivity and low water content materials such as bone, fat, and skin will in turn have low conductivity. This causes problems when surface plates or capacitive type heating is carried out since the current produced by the electrical field tends to flow from one layer to the next and skin and subcutaneous fat will heat more than the underlying muscle. This case is similar to a large (fat) and small (muscle) resistor in series with the same current existing in each resistor. More heat energy will be produced in the large resistor than the small. Microwave and UHF applicators are designed to produce an electrical field that is parallel to the surface thereby causing the current to be parallel to the interface between tissues. This arrangement allows the deep muscle layer to heat more than the subcutaneous fat and skin. A parallel high (fat) and low (muscle) resistance circuit is a simple analogy of the microwave and UHF applicator. Now the high resistance will develop less heat than the low resistance.

The author's experience in the field of cancer therapy by use of electromagnetic energy began in 1974 when he assembled a rf contact plate treatment unit at The Emory Clinic. Several years later a radiative applicator and wave guide system operating at 433 MHz were purchased. The first treatment with the contact electrodes was carried out using the method shown in Figure 8 to treat a tumor located in the neck of a patient. The patient had been treated with surgery and radiation in the past and tumor control had not been achieved. Heat treatment with additional radiation was the only option left for the patient. As depicted in Figure 8 the electrodes were placed adjacent to each other and the tumor volume was adequately covered by the high field region. Similar to Tesla's equipment, the oscillator was operated below 1MHz and 15 watt of output power was sufficient to elevate the temperature to a level where tumor destruction occurs(42-43° C). Heat treatments were given for one hour three times a week for three weeks. Radiation therapy was carried out after each heating cycle. During the course of the treatment the size of the tumor was observed to shrink. The patient was followed for five years and was free of disease at the end of the period.

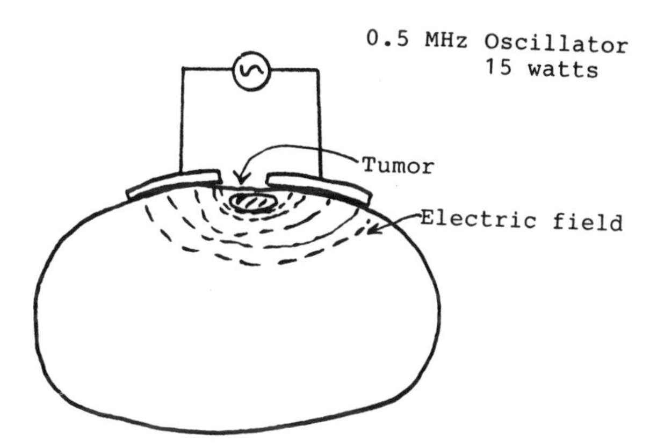

Figure 8. Electric field distribution in patient's neck by contact electrodes.

A total of over 500 patients were treated by 1987 when we decommissioned our heat therapy equipment which was by then outdated. At present we are planning the purchase of modern heat therapy units. A review of the clinical data shows that many patients who had been refused further treatment with radiation or surgery obtained benefit from the use of hyperthermia.

Conclusions and Summary

One can follow the development of the medical use of electromagnetic energy from Tesla's discoveries to the present time. Of course, the equipment employed for electrotherapy has changed greatly but the physical principles found by Tesla are still in use. There can be no doubt that his research led to the development of electrosurgery, diathermy, and hyperthermia. Possibly in the future some of the non-thermal effects found by Tesla will prove useful for the treatment and cure of diseases.

References

[1] Beebe and Van Alstyne. Surg., Gynec. and Obst.; vol. 18, 1914. p.438.

[2] Cheney, M. Tesla: Man Out of Time. New York: Laurel, 1981. (ch. 7; p. 73).

[3] Cheney, M. Tesla: Man Out of Time. New York: Laurel, 1981. (ch. 16; p. 166).

[4] Curtis, Thomas. High Frequency Apparatus. New York: Everday Mechanics, 1916. (ch. 11, pp. 111-136)

[5] D'Arsonval, A. "Sur les Effects Physiologiques de l'etat Variable et de Courants Alternatifs;" Bull. Soc. Internat. Electro, April, 1892.

[6] D'Arsonval, A. " Action Physiologique des Courants Alternatifs a Grande Frequence," Arch. Electrol. Med., vol. 6, 1897. (p. 133)

[7] De Keating-Hart; La Fulguration et ses Resultants dans le Traitmet du Cancer, d'Ares une Statistique personnelle de 247 cas. Paris: Malonie, 1909.

[8] Denier, A. "Les Ondes Hertziennes Ultracourtes de 80 cm;" J. Radiol. Electrol., vol. 20, 1936. (p. 193)

[9] Guy, Authur. "Biophysics of High Frequency Currents and Electromagnetic Radiation," Chapter 6 in Therapeutic Heat and Cold. Edited by Lehmann, J. Baltimore: Williams and Wilkins Pub, 1982.

[10] Licht, Sidney. "History of Therapeutic Heat and Cold," Chapter 1 in Therapeutic Heat and Cold. Edited by Lehmann, J. Baltimore: Williams and Wilkins Pub, 1982.

[11] Oudin, M. "Nouveau Mode de Transformation des Courants de Haute Frequence," J. Soc. Franc. Electrother, 1893.

[12] Schilephake, E. Les Ondes Courtes en Biologie. Paris, 1938.

[13] Strong, Frederick. "Electricity and Life," The Electrical Experimenter. March, 1917. (pp. 830-831.)

[14] Tesla, Nikola. "Experiments with Alternate Currents of Very High Frequency and Their Application to Methods of Artificial Illumination," a lecture delivered before the AIEE at Columbia College on May 20, 1891.

[15] Tesla, Nikola. "High Frequency Oscillators for Electortherapeutic and Other Purposes," a lecture delivered before the American Electro-Therapeutic Association in Buffalo, September 13, 1898.

About The Author:

Dr. Patton H. McGinley is an Associate Professor of Radiology at the Emory University. He received his PhD in 1971 from Georgia Institute of Technology. His thesis research was related to ***medical physics*** and he has been engaged in cancer treatment and research since 1973. Dr. McGinley's interest in Tesla developed from his work in hyperthermia and he developed a paper before the International Clinical Hyperthermia Society in 1986 in Sarajevo, Yugoslavia on Tesla's studies in electrotherapy.

Part Six
Tesla's FBI Files

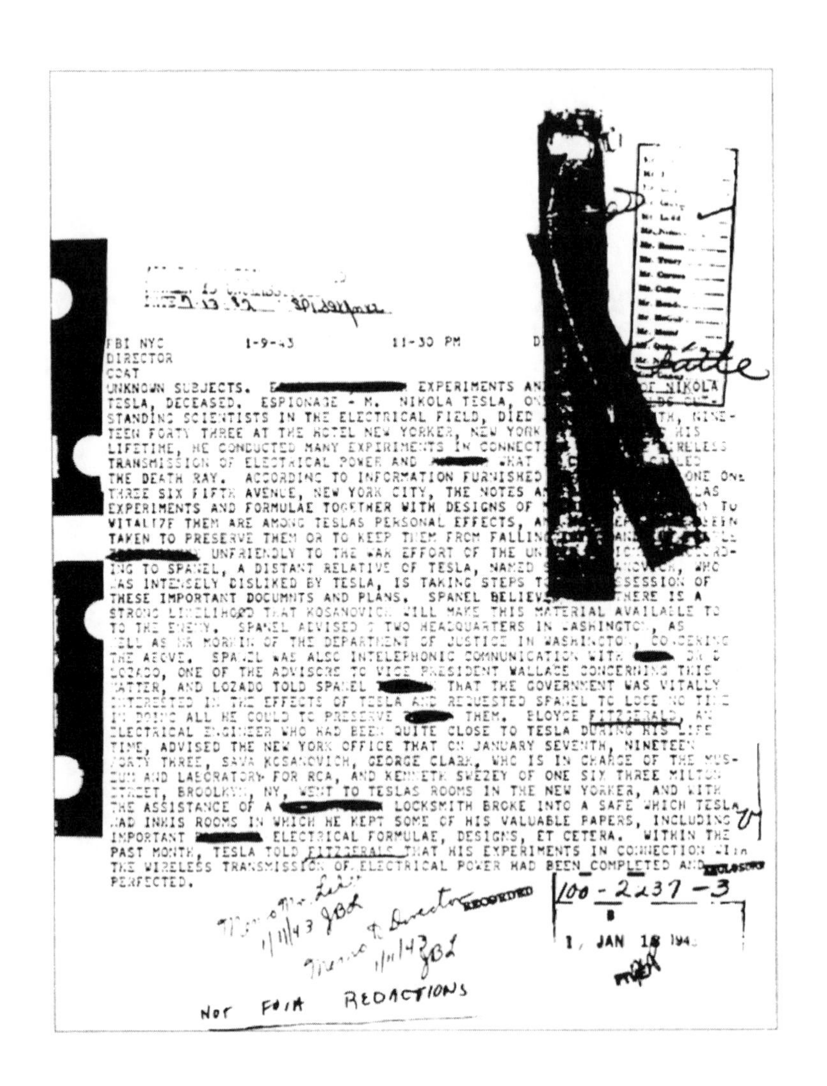

FBI NYC 1-9-43 11-30 PM

DIRECTOR

CODE

UNKNOWN SUBJECTS. [illegible] EXPERIMENTS AN[illegible] OF NIKOLA
TESLA, DECEASED. ESPIONAGE - M. NIKOLA TESLA, ON[illegible]
STANDING SCIENTISTS IN THE ELECTRICAL FIELD, DIED [illegible]TH, NINE-
TEEN FORTY THREE AT THE HOTEL NEW YORKER, NEW YORK [illegible] HIS
LIFETIME, HE CONDUCTED MANY EXPIRIMENTS IN CONNECT[illegible]IRELESS
TRANSMISSION OF ELECTRICAL POWER AND [illegible] WHAT [illegible]LED
THE DEATH RAY. ACCORDING TO INFORMATION FURNISHED [illegible] ONE ONE
THREE SIX FIFTH AVENUE, NEW YORK CITY, THE NOTES A[illegible]LAS
EXPERIMENTS AND FORMULAE TOGETHER WITH DESIGNS OF [illegible] TO
VITALIZE THEM ARE AMONG TESLAS PERSONAL EFFECTS, A[illegible]EEN
TAKEN TO PRESERVE THEM OR TO KEEP THEM FROM FALLIN[illegible]
[illegible] UNFRIENDLY TO THE WAR EFFORT OF THE UN[illegible]
ING TO SPANEL, A DISTANT RELATIVE OF TESLA, NAMED [illegible]ANOVICH, WHO
WAS INTENSELY DISLIKED BY TESLA, IS TAKING STEPS T[illegible]SSESSION OF
THESE IMPORTANT DOCUMNTS AND PLANS. SPANEL BELIEV[illegible] THERE IS A
STRONG LIKELIHOOD THAT KOSANOVICH WILL MAKE THIS MATERIAL AVAILABLE TO
TO THE ENEMY. SPANEL ADVISED [illegible] TWO HEADQUARTERS IN WASHINGTON, AS
WELL AS MR MORHIN OF THE DEPARTMENT OF JUSTICE IN WASHINGTON, CONCERNING
THE ABOVE. SPANEL WAS ALSO INTELEPHONIC COMMUNICATION WITH [illegible] DR
LOZADO, ONE OF THE ADVISORS TO VICE PRESIDENT WALLACE CONCERNING THIS
MATTER, AND LOZADO TOLD SPANEL [illegible] THAT THE GOVERNMENT WAS VITALLY
INTERESTED IN THE EFFECTS OF TESLA AND REQUESTED SPANEL TO LOSE NO TIME
IN DOING ALL HE COULD TO PRESERVE [illegible] THEM. BLOYCE FITZGERALD, AN
ELECTRICAL ENGINEER WHO HAD BEEN QUITE CLOSE TO TESLA DURING HIS LIFE
TIME, ADVISED THE NEW YORK OFFICE THAT ON JANUARY SEVENTH, NINETEEN
FORTY THREE, SAVA KOSANOVICH, GEORGE CLARK, WHO IS IN CHARGE OF THE MUS-
EUM AND LABORATORY FOR RCA, AND KENNETH SWEZEY OF ONE SIX THREE MILTON
STREET, BROOLKYN, NY, WENT TO TESLAS ROOMS IN THE NEW YORKER, AND WITH
THE ASSISTANCE OF A [illegible] LOCKSMITH BROKE INTO A SAFE WHICH TESLA
HAD INHIS ROOMS IN WHICH HE KEPT SOME OF HIS VALUABLE PAPERS, INCLUDING
IMPORTANT [illegible] ELECTRICAL FORMULAE, DESIGNS, ET CETERA. WITHIN THE
PAST MONTH, TESLA TOLD FITZGERALD THAT HIS EXPERIMENTS IN CONNECTION WITH
THE WIRELESS TRANSMISSION OF ELECTRICAL POWER HAD BEEN COMPLETED AND
PERFECTED.

RECORDED

100-2237-3

1, JAN 18 194[illegible]

NOT FOIA REDACTIONS

An FBI memorandum from agent P.E. Foxworth dated January 9. 1943 concerning Tesla's death. This is a typical page from Tesla's FBI file, with much of the text redacted (blacked-out).

NIKOLA TESLA'S FBI FILES

This anonymous summary of Tesla's FBI files was found on the internet:
(All redacted [blacked-out] text from the files appears as XXXXXX. Some pages were 90% redacted)

This is a complete summary of relevant information found in the 156 pages (158 including the FBI deleted page sheets) of Enclosure # I of a Freedom of Information Act (FOIA) request of the Nikola Tesla FBI files.
Page numbers are sequential from the title sheet to the end as received from the FBI.

Page #-|———DATE——|————FROM————|————TO————|

1- .....Title sheet.....Nikola Tesla....MAIN FILE....156 Pages
2- .....Apr 3, 1950 | J. Edgar Hoover | XXXXXX |
Summary> This is a letter to XXXXX form J.Edgar stating the FBI has never been in possession of Dr. Tesla's papers.
3- .....Apr xx, xxxx | J. edgar Hoover | XXXXX |
Summary> This is a very poor copy of another request for Tesla's papers. This time FBI refers them to the Office of Alien Property 4- .....Oct 1, 1940 | J. Edgar Hoover | XXXXXX |
Summary> Letter from Hoover to XXXXXX acknowledging receipt of a letter Sep 24, 1940.
5- .....Jan 12, 1943 | FBI NY Foxworth | Director FBI |
Summary> Dept of Justice teletype states Tesla died Jan 8, rather than Thursday, Jan 7 as XXXXX stated in teletype. On the night of Jan 8, Sava Kosanovic, George Clark, and Kenneth Sweezey visited Tesla's hotel room with a representative of Shaw Walker Co. In order to open the Tesla's safe. Kosanovich was looking for Tesla's will. Three asst. managers of the hotel New Yorker as well as representatives of Yugoslavian Consulate present. Sweezey took a book. safe closed and a new combination in the possession of Kosanovich. Saturday, Jan 9, Gorsuch and XXXXXXX Fitzgerald of Alien Property Control went to the hotel and seized the property of Tesla, consisting of about two truckloads of material, sealed all articles and transferred them to the Manhattan Storage and Warehouse Co. NY, where they are now located.
At that time there were also thirty barrels and bundles belonging to Tesla which had been there since 1934. These also have been sealed.
6- .....Cont.5- Tesla also had some property allegded by INFORMANT FITZGEARLD in this case Tesla had an invention in a safe deposit box at the Govonor Clinton Hotel. He placed it there in 1932. The hotel will not release it till a bill is paid. They will inform us if anyone attempts to get it. Sweezey is a writer for Popular Mechanics and is desirous to publish Tesla's biography. Clark is employed by RCA. A trained person should review Tesla's notes.
The name Foxworth is typed at the bottom of this page.
7- .....Jan 9, 1943 | FBI NY Foxworth | Director FBI |
Summary> This looks like the teletype that proceeded the one of page 5 & 6. States Sava Kosanovich was intensley disliked by Tesal is trying to gain posession of Tesla's effects. Spanel thinks SAva will make this information available to the enemy.
Spanel was in telephonic communication with Dr. Dlozado, an advisor to Vice Presi-

dent Wallace, and was told to lose no time in doing what he could to preserve Tesla's effects. Bloyce Fitzgearld who had been quite close to Tesla, advised the NY office about Tesla's death and of Sava Kosanovick, George Clark and Sweezey breaking into Tesla's safe. Tesla told Fitzgearld he had completed and perfected his experiments in connection with the wireless transmission of power.

8- .....Cont.7- Fitzgearld talks of a new torpedo and knows the where abouts of plans. He also says that a working model that cost $10,000 to build, is in the safety deposit box of the Governor Clinton hotel. He thinks the model has to do with Tesla's Death Ray or the wireless transmission of electrical current.

Bureau is requested to advise immediately what action to take.

9- .....Jan 11, 1943 | D.M. Ladd | Agents |

Summary> This looks like a memo sent to all agents from Hoover basically explaining everything in the two preceeding teletypes.

10- .....Cont.9- The NY office was instructed to discuss and take the matter up with the State's Attorney in NY City with possibility of taking Kosanovich into custody on a burglary charge and obtain the papers he took from the safe.It is pointed out that any activities be handled in a MOST SECRET FASHION IN ORDER TO AVOID ANY PUBLICITY IN RESPECT TO TESLA'S INVENTIONS.

EVERY PRECAUTION TAKEN TO PRESERVE THE SECRECY OF TESLA'S INVENTIONS.

There is a hand written note at the bottom of this page that says L.M.C. Smith is handling this with Alien Property Custodian so there appears to be no need for us to mess around in it.(initals appears to be EH.

11- .....Jan 12, 1943 | Edw. A. Tamm | Mr. Ladd |

Summary> A memo, Mr. Tamm tells Mr. Ladd that the Alien Property Custodian's Office will handel the matter and the Bureau should take no action.

12- .....Apr 16, 1948 | XXXXXX | J. Edgar Hoover |

Summary> A letter from a student at Milwaukee School of Engineering requesting Tesla's papers so he can complete a thesis on 13- .....Cont.12- Letter states he knows some of Tesla's stuff is under National Security.

14- .....Apr 17, 1950 | SAC NY | Director, FBI |

Summary> April 7, 1950 agents interviewed Mr. L.V. Potts, vice president of the Mahattan Warehouse and Storage Co. He relates process necessary to gain access to things stored there.In review of Tesla's effects, only one visit had been made. This one session took place on Jan 26 and 27, 1943 by agents of Alien Property to review the effects.

The Tesla effects are stored in rooms 5J and 5L of the storage Co.

Mr. Michael King the floor supervisor for 10 years states the only time anyone examined Tesla's effects was in early 1943. He states that at that time numerious photographs were taken with what would tend to indicate microfilm equipment. Mr. King added that several of group wore U.S. Navy uniforms, the civilians were referred to as federal authorities.

15- .....Cont.14- Intervewing agents explained to Mr. Potts that the examination made as mentioned, was not instigated by the Bureau, nor had the Bureau taken part in the examination. CLOSED 16- .....Aug 18, 1953 | XXXXXX | FBI Washington D.C. |

Summary> A letter to the FBI requestingTesla's effects including personal letters, periodicals,rare books and a collection of patents issued to Tesla. This person has contacted the Office of Alien Property and the Bureau of Naval Research and the Dept. of Commerce with no results....

The FBI special agent in charge of Tesla's file, P.E. Foxworth.

" -I am especially interested in the research work in which Tesla was engaged in his later years. There are various unpublished works, such as a 10-page typewritten statement presented in 1937 at a meeting of several well-known editors outlining his discoveries and giving a resume of his work in the fields of gravity and cosmic ray research, ect. Also, Tesla prepared various papers, one of which was in effect to secure the Pierre Gutzman Prize from the Institute of France. My inquiry is in effort to determine whether any of these documents, as well as others, are at this time available. " 17- .....Aug 26, 1952 | John Edgar Hoover | XXXXXXX| Summary> This is the standard reply used by the FBI to the request for Tesla infromation to the page 16 request. It tells the requester to contact Roland F. Kirks, Assistant Attorney General for help.

It also shows a cc. ,carbon copy to Kirks...incoming.

18- ......May 5, 1953 | L.L. Laughlin | A.H. Belmont |

Summary> Memo about information in the book /Prodigal Genius/ which states the FBI took Tesla's effects after his death. The memo also refers to the request made by the student from Milwaukee (page-12) 19- .....Cont.18- Mr Laughlin states he called XXXXX and told him to contact the Alien Property office.

The book Prodical Genius is not in the Bureau library but a copy will be obtained and a determination can be made as to what futher action should be taken.

Addendum..5-14-53..page 277 of the /Prodigal Genius/ states that the FBI took papers from Tesla's safe. Since this work was published inn 1944 no objection would be raised with the publisher at this time.

The individual that raised the question has been set stright.No action.

20- .....Feb 19, 1954 | SCA New York | Director, FBI |

Summary> This is a memo about letters in the possession of Mr. George H. Scherff, Jr. which he recieved from Leland I. Anderson. Mr. Anderson was requesting any writings that Mr. Scherff may have from Tesla.

Scherff was an associate of Dr. Tesla in 1914 and that his father was Dr. Tesla's private secretary for years. Mr. Scherff says he has quite a bit of Tesla's writings and he does not know Mr. Anderson and wants to if Tesla's papers would be of value to a foreign government.

21- .....Feb 3, 1954 | Leland I. Anderson | Mr. George H. Scherff, JR.

Summary> This is a copy of the letter mentioned in the previous memo.

In the letter Mr. Anderson is asking Mr.Schreff for any parers which he may have regarding Tesla. He states he has been collecting Tesla's papers for some time to publish them.

22- .....Cont.21- This was an enclosure with Mr. Anderson's letter.

It is an outline of /Tesla-International/. An organization which Mr. Anderson founded to bring Tesla's name back to life.

23- .....Cont.23- This is a photocopy of the envelop this letter came to Mr. Schreff in. It was registered.

24- .....Feb 12, 1953 | Leland I. Anderson | Mr. George H. Scherff, JR.

Summary> This appears to be a second letter to Mr. Schreff. Mr.

Anderson says he is glad to get a reply and goes into detail about the people and places he been to to get all the Tesla effects he can. He is sending the first two Tesla journals to him. He is looking for information as to the where abouts of a lady mentioned in O'neill's book that has many Tesla documents. She feared a run in with the Army and moved to the mid-west. Anderson writes of the sad death of O'Neill and Edwin Armstrong. He mentions the purchase of 70 pieces of Tesla related material of Robert U. Johnson.

25- .....Cont.24- States James.A. McChesney of Long Island has many correspondence between Tesla and George Sylvester Viereck.

Anderson ask if Schreff has one of the brochures that Tesla issued Feb 1904. It came in a large square envelope bearing a large red wax seal with the initials /N.T./ stamped thereon.

He wants to reproduce it for the orginazation. Anderson writes about the upcoming 1956 Tesla Centenary. Wants help organizing something here in the US.

26- .....Cont.25- This is a REAL bad photocopy of an announcement from /Tesla-International/ about the avalibility of a reprinting of literature of somesort. Thats all I can makeout.

27- .....Cont.26- This is a photocopy of the envelope that contained the previous correspondence. It was Special Delivery.

28- .....Sep 24, 1940 | XXXXXX | J. Edgar Hoover |

Summary> A letter from someone concerned about an article in the 9/22/1940 N.Y. Times that made references to Tesla's Teleforce and /Death Ray/ for planes. They were concerned someone may kidnap and torture Tesla for this information.

29- .....Cont.28- This is a photocopy of the NY Times article. The author states he has known Tesla for many years and Tesla who is 84 now, still retains full intellectual vigor. The article goes on to explain the /Teleforce/ and seems amazingly similar to present day particle beam devices.

30- .....Mar 10, 1954 | SAC New York | Director, FBI |

Summary> A memo about two letters from Leland I. Anderson which George Schreff orginally sent to the Bureau.

31- .....Jun 25, 1955 | Kenneth M. Swezey | J. Edgar Hoover |

Summary> This letter from Swezey states that when he and Savy Kosanovic, and others opened Tesla's safe, the only article removed was a volumn of greetings from prominent people wishing Tesla well on his 75th birthday. A set of keys and the Edison medal were left in the safe.

The safe was never opened again.

32- .....Cont.31- Until it arrived in Belgrade at Tesla's Museum.

Mr. Kosanovic opened the safe to find the keys and the Edison medal were missing. The keys were later found in a box outside the safe but the Edison Medal was never found. A duplicate for the museum is in the works.Swezey ask the FBI to help strighten out the dis-information about the safe opening and help find or replace the Edison Medal.

33- .....Cont.32- This is a photocopy of an article from *Power* magazine May 1955 which Swezey included with his letter. It traces Tesla's inventions and his forgotten inportance in our society.

34- .....Page Missing?

35- .....Unknown | Unknown | Unknown |

Summary> Page two of a documment marked Secret. I lost, or the FBI didn't send page one. So I'll start here...

This documment is referring to Tesla's death Jan 7, 1943 and goes into Tesla's life and inportance. States he was naturalized in 1889.

Mentiones his patents on Neon and flourescent lights, but made little money on them, although continued to do experiments leading to devices of great potential worth, which he never patented. He donated papers to the Tesla Institute in Belgrade in 1930's. There were reports copies of Tesla's effects existed in the US.

36- .....Cont.35- Also, the Soviet Union has had access to Tesla's papers which influenced their research in energy weapons. Butler's expertise is in Beam weapons and has searched libraries at WPAFB.

and is unable to find Tesla's papers. Reports have the Alien Property Custodian shipping some of Tesla's papers to a research lab at WPAFB. XXXXX travels to Washington DC. on FTD business and can review FBI files at FBIHQ relating to Tesla and Kosanovic.Bureau should consider granting XXXXX of FTD offical access to files.

37- .....Cont.36- Leads/ New York / at New York / Cincinnatti / at Dayton Ohio. Remain in contact XXXXX.

38- .....Arp 1, 1981 | Allen J. MacLaren USAF | Roger S. Young FBI |

Summary> This is a letter to Mr. Young thanking him for the effort that he went to in getting Tesla's data.

This letter has FOIA # 356,608 on it. That was assigned to my request.

39- .....Jul 21, 1981 | XXXXX | Director, FBI |

Summary> This is a request for Tesla's papers the FBI may have. The requester's name is blacked out.

40- ..... Cont.39- This is page two of the lengthy request.

41- .....Aug 7, 1981 | Roger S. Young FBI | XXXXX |

Summary> This is the answer to the request for Tesla's effects that the FBI may have. FBI says they never had them and was forwarding the communication to the Dept. of Justice.

42- .....Aug 18, 1983 | SAC, Cincinnati | Director, FBI |
Summary>********. This is a HOOT!..IT'S 1983.****** .... Verbatim...

This communication is classified / Secret / in it's entirety.

Re telephone call of SA XXXXXXXX, Cincinnati Division, to Supervisor XXXXXXXX FBIHQ, on 8/11/83.

Enclosed for Bureau and New York is one copy each of pertinent pages from the 1981 book titled /Tesla: Man out of Time/ by Margaret Cheney, with important passages underlined.

For information of Bureau and New York XXX-3 lines-XXX at Wright-Patterson Air Force Base (WPAFB) and XXX-2 lines-XXX also at WPAFB, have both been in contact with SA XXX at Dayton, Ohio RA regarding possible FBI.

****Can you believe this...The FBI photocopied Cheney's book and UNDERLINED the important parts!!!**** 43-63.....These are photocopies of pages of/ Tesla: Man out of time/.

there are two pages of the book on each one. The main thing underlined is relating to Sava Kosanovic's actions.

64- .....Nov 10, 1975 | Clarence M. Kelley | XXXXX |

Summary> This is a letter from Director Kelley acknowledging receipt of an FOIA request for Tesla's files. The letter was forwarded to the FBI from the Dept. of Justice Oct 22.

Kelley writes the person XXXXXX that the FBI will not indicate wheather or not they have such files. Then he says a notarized authorization from Tesla's next of kin directing the FBI to release any information their files may contain. This does not infer that we do or do not have the info you requested.

65- .....Oct 17, 1975 | XXXXX | Deputy Attorney General |

Summary> This is the orginal FOIA request for Tesla's files noted in the above page. It is to the Dept of Justice. The person wants to see the Tesla files if the Dept of Justice has them and if they don't, where are they! 66- .....Apr 30, 1976 | C. M. Kelley | XXXXX |

Summary> A letter from Kelley thanking someone for a letter with enclosures. A note at the bottom says the FBI never had Tesla's files and again points to the Dept of Justice, Alien Property.

67-68.....These are copies of the enclosures noted it the previous letter They are from / Design News/. The article is about project Tesla headed by Robert K. Golka and Dr. Robert W. Bass. The project is about ball lightning.

69- .....Apr 20, 1976 | XXXXX | C.M. Kelley FBI Director |

Summary> This letter states that Mr. Allen and MR. Ruchlehaus, former acting director of the FBI, contacted XX in 1973 regarding the unavialability of American microfilm records of Nikola Tesla's unpublished diary. Below quoted...from letter...

At the time I discounted the possibility that these unpublished discoveries had military significance. But because of experiments now underway at Hill AFB, I now suspect such military applications exist and feel it imperative that you be notified, particularly in view of the fact that the Soviets have primary access to the ENTIRE collection Two photos of each page exist.

After Tesla's death, scientist from the Navy and OSS performed a cursory examination of the diary and notes, which if my memory serves me correctly, was one month long, hardly enough time to decipher Tesla's torturous hadrwriting. Though in English.

According to the museum director (1971), the Soviets had made copies of some portions, but not the Colorado Springs diary, which number 500 pages ,20 that directly pertain to ball lightning, and 20 or so relevant to the equipment construction. (we copied the most significant portions, but feel more exists.

XXXXXXXX an article XXX magazine, EDN (an electrical engineering magazine , but only with the very recent reciept of an unpublished manuscript from John J. O'Neill's book /Prodigal Genius/ did I place credence on Tesla's later calim to military applications. Incidentally, some of O'Neill's discriptions were inaccurate and exagerated, as we have exceeded Tesla's results and are familiar with the experiments. At any rate, there are three possible military applicatoins.

70- .....Cont.69- First, Tesla claimed that the lightning balls (which destroyed his equipment) could be used to destroy aircraft.

XXX saw one inside his plane and most AF personal fear these / rf-balls/ as they call them.

Second, it is a suspicion of mine that ball lightning, if injected with lithium, could produce a cheap fusion bomb.

Third— and this may be no more than a suspicion— the propulsion mode of ball lightning involves electro-gravitic interaction by which means air vehicles of revolutionary configuration may be constructed.

There are no presently-known laws of physics that can account for the propulsion (400 mph or so when following an airliner). Other hitherto unsuspected applications may exist.

None of these applications were the goal of Project Tesla, which centered on producing ball lightning as Tesla did and studying it as a plasma confinement technique for fusion reactors. Incidentally, Tesla's claim to setting up standing waves on the earth's surface (wireless power) was erroneous and involved techniques similar to Project Sanguine, that is, using the earth's atmosphere as a wave guide XXX is aware of our research.

Signed XXXX. P.S. I am notifying the CIA.

71-74.....Photo copies of Chapter 34 from John J. O'Neill's book.

I do not know if these are the unpublished pages of same or if they are the same that appear in the published book. They mention /lightning balls or fire balls/ and Tesla's plan to build a defence system for Great Britain.

75- .....Jun 16, ???? | United States Senate | XXXXXX |

Summary> A note...Please return all correspondence to the Attention of XXXXXX FBI.

76-77.....Jun ?, 1976 | XXXXX | XXXXX |

Summary> This is a hand written letter to someone in the Senate Office Building, room 5241. It is a request for Tesla's papers from a substitute Milwaukee Public School and scientist. It says the FBI took the papers after his death.

78- .....Jun 23, 1976 | C.M. Kelley | Senator XXXXX |

Summary> This is a letter of responce to the handwritten letteron the previous page. Tells the Senator the FBI was not involved.

79- .....Jul 26, 1979 | XXXXX | Mr. Webster |

Summary> Another handwritten letter wanting Tesla information.

This person says he has almost everyting Tesla has written and to futher his studies want's what the FBI has on Tesla.

80- .....Aug 6, 1979 | William L. Bailey | XXXXX |

This looks like a responce to the previous letter. It says, again the FBI has NO Tesla papers. Also goes on to say that the person will have to make a seprate FOIA request if he wants to get the FBI documents. The letter must request documents regarding the specific topic of interest.

81- .....Feb 2, 1980 | XXXXX | Mr. Bresson |

Summary> A memo about disclosing Nikola Tesla information. It states Bufile 100-2237 is mostly inquiry type correspondence. Copies of the remaining material in the file totaling 29 pages , has been processed for disclosure. Bufile 190-16504-4 be considered the preprocessed release appropriate for futher request.

82- .....Date form was printed 10/10/79/ | Thomas H. Bresson | ????? |

Summary> This is a photocopy of an FBI Internal Routing Action Slip.

There is a handwritten note...I would prefer IPU being altered that any request for Tesla file be refered directly to XXXX where upon she would see that appropriate release is made.

83- .....NO date | XXXXX | Tom |

Summary> This is a handwritten note... Tom- It is my understanding that EVERY case which is completed through disclosure is /preprocessed/ IPU reading room maintains ONLY these /preprocessed/ cases which have been deemed to have sufficient public interest to warrent inclusion in the Reading Room. The question therefore, is : Does the Tesla material fit the criteria for inclusion in the FOIA Reading Room? I Think not. XXX

84- .....Feb 9, 1981 | A. J. MacLaren USAF | Director, FBI |

Summary> This is a memo from LtColonel MacLaren at the Strategic and Space Systems requesting Tesla's papers." We believe that certain of Tesla's papers may contain basic principles which would be of considerable value to certain ongoing research within the DoD. It would be very helpful to have access to his papers." At the bottonm of the page are hand written notes stating all information contained herein is unclassified 4-8-93.

It also has my FOIPA No. 356,608?

85- .....Mar 10, 1981 | Robert S. Young | Colonel MacLaren |

Summary> This is a responce to MacLaren's letter. Young's title is ..Assistant Director in Charge..Office of Congressional and Public Affairs. Young states that the FBI did not participate in the handling of Tesla's papers. His papers were exhimined by representatives of the Office of Alien Property, the Navy Department and the Office of Scientific Research and Development.A complete search of our indices is being made to determine if we have any information that might be useful to you. A note at the bottom of the page states..IN previous request we have said that the Office of Alien Property of the Dept. of Justice impounded Tesla's papers. However the Office of Foreign Litigation, Civil Division, indicats that Dr. Tesla'spapers are not in their possession.

86- .....Mar 20,1981 | Robert S. Young | LtColonel MacLeram | Summary> This is another letter to MacLearn futher explaining that the FBI didn't have Tesla's papers. He says the Unites States Naval Researce and the Office of Naval Intelligance were at the Jan 26,27 1943 exhimination of Tesla's papers in Manhattan. It was the opinion of all there that there was NO information in Tesla's papers of scientific or millitary importance.

87- .....Cont.86-more notes to MacLearn telling him once in Oct 26, 1945 XXXXXX(I am Sure the blanked out name is Bloyce Fitzgearld) a young scientist who had been Tesla's protege, called in person at our New York Field Office. With him were XXXXX (I have think this states three army officers were with him.) and XXXXX from the research development unit at Wright Field, Dayton, Ohio. The men carried a letter signed by XXXXX requesting the FBI allow the bearers of the letter access to the effects of Nikola Tasla (65-47943-15). Young tells Maclearn to contact the other agencies for more information.

88- .....Cont.87-last note from Young. He states that from 1973-1980 the FBI has recieved 20 FOIA request for Tesla's papers.

89- .....XXX 29, 1955 | XXXXX | XXXXX |

Summary> This is a photocopy of a page from the American Srbobran It tells of the progress of Leland I Anderson, head of the Tesla Society headquartered at University of Minnesota, getting Dr. Tesla proper recognition. Talks of the Teslian publication that Anderson puts out. It mentions an article in POWER magazine.

90- .....Jun 29,1955 | A.H. Belmont | L.V. Boardman |

Summary> Mr. Belmont relates to a letter the bureau received dated 6-25-1955 from Kenneth M. Swezey which ask the FBI to help locate the gold Edison metal which was missing from Tesla's safe when it arrived at the Belgrade museum.Swezey mentions two publications, The Prodigal Genius, and The Genius who Walked Alone, that state the FBI was present at the opening of Mr. Tesla's safe the day he died 1-8-43.

91- .....Cont.90-Recommendation that Swezey be advised bureau did not participate in the handeling of Tesla's effects.

92- .....Jun 30,1955 | J. Edgar Hoover | Mr. Kenneth Swezey| Summary> This is the letter to Mr. Swezey which was recommended in page 91. It states the FBI was not involved in Tesla's effects.

93- .....Aug 1, 1955 | XXX | J. Edgar Hoover |

Summary> This is another letter requesting whatever material the FBI has concerning Tesla.

94- .....Aug 11, 1955 | John Edgar Hoover | XXX |

Summary> Answer to the letter of page 93. It states the FBI has no Tesla effects.

95- .....Sep 10, 1955 | XXX | John Edgar Hoover |

Summary> This is kind of interesting...This letter was written by the same person requesting Tesla information back on page 93.

It was from a typewriter with a bad capital (M) and is easily seen to be from the same typewriter. The letter shows XXX wrote the office of Alien Property seeking the where abouts of Tesla's effects after the FBI said they had no information. Mr. XXX states he received a letter from Mr. Henry G. Hilken-that department-File Number HGH:N SM: elk 017-3566, in which I am told that office "NEVER HAD CUSTODY, NOR HAS VESTED, any property of Nikola Tesla".

96- .....Sep 20, 1955 | John Edgar Hoover | XXX |

Summary> This is a short reply to the previous letter of page 95.

Restates the fact that the FBI can furnish additional information on Tesla.

97- .....Jun 29, 1955 | The American Srbobran | XXX |

Summary> This looks like a copy of a form that was printed in the June 29, 1955 American Srbobran, Pittsburg, Penn. It is a request to Postmaster General Hon. Auther E. Summerfield, to issue a 1956 commemorative stamp on the centennial of Tesla's birthday.

It is sad it wasn't until September 21, 1983 such a stamp was issued.

98-102...This is a photocopy of an article in June 1955 Coronet magazine It is by Alfred Sinks and is quite good. It seems to be most of what we now know of Tesla's life.

103-.....Sep 28, 1955 | The American Srbobran | XXX |

Summary> This is a photocopy from The American Srbobran which states a bust of Tesla will be placed at the Monastery at Libertyville Ill. If the $3000.00 can be raised.

104-.....Oct 24, 1955 | The American Srbobran | XXX |

Summary> This is a photocopy from The American Srbobran which states a portrait of Tesla which was used on the cover of Time magazine has been sold at auction and it's location is unknown.

It is 53"x48" and was painted by Princess Vilma Lwoff-Parlaghy.

105-106..Nov 14, XXXX | The American Srbobran | XXX |

Summary> This is a photocopy from The American Srbobran which updates the status of a bust of Tesla to be placed at the Monastery at Libertyville, Ill. It list names and ammounts of contribution.

107-.....Jun 23,1956 | XXX | FBI |

Summary> A letter requesting information the FBI holds about Tesla.

108-.....Jun 29, 1956 |John Edgar Hoover | XXX |

Summary> Answer to the previous letter of page 107. Again the FBI claims to have NO Tesla information.

109-112 ...July 25, 1957 | SAC, New York | Director FBI |

Summary> A very interesting 4 page memorandum about an informant's story telling of a Tesla set which was finished by Tesla engineers in 1950 which enabled interplanetary communication. The set was in the posession of a Mr.John Storm and his wife Margaret. It also tells of a 9-22-40 New York Times article about Tesla's "Death Ray".

You will want copies of this! 113-.....Mar 7,1958 | XXX | J. Edgar Hoover |

Summary> A letter requesting Tesla's FBI files.

114-.....Mar 14,1958 | J. Edgar Hoover | XXX |

Summary> Reply to previous letter of page 113. The FBI has no Tesla files.

There is a note at the bottom of the page of interest! "NOTE: Bufiles reflect no record on correspondent".

How about that...the FBI looks up the people that make a Tesla request to see if they have a file on THEM! I have seen this note on other request but just now realized the importance of it.

115-158 ..The last 43 pages are 19 request for Tesla information from June 24, 1959 - Nov 13, 1975. They are of moderate interest and are all about the same with the same responce...no files on Tesla.

FRUITS OF GENIUS WERE SWEPT AWAY.

By a Fire the Noted Electrician, Nicola Tesla, Loses Mechanisms of Inestimable Value.

INVENTIONS IN THE RUINS.

The Workshop Where He Evolved Ideas That Startled Electricians Entirely Destroyed.

YEARS OF LABOR LOST.

New York Herald, March 14, 1895.

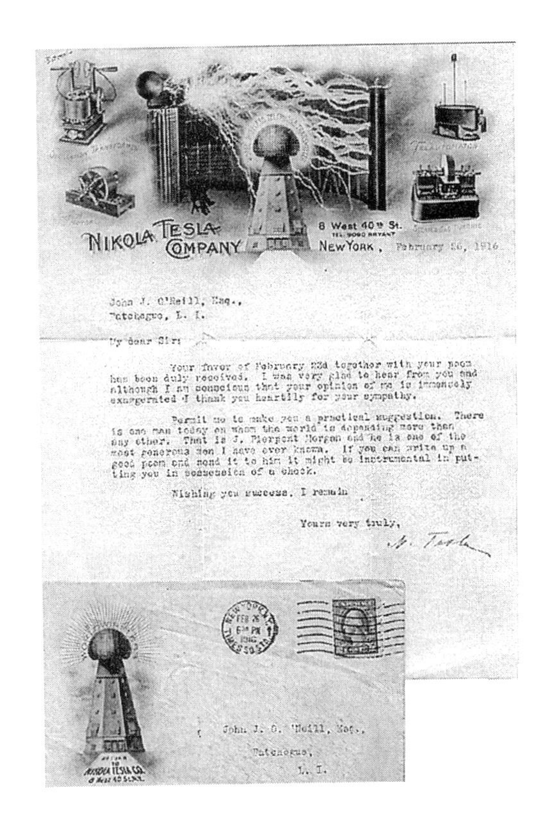

NIKOLA TESLA COMPANY
8 West 40th St.
NEW YORK, February 26, 1916

John J. O'Neill, Esq.,
Patchogue, L. I.

My dear Sir:

Your favor of February 23d together with your poem has been duly received. I was very glad to hear from you and although I am conscious that your opinion of me is immensely exaggerated I thank you heartily for your sympathy.

Permit me to make you a practical suggestion. There is one man today on whom the world is depending more than any other. That is J. Pierpont Morgan and he is one of the most generous men I have ever known. If you can write up a good poem and send it to him it might be instrumental in putting you in possession of a check.

Wishing you success, I remain

Yours very truly,

N. Tesla

John J. O'Neill, Esq.,
Patchogue,
L. I.

Guglielmo Marconi

Part Seven
The Marconi—Tesla Transcripts

Case Name: MARCONI WIRELESS CO. V. U.S., 320 U.S. 1 NO. 369.
ARGUED APRIL 9, 12, 1943.
DECIDED JUNE 21, 1943.
99 CT.CLS. 1, AFFIRMED IN PART.

1. THE BROAD CLAIMS OF THE MARCONI PATENT NO. 763,772, FOR IMPROVEMENTS IN APPARATUS FOR WIRELESS TELEGRAPHY - BRIEFLY, FOR A STRUCTURE AND ARRANGEMENT OF FOUR HIGH-FREQUENCY CIRCUITS WITH MEANS OF INDEPENDENTLY ADJUSTING EACH SO THAT ALL FOUR MAY BE BROUGHT INTO ELECTRICAL RESONANCE WITH ONE ANOTHER - HELD INVALID BECAUSE ANTICIPATED. P. 38.

MARCONI SHOWED NO INVENTION OVER STONE (PATENT NO. 714,756) BY MAKING THE TUNING OF HIS ANTENNA CIRCUIT ADJUSTABLE, OR BY USING LODGE'S (PATENT NO. 609,154) VARIABLE INDUCTANCE FOR THAT PURPOSE. WHETHER STONE'S PATENT INVOLVED INVENTION IS NOT HERE DETERMINED.

2. MERELY MAKING A KNOWN ELEMENT OF A KNOWN COMBINATION ADJUSTABLE BY A MEANS OF ADJUSTMENT KNOWN TO THE ART, WHEN NO NEW OR UNEXPECTED RESULT IS OBTAINED, IS NOT INVENTION. P. 32.

3. AS BETWEEN TWO INVENTORS, PRIORITY OF INVENTION WILL BE AWARDED TO THE ONE WHO BY SATISFYING PROOF CAN SHOW THAT HE FIRST CONCEIVED OF THE INVENTION. P. 34.

4. COMMERCIAL SUCCESS ACHIEVED BY THE LATER INVENTOR AND PATENTEE CANNOT SAVE HIS PATENT FROM THE DEFENSE OF ANTICIPATION BY A PRIOR INVENTOR. P. 35.

5. IN THE EXERCISE OF ITS APPELLATE POWER, THIS COURT MAY CONSIDER ANY EVIDENCE OF RECORD WHICH, WHETHER OR NOT CALLED TO THE ATTENTION OF THE COURT BELOW, IS RELEVANT TO AND MAY AFFECT THE CORRECTNESS OF ITS DECISION SUSTAINING OR DENYING ANY CONTENTION WHICH A PARTY HAS MADE BEFORE IT. P. 44.

6. ALTHOUGH THE INTERLOCUTORY DECISION OF THE COURT OF CLAIMS IN THIS CASE THAT CLAIM 16 OF MARCONI PATENT NO. 763,772 WAS VALID AND INFRINGED WAS APPEALABLE, THE DECISION WAS NOT FINAL UNTIL THE CONCLUSION OF THE ACCOUNTING; HENCE, THE COURT DID NOT LACK POWER AT ANY TIME PRIOR TO ENTRY OF ITS FINAL JUDGMENT AT THE CLOSE OF THE ACCOUNTING TO RECONSIDER ANY PORTION OF ITS DECISION AND REOPEN ANY PART OF THE CASE, AND IT WAS FREE IN ITS DISCRETION TO GRANT A REARGUMENT BASED EITHER ON ALL THE EVIDENCE THEN OF RECORD OR ONLY THE EVIDENCE BEFORE THE COURT WHEN IT RENDERED ITS INTERLOCUTORY DECISION, OR TO REOPEN THE CASE FOR FURTHER EVIDENCE. P. 47.

7. THE JUDGMENT OF THE COURT OF CLAIMS HOLDING VALID AND INFRINGED CLAIM 16 OF MARCONI PATENT NO. 763,772 IS VACATED AND REMANDED IN ORDER THAT THAT COURT MAY DETERMINE WHETHER TO RECONSIDER ITS DECISION IN THE LIGHT OF THE GOVERNMENT'S PRESENT CONTENTION THAT CLAIM 16, AS CONSTRUED BY THE COURT OF CLAIMS, WAS ANTICIPATED BY THE PATENTS TO PUPIN, NO. 640,516, AND FESSENDEN, NO. 706,735. P. 48.

8. A DEFENDANT IN A PATENT INFRINGEMENT SUIT WHO HAS ADDED NON INFRINGING AND VALUABLE IMPROVEMENTS WHICH CONTRIBUTED TO THE MAKING OF THE PROFITS IS NOT LIABLE FOR BENEFITS RESULTING FROM SUCH IMPROVEMENTS. P. 50.

9. DISCLOSURE BY PUBLICATION MORE THAN TWO YEARS BEFORE APPLICATION FOR A PATENT BARS ANY CLAIM FOR A PATENT FOR AN INVENTION EMBODYING THE PUBLISHED DISCLOSURE. P. 57.

10. INVALIDITY IN PART OF A PATENT DEFEATS THE ENTIRE PATENT UNLESS THE INVALID PORTION WAS CLAIMED THROUGH INADVERTENCE, ACCIDENT, OR MISTAKE, AND WITHOUT ANY FRAUDULENT OR DECEPTIVE INTENTION, AND IS DISCLAIMED WITHOUT UNREASONABLE NEGLECT OR DELAY. P. 57.

11. FLEMING PATENT NO. 803,864 HELD INVALID BY REASON OF AN IMPROPER DISCLAIMER. P. 58.

THE SPECIFICATIONS PLAINLY CONTEMPLATED THE USE OF THE CLAIMED DEVICE WITH LOW AS WELL AS HIGH FREQUENCY CURRENTS, AND THE PATENT WAS INVALID FOR WANT OF INVENTION SO FAR AS APPLICABLE TO USE WITH LOW FREQUENCY CURRENTS; THE CLAIM WAS NOT INADVERTENT, AND THE DELAY OF TEN YEARS IN MAKING THE DISCLAIMER WAS UNREASONABLE.

12. THAT THE PATENTEE'S CLAIM FOR MORE THAN HE HAD INVENTED WAS NOT INADVERTANT, AND THAT HIS DELAY IN MAKING DISCLAIMER WAS UNREASONABLE, WERE QUESTIONS OF FACT; BUT, SINCE THE COURT OF CLAIMS IN ITS OPINION IN THIS CASE PLAINLY STATES ITS CONCLUSIONS AS TO THEM, AND THOSE CONCLUSIONS ARE SUPPORTED BY SUBSTANTIAL EVIDENCE, ITS OMISSION TO MAKE FORMAL FINDINGS OF FACT IS IMMATERIAL. P. 58.

13. THE DISCLAIMER STATUTES ARE APPLICABLE TO ONE WHO ACQUIRES A PATENT UNDER AN ASSIGNMENT OF THE APPLICATION. P. 59.

MARCONI WIRELESS TELEGRAPH COMPANY OF AMERICA V. UNITED STATES* .

*TOGETHER WITH NO. 373, UNITED STATES V. MARCONI WIRELESS TELEGRAPH COMPANY OF AMERICA, ALSO ON WRIT OF CERTIORARI, 317 U.S. 620, TO THE COURT OF CLAIMS.

CERTIORARI TO THE COURT OF CLAIMS.

WRITS OF CERTIORARI, 317 U.S. 620, ON CROSS-PETITIONS TO REVIEW A JUDGMENT IN A SUIT AGAINST THE UNITED STATES TO RECOVER DAMAGES FOR INFRINGEMENT OF PATENTS. SEE 81 CT. CLS. 741.

MR. CHIEF JUSTICE STONE DELIVERED THE OPINION OF THE COURT.

THE MARCONI COMPANY BROUGHT THIS SUIT IN THE COURT OF CLAIMS PURSUANT TO 35 U.S.C. SEC. 68, TO RECOVER DAMAGES FOR INFRINGEMENT OF FOUR UNITED STATES PATENTS. TWO, NO. 763,772, AND REISSUE NO. 11,913, WERE ISSUED TO MARCONI, A THIRD, NO. 609,154, TO LODGE, AND A FOURTH, NO. 803,684, TO FLEMING. THE COURT HELD THAT THE MARCONI REISSUE PATENT WAS NOT INFRINGED. IT HELD ALSO THAT THE CLAIMS IN SUIT, OTHER THAN CLAIM 16, OF THE MARCONI PATENT NO. 763,772, ARE INVALID; AND THAT CLAIM 16 OF THE PATENT IS VALID AND WAS INFRINGED. IT GAVE JUDGMENT FOR PETITIONER ON THIS CLAIM IN THE SUM OF $42,984.93 WITH INTEREST. IT HELD THAT THE LODGE PATENT WAS VALID AND INFRINGED, AND THAT THE FLEMING PATENT WAS NOT INFRINGED AND WAS RENDERED VOID BY AN IMPROPER DISCLAIMER. THE CASE COMES HERE ON CERTIORARI, 317 U.S. 620, 28 U.S.C. SEC. 288(B), ON PETITION OF THE MARCONI COMPANY IN NO. 369, TO REVIEW THE JUDGMENT OF THE COURT OF CLAIMS HOLDING INVALID THE CLAIMS IN SUIT, OTHER THAN CLAIM 16, OF THE MARCONI PATENT, AND HOLDING THE FLEMING PATENT INVALID AND NOT INFRINGED, AND ON PETITION OF THE GOVERNMENT IN NO. 373, TO REVIEW THE DECISION ALLOWING RECOVERY FOR INFRINGEMENT OF CLAIM 16 OF THE MARCONI PATENT. NO REVIEW WAS SOUGHT BY EITHER PARTY OF SO MUCH OF THE COURT'S JUDGMENT AS SUSTAINED THE LODGE PATENT AND HELD THE FIRST MARCONI REISSUE PATENT NOT INFRINGED.

MARCONI PATENT NO. 763,772.

THIS PATENT, GRANTED JUNE 28, 1904, ON AN APPLICATION FILED NOVEMBER 10, 1900, AND ASSIGNED TO THE MARCONI COMPANY ON MARCH 6, 1905, FN1 IS FOR IMPROVEMENTS IN APPARATUS FOR WIRELESS TELEGRAPHY BY MEANS OF HERTZIAN OSCILLATIONS OR ELECTRICAL WAVES. IN WIRELESS TELEGRAPHY, SIGNALS GIVEN BY MEANS OF CONTROLLED ELECTRICAL PULSATIONS ARE TRANSMITTED THROUGH THE ETHER BY MEANS OF THE SOCALLED HERTZIAN OR RADIO WAVES. HERTZIAN WAVES ARE ELECTRICAL OSCILLATIONS WHICH TRAVEL WITH THE SPEED OF LIGHT AND HAVE VARYING WAVE LENGTHS AND CONSEQUENT FREQUENCIES INTERMEDIATE BETWEEN THE FREQUENCY RANGES OF LIGHT AND SOUND WAVES. THE TRANSMITTING APPARATUS USED FOR SENDING THE SIGNALS IS CAPABLE, WHEN ACTUATED BY A TELEGRAPH KEY OR OTHER SIGNALLING DEVICE, OF PRODUCING, FOR SHORT PERIODS OF VARIABLE LENGTHS, ELECTRICAL OSCILLATIONS OF RADIO FREQUENCY (OVER 10,000 CYCLES PER SECOND) IN AN ANTENNA OR OPEN CIRCUIT

FROM WHICH THE OSCILLATIONS ARE RADIATED TO A DISTANT RECEIVING APPARATUS. THE RECEIVER HAS AN OPEN ANTENNA CIRCUIT WHICH IS ELECTRICALLY RESPONSIVE TO THE TRANSMITTED WAVES AND IS CAPABLE OF USING THOSE RESPONSES TO ACTUATE BY MEANS OF A RELAY OR AMPLIFIER ANY CONVENIENT FORM OF SIGNALLING APPARATUS FOR MAKING AUDIBLE AN ELECTRICALLY TRANSMITTED SIGNAL, SUCH AS A TELEGRAPH SOUNDER OR A LOUD SPEAKER. IN BRIEF, SIGNALS AT THE TRANSMITTER ARE UTILIZED TO CONTROL HIGH FREQUENCY ELECTRICAL OSCILLATIONS WHICH ARE RADIATED BY AN ANTENNA THROUGH THE ETHER TO THE DISTANT RECEIVER AND THERE PRODUCE AN AUDIBLE OR VISIBLE SIGNAL.

ALL OF THESE WERE FAMILIAR DEVICES AT THE TIME OF MARCONI'S APPLICATION FOR THE PATENT NOW IN SUIT. BY THAT TIME RADIO HAD PASSED FROM THE THEORETICAL TO THE PRACTICAL AND COMMERCIALLY SUCCESSFUL. FOUR YEARS BEFORE, MARCONI HAD APPLIED FOR HIS ORIGINAL AND BASIC PATENT, WHICH WAS GRANTED AS NO. 586,193, JULY 13, 1897 AND REISSUED JUNE 4, 1901 AS REISSUE NO. 11,913. HE APPLIED FOR HIS CORRESPONDING BRITISH PATENT, NO. 12039 OF 1896, ON JUNE 2, 1896. MARCONI'S ORIGINAL PATENT SHOWED A TWO-CIRCUIT SYSTEM, IN WHICH THE HIGH FREQUENCY OSCILLATIONS ORIGINATED IN THE TRANSMITTER ANTENNA CIRCUIT AND THE DETECTING DEVICE WAS CONNECTED DIRECTLY IN THE RECEIVER ANTENNA CIRCUIT. BETWEEN 1896 AND 1900 HE DEMONSTRATED ON NUMEROUS OCCASIONS THE PRACTICAL SUCCESS OF HIS APPARATUS, ATTAINING SUCCESSFUL TRANSMISSION AT DISTANCES OF 70 AND 80 MILES. DURING THOSE YEARS HE APPLIED FOR A LARGE NUMBER OF PATENTS IN THIS AND OTHER COUNTRIES FOR IMPROVEMENTS ON HIS SYSTEM OF RADIO COMMUNICATION. FN2 THE PARTICULAR ADVANCE SAID TO HAVE BEEN ACHIEVED BY THE MARCONI PATENT WITH WHICH WE ARE HERE CONCERNED WAS THE USE OF TWO HIGH FREQUENCY CIRCUITS IN THE TRANSMITTER AND TWO IN THE RECEIVER, ALL FOUR SO ADJUSTED AS TO BE RESONANT TO THE SAME FREQUENCY OR MULTIPLES OF IT. THE CIRCUITS ARE SO CONSTRUCTED THAT THE ELECTRICAL IMPULSES IN THE ANTENNA CIRCUIT OF THE TRANSMITTER VIBRATE LONGER WITH THE APPLICATION TO THE TRANSMITTER OF A GIVEN AMOUNT OF ELECTRICAL ENERGY THAN HAD BEEN THE CASE IN THE PREVIOUS STRUCTURES KNOWN TO THE ART, AND THE SELECTIVITY AND SENSITIVITY OF THE RECEIVER IS LIKEWISE ENHANCED. THUS INCREASED EFFICIENCY IN THE TRANSMISSION AND RECEPTION OF SIGNALS IS OBTAINED. THE SPECIFICATIONS OF THE MARCONI PATENT STATE THAT ITS OBJECT IS "TO INCREASE THE EFFICIENCY OF THE SYSTEM AND TO PROVIDE NEW AND SIMPLE MEANS WHEREBY OSCILLATIONS OF ELECTRICAL WAVES FROM A TRANSMITTING STATION MAY BE LOCALIZED WHEN DESIRED AT ANY ONE SELECTED RECEIVING STATION OR STATIONS OUT OF A GROUP OF SEVERAL RECEIVING STATIONS." THE SPECIFICATIONS DESCRIBE AN ARRANGEMENT OF FOUR HIGH FREQUENCY CIRCUITS TUNED TO ONE ANOTHER - TWO AT THE SENDING STATION ASSOCIATED WITH A SOURCE OF LOW FREQUENCY OSCILLATIONS, AND TWO AT THE RECEIVING STATION ASSOCIATED WITH A RELAY OR AMPLIFIER OPERATING A SIGNALLING DEVICE. AT THE SENDING STATION THERE IS AN OPEN ANTENNA CIRCUIT WHICH IS "A GOOD RADIATOR," CONNECTED WITH THE SECONDARY COIL OF A TRANSFORMER, AND THROUGH IT INDUCTIVELY COUPLED WITH A CLOSED CIRCUIT, WHICH IS CONNECTED WITH THE PRIMARY COIL OF THE TRANSFORMER, THIS CLOSED CIRCUIT BEING A "PERSISTENT OSCILLATOR." AT THE RECEIVING STATION THERE IS AN OPEN ANTENNA CIRCUIT CONSTITUTING A "GOOD ABSORBER" INDUCTIVELY COUPLED WITH A CLOSED CIRCUIT CAPABLE OF ACCUMULATING THE RECEIVED OSCILLATIONS.

THE PATENT, IN DESCRIBING THE ARRANGEMENT OF THE APPARATUS SO AS TO SECURE THE DESIRED RESONANCE OR TUNING, SPECIFIES: "THE CAPACITY AND SELF-INDUCTION OF THE FOUR CIRCUITS - I.E., THE PRIMARY AND SECONDARY CIRCUITS AT THE TRANSMITTING-STATION AND THE PRIMARY AND SECONDARY CIRCUITS AT ANY ONE OF THE RECEIVING-STATIONS IN A COMMUNICATING SYSTEM ARE EACH AND ALL TO BE SO INDEPENDENTLY ADJUSTED AS TO MAKE THE PRODUCT OF THE SELF-INDUCTION MULTIPLIED BY THE CAPACITY THE SAME IN EACH CASE OR MULTIPLES OF EACH OTHER - THAT IS TO SAY, THE ELECTRICAL

TIME PERIODS OF THE FOUR CIRCUITS ARE TO BE THE SAME OR OCTAVES OF EACH OTHER." FN3 AND AGAIN, "IN EMPLOYING THIS INVENTION TO LOCALIZE THE TRANSMISSION OF INTELLIGENCE AT ONE OF SEVERAL RECEIVING-STATIONS THE TIME PERIOD OF THE CIRCUITS AT EACH OF THE RECEIVING-STATIONS IS SO ARRANGED AS TO BE DIFFERENT FROM THOSE OF THE OTHER STATIONS. IF THE TIME PERIODS OF THE CIRCUITS OF THE TRANSMITTING-STATION ARE VARIED UNTIL THEY ARE IN RESONANCE WITH THOSE OF ONE OF THE RECEIVING STATIONS, THAT ONE ALONE OF ALL THE RECEIVING-STATIONS WILL RESPOND, PROVIDED THAT THE DISTANCE BETWEEN THE TRANSMITTING AND RECEIVING STATIONS IS NOT TOO SMALL." THE DRAWINGS AND SPECIFICATIONS SHOW A CLOSED CIRCUIT AT THE TRANSMITTING STATION CONNECTED WITH THE PRIMARY OF AN INDUCTION COIL, AND EMBRACING A SOURCE OF ELECTRICAL CURRENT AND A CIRCUIT-CLOSING KEY OR OTHER SIGNALLING DEVICE. THE SECONDARY OF THE INDUCTION COIL IS CONNECTED IN A CIRCUIT WHICH INCLUDES A SPARK GAP OR OTHER PRODUCER OF HIGH FREQUENCY OSCILLATIONS AND, IN A SHUNT AROUND THE SPARK GAP, THE PRIMARY COIL OF AN OSCILLATION TRANSFORMER AND A CONDENSER, PREFERABLY SO ARRANGED THAT ITS CAPACITY CAN READILY BE VARIED. THIS SHUNT CIRCUIT CONSTITUTES ONE OF THE TWO TUNED CIRCUITS OF THE TRANSMITTER, AND IS OFTEN REFERRED TO AS THE CLOSED OR CHARGING CIRCUIT. THE SECONDARY COIL OF THE TRANSFORMER IS CONNECTED IN THE OPEN OR ANTENNA CIRCUIT, ONE END OF WHICH IS CONNECTED WITH THE EARTH, THE OTHER TO A VERTICAL WIRE ANTENNA OR AN ELEVATED PLATE. THIS ANTENNA CIRCUIT ALSO INCLUDES AN INDUCTION COIL, PREFERABLY ONE WHOSE INDUCTANCE IS READILY VARIABLE, LOCATED BETWEEN THE ANTENNA OR PLATE AND THE TRANSFORMER.

THE RECEIVER CONSISTS OF A SIMILAR ANTENNA CIRCUIT CONNECTED WITH THE PRIMARY COIL OF A TRANSFORMER, AND HAVING A VARIABLE INDUCTION COIL LOCATED BETWEEN THE ANTENNA OR PLATE AND THE TRANSFORMER. A SHUNT CIRCUIT BRIDGING THE TRANSFORMER AND CONTAINING A CONDENSER WHICH IS PREFERABLY ADJUSTABLE MAY ALSO BE ADDED. THE SECONDARY COIL OF THE TRANSFORMER IS CONNECTED THROUGH ONE OR MORE INTERPOSED INDUCTANCE COILS, "PREFERABLY OF VARIABLE INDUCTANCE," WITH THE TERMINALS OF A COHERER FN4 OR OTHER SUITABLE DETECTOR OF ELECTRICAL OSCILLATIONS. THE CLOSED RECEIVER CIRCUIT ALSO CONTAINED ONE OR MORE CONDENSERS.

THE DEVICES AND ARRANGEMENTS SPECIFIED ARE SUITABLE FOR EFFECTING THE ELECTRICAL TRANSMISSION OF SIGNALS IN THE MANNER ALREADY INDICATED. BY THE MAINTENANCE OF THE SAME HIGH FREQUENCY THROUGHOUT THE FOUR-CIRCUIT SYSTEM THE CUMULATIVE RESONANCE IS ATTAINED WHICH GIVES THE DESIRED INCREASED EFFICIENCY IN TRANSMISSION AND INCREASED SELECTIVITY AT THE RECEIVING STATION.

THE PATENT DESCRIBES THE OPERATION OF THE FOUR CIRCUITS AS FOLLOWS, BEGINNING WITH THE TRANSMITTER: "IN OPERATION THE SIGNALLING-KEY B IS PRESSED, AND THIS CLOSES THE PRIMARY OF THE INDUCTION-COIL. CURRENT THEN RUSHES THROUGH THE TRANSFORMER-CIRCUIT AND THE CONDENSER E IS CHARGED AND SUBSEQUENTLY DISCHARGES THROUGH THE SPARK-GAP. IF THE CAPACITY, THE INDUCTANCE, AND THE RESISTANCE OF THE CIRCUIT ARE OF SUITABLE VALUES, THE DISCHARGE IS OSCILLATORY, WITH THE RESULT THAT ALTERNATING CURRENTS OF HIGH FREQUENCY PASS THROUGH THE PRIMARY OF THE TRANSFORMER AND INDUCE SIMILAR OSCILLATIONS IN THE SECONDARY, THESE OSCILLATIONS BEING RAPIDLY RADIATED IN THE FORM OF ELECTRIC WAVES BY THE ELEVATED CONDUCTOR (ANTENNA).

"FOR THE BEST RESULTS AND IN ORDER TO EFFECT THE SELECTION OF THE STATION OR STATIONS WHEREAT THE TRANSMITTED OSCILLATIONS ARE TO BE LOCALIZED I INCLUDE IN THE OPEN SECONDARY CIRCUIT OF THE TRANSFORMER, AND PREFERABLY BETWEEN THE RADIATOR F AND THE SECONDARY COIL D', AN INDUCTANCE-COIL G, FIG. 1, HAVING NUMEROUS COILS, AND THE CONNECTION IS SUCH THAT A GREATER OR LESS NUMBER OF TURNS OF THE COIL CAN BE PUT IN USE, THE PROPER NUMBER BEING ASCERTAINED BY

EXPERIMENT." THE INVENTION THUS DESCRIBED MAY SUMMARILY BE STATED TO BE A STRUCTURE AND ARRANGEMENT OF FOUR HIGH FREQUENCY CIRCUITS, WITH MEANS OF INDEPENDENTLY ADJUSTING EACH SO THAT ALL FOUR MAY BE BROUGHT INTO ELECTRICAL RESONANCE WITH ONE ANOTHER. THIS IS THE BROAD INVENTION COVERED BY CLAIM 20. COMBINATIONS COVERING SO MUCH OF THE INVENTION AS IS EMBODIED IN THE TRANSMITTER AND THE RECEIVER RESPECTIVELY ARE SEPARATELY CLAIMED. FN5 LONG BEFORE MARCONI'S APPLICATION FOR THIS PATENT THE SCIENTIFIC PRINCIPLES OF WHICH HE MADE USE WERE WELL UNDERSTOOD AND THE PARTICULAR APPLIANCES CONSTITUTING ELEMENTS IN THE APPARATUS COMBINATION WHICH HE CLAIMED WERE WELL KNOWN. ABOUT SEVENTY YEARS AGO CLERK MAXWELL DESCRIBED THE SCIENTIFIC THEORY OF WIRELESS COMMUNICATION THROUGH THE TRANSMISSION OF ELECTRICAL ENERGY BY ETHER WAVES. FN6 BETWEEN 1878 AND 1890 HERTZ DEVISED APPARATUS FOR ACHIEVING THAT RESULT WHICH WAS DESCRIBED BY DE TUNZELMANN IN A SERIES OF ARTICLES PUBLISHED IN THE LONDON ELECTRICIAN IN 1888. ONE, OF SEPTEMBER 21, 1888, SHOWED A TRANSMITTER COMPRISING A CLOSED CIRCUIT INDUCTIVELY COUPLED WITH AN OPEN CIRCUIT. THE CLOSED CIRCUIT INCLUDED A SWITCH OR CIRCUIT BREAKER CAPABLE OF USE FOR SENDING SIGNALS, AND AN AUTOMATIC CIRCUIT BREAKER CAPABLE, WHEN THE SWITCH WAS CLOSED, OF SETTING UP AN INTERMITTENT CURRENT IN THE CLOSED CIRCUIT WHICH IN TURN INDUCED THROUGH A TRANSFORMER AN INTERMITTENT CURRENT OF HIGHER VOLTAGE IN THE OPEN CIRCUIT. THE OPEN CIRCUIT INCLUDED A SPARK GAP ACROSS WHICH A SUCCESSION OF SPARKS WERE CAUSED TO LEAP WHENEVER THE SIGNAL SWITCH WAS CLOSED, EACH SPARK PRODUCING A SERIES OF HIGH FREQUENCY OSCILLATIONS IN THE OPEN CIRCUIT.

BY CONNECTING THE SPARK GAP TO LARGE AREA PLATES IN THE OPEN CIRCUIT HERTZ INCREASED THE CAPACITY AND THUS NOT ONLY INCREASED THE FORCE OF THE SPARKS BUT ALSO CHANGED ONE OF THE TWO FACTORS DETERMINING THE FREQUENCY OF THE OSCILLATIONS IN THE CIRCUIT, AND HENCE THE WAVE LENGTH OF THE OSCILLATIONS TRANSMITTED. HERTZ'S RECEIVER WAS SHOWN AS A RECTANGLE OF WIRE CONNECTED TO THE KNOBS OF A SPARK GAP, BOTH THE WIRE AND THE SPARK GAP BEING OF SPECIFIED LENGTHS OF SUCH RELATIONSHIP AS TO RENDER THE CIRCUIT RESONANT TO THE WAVE LENGTHS IN THE TRANSMITTER. AT TIMES HERTZ ATTACHED TO THE RECTANGLE ADDITIONAL VERTICAL WIRES WHICH PROVIDED ADDITIONAL CAPACITY, AND WHOSE LENGTH COULD READILY BE VARIED SO AS TO VARY THE WAVE LENGTHS TO WHICH THE RECEIVER WAS RESPONSIVE, THUS PROVIDING A "METHOD OF ADJUSTING THE CAPACITY" OF THE RECEIVER. FN7 THUS HERTZ AT THE OUTSET OF RADIO COMMUNICATION RECOGNIZED THE IMPORTANCE OF RESONANCE AND PROVIDED MEANS FOR SECURING IT BY TUNING BOTH HIS TRANSMITTING AND RECEIVING CIRCUITS TO THE SAME FREQUENCY, BY ADJUSTING THE CAPACITY OF EACH. FN8 LODGE, WRITING IN THE LONDON ELECTRICIAN IN 1894, ELABORATED FURTHER ON THE DISCOVERIES OF HERTZ AND ON HIS OWN EXPERIMENTS ALONG THE SAME LINES. IN ONE ARTICLE, OF JUNE 8, 1894, HE DISCUSSED PHENOMENA OF RESONANCE AND MADE AN OBSERVATION WHICH UNDERLIES SEVERAL OF THE DISCLOSURES IN MARCONI'S PATENT. LODGE POINTED OUT THAT SOME CIRCUITS WERE BY THEIR NATURE PERSISTENT VIBRATORS, I.E., WERE ABLE TO SUSTAIN FOR A LONG PERIOD OSCILLATIONS SET UP IN THEM, WHILE OTHERS WERE SO CONSTRUCTED THAT THEIR OSCILLATIONS WERE RAPIDLY DAMPED. HE SAID THAT A RECEIVER SO CONSTRUCTED AS TO BE RAPIDLY DAMPED WOULD RESPOND TO WAVES OF ALMOST ANY FREQUENCY, WHILE ONE THAT WAS A PERSISTENT VIBRATOR WOULD RESPOND ONLY TO WAVES OF ITS OWN NATURAL PERIODICITY. LODGE POINTED OUT FURTHER THAT HERTZ'S TRANSMITTER "RADIATES VERY POWERFULLY" BUT THAT "IN CONSEQUENCE OF ITS RADIATION OF ENERGY, ITS VIBRATIONS ARE RAPIDLY DAMPED, AND IT ONLY GIVES SOME THREE OR FOUR GOOD STRONG SWINGS. HENCE IT FOLLOWS THAT IT HAS A WIDE RANGE OF EXCITATION, I.E., IT CAN EXCITE SPARKS IN CONDUCTORS BARELY AT ALL IN TUNE WITH IT." ON THE OTHER HAND HERTZ'S RECEIVER WAS "NOT A GOOD ABSORBER BUT A PERSISTENT

VIBRATOR, WELL ADAPTED FOR PICKING UP DISTURBANCES OF PRECISE AND MEASURABLE WAVE LENGTH." LODGE CONCLUDED THAT "THE TWO CONDITIONS, CONSPICUOUS ENERGY OF RADIATION AND PERSISTENT VIBRATION ELECTRICALLY PRODUCED, ARE AT PRESENT INCOMPATIBLE." (PP. 154-5.) IN 1892, CROOKES PUBLISHED AN ARTICLE IN THE FORTNIGHTLY REVIEW IN WHICH HE DEFINITELY SUGGESTED THE USE OF HERTZIAN WAVES FOR WIRELESS TELEGRAPHY AND POINTED OUT THAT THE METHOD OF ACHIEVING THAT RESULT WAS TO BE FOUND IN THE USE AND IMPROVEMENT OF THEN KNOWN MEANS OF GENERATING ELECTRICAL WAVES OF ANY DESIRED WAVE LENGTH, TO BE TRANSMITTED THROUGH THE ETHER TO A RECEIVER, BOTH SENDING AND RECEIVING INSTRUMENTS BEING ATTUNED TO A DEFINITE WAVE LENGTH. FN9 A YEAR LATER TESLA, WHO WAS THEN PREOCCUPIED WITH THE WIRELESS TRANSMISSION OF POWER FOR USE IN LIGHTING OR FOR THE OPERATION OF DYNAMOS, PROPOSED, IN A LECTURE BEFORE THE FRANKLIN INSTITUTE IN PHILADELPHIA, THE USE OF ADJUSTABLE HIGH FREQUENCY OSCILLATIONS FOR WIRELESS TRANSMISSION OF SIGNALS. FN10 MARCONI'S ORIGINAL PATENT NO. 586,193, WHICH WAS GRANTED JULY 13, 1897, AND BECAME REISSUE NO. 11,913, DISCLOSED A TWO-CIRCUIT SYSTEM FOR THE TRANSMISSION AND RECEPTION OF HERTZIAN WAVES. THE TRANSMITTER COMPRISED AN ANTENNA CIRCUIT CONNECTED AT ONE END TO AN AERIAL PLATE AND AT THE OTHER TO THE GROUND, AND CONTAINING A SPARK GAP. TO THE KNOBS OF THE SPARK GAP WAS CONNECTED A TRANSFORMER WHOSE SECONDARY WAS CONNECTED WITH A SOURCE OF CURRENT AND A SIGNALLING KEY. THE LOW FREQUENCY CURRENT THEREBY INDUCED IN THE ANTENNA CIRCUIT WAS CAUSED TO DISCHARGE THROUGH THE SPARK GAP, PRODUCING THE HIGH FREQUENCY OSCILLATIONS WHICH WERE RADIATED BY THE ANTENNA. THE RECEIVER SIMILIARLY CONTAINED AN ANTENNA CIRCUIT BETWEEN AN ELEVATED PLATE AND THE GROUND, IN WHICH A COHERER WAS DIRECTLY CONNECTED. MARCONI CLAIMED THE CONSTRUCTION OF TRANSMITTER AND RECEIVER SO AS TO BE RESONANT TO THE SAME FREQUENCY, AND DESCRIBED MEANS OF DOING SO BY CAREFUL DETERMINATION OF THE SIZE OF THE AERIAL PLATES.

THE TESLA PATENT NO. 645,576, APPLIED FOR SEPTEMBER 2, 1897 AND ALLOWED MARCH 20, 1900, DISCLOSED A FOUR-CIRCUIT SYSTEM, HAVING TWO CIRCUITS EACH AT TRANSMITTER AND RECEIVER, AND RECOMMENDED THAT ALL FOUR CIRCUITS BE TUNED TO THE SAME FREQUENCY. TESLA'S APPARATUS WAS DEVISED PRIMARILY FOR THE TRANSMISSION OF ENERGY TO ANY FORM OF ENERGY CONSUMING DEVICE BY USING THE RARIFIED ATMOSPHERE AT HIGH ELEVATIONS AS A CONDUCTOR WHEN SUBJECTED TO THE ELECTRICAL PRESSURE OF A VERY HIGH VOLTAGE. BUT HE ALSO RECOGNIZED THAT HIS APPARATUS COULD, WITHOUT CHANGE, BE USED FOR WIRELESS COMMUNICATION, WHICH IS DEPENDENT UPON THE TRANSMISSION OF ELECTRICAL ENERGY. HIS SPECIFICATIONS DECLARE: "THE APPARATUS WHICH I HAVE SHOWN WILL OBVIOUSLY HAVE MANY OTHER VALUABLE USES - AS, FOR INSTANCE, WHEN IT IS DESIRED TO TRANSMIT INTELLIGIBLE MESSAGES TO GREAT DISTANCES ... " FN11 TESLA'S SPECIFICATIONS DISCLOSED AN ARRANGEMENT OF FOUR CIRCUITS, AN OPEN ANTENNA CIRCUIT COUPLED, THROUGH A TRANSFORMER, TO A CLOSED CHARGING CIRCUIT AT THE TRANSMITTER, AND AN OPEN ANTENNA CIRCUIT AT THE RECEIVER SIMILARLY COUPLED TO A CLOSED DETECTOR CIRCUIT. HIS PATENT ALSO INSTRUCTED THOSE SKILLED IN THE ART THAT THE OPEN AND CLOSED CIRCUITS IN THE TRANSMITTING SYSTEM AND IN THE RECEIVING SYSTEM SHOULD BE IN ELECTRICAL RESONANCE WITH EACH OTHER. HIS SPECIFICATIONS STATE THAT THE "PRIMARY AND SECONDARY CIRCUITS IN THE TRANSMITTING APPARATUS" ARE "CAREFULLY SYNCHRONIZED." THEY DESCRIBE THE METHOD OF ACHIEVING THIS BY ADJUSTING THE LENGTH OF WIRE IN THE SECONDARY WINDING OF THE OSCILLATION TRANSFORMER IN THE TRANSMITTER, AND SIMILARLY IN THE RECEIVER, SO THAT "THE POINTS OF HIGHEST POTENTIAL ARE MADE TO COINCIDE WITH THE ELEVATED TERMINALS" OF THE ANTENNA, I.E., SO THAT THE ANTENNA CIRCUIT WILL BE RESONANT TO THE FREQUENCY DEVELOPED IN THE CHARGING CIRCUIT OF THE TRANSMITTER. THE

SPECIFICATIONS FURTHER STATE THAT "THE RESULTS WERE PARTICULARLY SATISFACTORY WHEN THE PRIMARY COIL OR SYSTEM A' WITH ITS SECONDARY C' (OF THE RECEIVER) WAS CAREFULLY ADJUSTED, SO AS TO VIBRATE IN SYNCHRONISM WITH THE TRANSMITTING COIL OR SYSTEM AC." TESLA THUS ANTICIPATED THE FOLLOWING FEATURES OF THE MARCONI PATENT: A CHARGING CIRCUIT IN THE TRANSMITTER FOR CAUSING OSCILLATIONS OF THE DESIRED FREQUENCY, COUPLED, THROUGH A TRANSFORMER, WITH THE OPEN ANTENNA CIRCUIT, AND THE SYNCHRONIZATION OF THE TWO CIRCUITS BY THE PROPER DISPOSITION OF THE INDUCTANCE IN EITHER THE CLOSED OR THE ANTENNA CIRCUIT OR BOTH. BY THIS AND THE ADDED DISCLOSURE OF THE TWO CIRCUIT ARRANGEMENT IN THE RECEIVER WITH SIMILAR ADJUSTMENT, HE ANTICIPATED THE FOUR-CIRCUIT TUNED COMBINATION OF MARCONI. A FEATURE OF THE MARCONI COMBINATION NOT SHOWN BY TESLA WAS THE USE OF A VARIABLE INDUCTANCE AS A MEANS OF ADJUSTING THE TUNING OF THE ANTENNA CIRCUIT OF TRANSMITTER AND RECEIVER. THIS WAS DEVELOPED BY LODGE AFTER TESLA'S PATENT BUT BEFORE THE MARCONI PATENT IN SUIT.

IN PATENT NO. 609,154, APPLIED FOR FEBRUARY 1, 1898 AND ALLOWED AUGUST 16, 1898, BEFORE MARCONI'S APPLICATION, LODGE DISCLOSED AN ADJUSTABLE INDUCTION COIL IN THE OPEN OR ANTENNA CIRCUIT IN A WIRELESS TRANSMITTER OR RECEIVER OR BOTH TO ENABLE TRANSMITTER AND RECEIVER TO BE TUNED TOGETHER. HIS PATENT PROVIDED FOR THE USE, IN THE OPEN CIRCUITS OF A TRANSMITTER AND A RECEIVER OF HERTZIAN WAVES, OF A SELF INDUCTION COIL BETWEEN A PAIR OF CAPACITY AREAS WHICH HE STATED MIGHT BE ANTENNA AND EARTH. HIS SPECIFICATIONS STATE THAT A COIL LOCATED AS DESCRIBED COULD BE MADE ADJUSTABLE AT WILL SO AS TO VARY THE VALUE OF ITS SELF-INDUCTANCE; THAT THE ADJUSTMENT, TO SECURE THE "DESIRED FREQUENCY OF VIBRATION OR SYNTONY WITH A PARTICULAR DISTANT STATION," MAY BE ATTAINED EITHER "BY REPLACING ONE COIL BY ANOTHER" OR BY THE USE OF A COIL CONSTRUCTED WITH A MOVABLE SWITCH SO RELATED TO THE COIL AS TO SHORT CIRCUIT, WHEN CLOSED, ANY DESIRED NUMBER OF TURNS OF THE WIRE, "SO THAT THE WHOLE OR ANY SMALLER PORTION OF THE INDUCTANCE AVAILABLE MAY BE USED IN ACCORDANCE WITH THE CORRESPONDINGLY-ATTUNED RECEIVER AT THE PARTICULAR STATION TO WHICH IT IS DESIRED TO SIGNAL." THUS LODGE ADJUSTED HIS TUNING BY VARYING THE SELF-INDUCTANCE OF THE ANTENNA CIRCUITS, FOR, AS HE EXPLAINED, THE ADJUSTMENT OF WAVE LENGTHS, AND HENCE OF FREQUENCY IN THE CIRCUITS, COULD BE MADE BY VARYING EITHER OR BOTH THE INDUCTANCE AND CAPACITY, WHICH ARE THE FACTORS CONTROLLING WAVE LENGTH AND HENCE FREQUENCY IN THE ANTENNA CIRCUITS.

LODGE THUS BROADLY CLAIMED THE TUNING, BY MEANS OF A VARIABLE INDUCTANCE, OF THE ANTENNA CIRCUITS IN A SYSTEM OF RADIO COMMUNICATION. HIS SPECIFICATIONS DISCLOSE WHAT IS SUBSTANTIALLY A TWO CIRCUIT SYSTEM, WITH ONE HIGH FREQUENCY CIRCUIT AT THE TRANSMITTER AND ONE AT THE RECEIVER. HE ALSO SHOWED A TWO-CIRCUIT RECEIVER WITH A TUNED ANTENNA CIRCUIT, HIS DETECTOR CIRCUIT AT THE RECEIVER BEING CONNECTED WITH THE TERMINALS OF A SECONDARY COIL WOUND AROUND THE VARIABLE INDUCTANCE COIL IN THE ANTENNA CIRCUIT AND THUS INDUCTIVELY COUPLED THROUGH A TRANSFORMER WITH THE ANTENNA CIRCUIT. FN12 LODGE THUS SUPPLIED THE MEANS OF VARYING INDUCTANCE AND HENCE TUNING WHICH WAS LACKING IN THE TESLA PATENT. HE ALSO SHOWED A RECEIVER WHICH COMPLETELY ANTICIPATED THOSE OF THE MARCONI RECEIVER CLAIMS WHICH PRESCRIBE ADJUSTABLE MEANS OF TUNING ONLY IN THE ANTENNA CIRCUIT (CLAIMS 2, 13 AND 18) AND PARTIALLY ANTICIPATED THE OTHER RECEIVER CLAIMS.

THE STONE PATENT NO. 714,756, APPLIED FOR FEBRUARY 8, 1900, NINE MONTHS BEFORE MARCONI'S APPLICATION, AND ALLOWED DECEMBER 2, 1902, A YEAR AND A HALF BEFORE THE GRANT OF MARCONI'S PATENT, SHOWED A FOUR CIRCUIT WIRELESS TELEGRAPH APPARATUS SUBSTANTIALLY LIKE THAT LATER SPECIFIED AND PATENTED BY MARCONI. IT DESCRIBED ADJUSTABLE TUNING, BY MEANS OF A VARIABLE INDUCTANCE, OF THE CLOSED CIRCUITS OF

BOTH TRANSMITTER AND RECEIVER. IT ALSO RECOMMENDED THAT THE TWO ANTENNA CIRCUITS BE SO CONSTRUCTED AS TO BE RESONANT TO THE SAME FREQUENCIES AS THE CLOSED CIRCUITS. THIS RECOMMENDATION WAS ADDED BY AMENDMENT TO THE SPECIFICATIONS MADE AFTER MARCONI HAD FILED HIS APPLICATION, AND THE PRINCIPAL QUESTION IS WHETHER THE AMENDMENTS WERE IN POINT OF SUBSTANCE A DEPARTURE FROM STONE'S INVENTION AS DISCLOSED BY HIS APPLICATION.

STONE'S APPLICATION SHOWS AN INTIMATE UNDERSTANDING OF THE MATHEMATICAL AND PHYSICAL PRINCIPLES UNDERLYING RADIO COMMUNICATION AND ELECTRICAL CIRCUITS IN GENERAL. IT CONTAINS A CRITICAL ANALYSIS OF THE STATE OF THE ART OF RADIO TRANSMISSION AND RECEPTION. HE SAID THAT AS YET IT HAD NOT BEEN FOUND POSSIBLE SO TO TUNE STATIONS USING A VERTICAL ANTENNA AS TO MAKE POSSIBLE SELECTIVE RECEPTION BY A PARTICULAR STATION TO THE EXCLUSION OF OTHERS. HIS EFFORT, ACCORDINGLY, WAS TO TRANSMIT A "SIMPLE HARMONIC WAVE" OF WELL DEFINED PERIODICITY TO A RECEIVER WHICH WOULD BE SELECTIVELY RESPONSIVE TO THE PARTICULAR FREQUENCY TRANSMITTED, AND THEREBY TO ACHIEVE GREATER PRECISION OF TUNING AND A HIGHER DEGREE OF SELECTIVITY.

STONE DISCUSSES IN SOME DETAIL THE DIFFERENCE BETWEEN "NATURAL" AND "FORCED" OSCILLATIONS. HE SAYS "IF THE ELECTRICAL EQUILIBRIUM OF A CONDUCTOR BE ABRUPTLY DISTURBED AND THE CONDUCTOR THEREAFTER BE LEFT TO ITSELF, ELECTRIC CURRENTS WILL FLOW IN THE CONDUCTOR, WHICH TEND TO ULTIMATELY RESTORE THE CONDITION OF ELECTRICAL EQUILIBRIUM." HE POINTS OUT THAT A CLOSED CIRCUIT CONTAINING A CONDENSER AND A COIL IS "CAPABLE OF OSCILLATORY RESTORATION OF EQUILIBRIUM UPON THE SUDDEN DISCHARGE OF THE CONDENSER" AND THAT "THE ELECTRICAL OSCILLATIONS WHICH IT SUPPORTS WHEN ITS EQUILIBRIUM IS ABRUPTLY DISTURBED AND IT IS THEN LEFT TO ITSELF ARE KNOWN AS THE NATURAL VIBRATIONS OR OSCILLATIONS OF THE SYSTEM." IN ADDITION TO ITS ABILITY TO ORIGINATE "NATURAL VIBRATIONS" WHEN ITS ELECTRICAL EQUILIBRIUM IS DISTURBED, STONE SAYS THAT AN ELECTRICAL CIRCUIT IS ALSO "CAPABLE OF SUPPORTING WHAT ARE TERMED FORCED VIBRATIONS" WHEN ELECTRICAL OSCILLATIONS ELSEWHERE CREATED ARE IMPRESSED UPON IT. IN CONTRAST TO THE "NATURAL" VIBRATIONS OF A CIRCUIT, WHOSE FREQUENCY DEPENDS UPON "THE RELATION BETWEEN THE ELECTROMAGNETIC CONSTANTS (CAPACITY AND SELF-INDUCTANCE) OF THE CIRCUIT," THE FREQUENCY OF THE "FORCED" VIBRATIONS IS "INDEPENDENT OF THE CONSTANTS OF THE CIRCUIT" ON WHICH THEY ARE IMPRESSED AND "DEPENDS ONLY UPON THE PERIOD (FREQUENCY) OF THE IMPRESSED FORCE." IN OTHER WORDS, STONE FOUND THAT IT WAS POSSIBLE NOT ONLY TO ORIGINATE HIGH FREQUENCY OSCILLATIONS IN A CIRCUIT, AND TO DETERMINE THEIR FREQUENCY BY PROPER DISTRIBUTION OF CAPACITY AND SELF-INDUCTANCE IN THE CIRCUIT, BUT ALSO TO TRANSFER THOSE OSCILLATIONS TO ANOTHER CIRCUIT AND RETAIN THEIR ORIGINAL FREQUENCY.

STONE POINTS OUT THAT IN THE EXISTING SYSTEMS OF RADIO TRANSMISSION THE ELECTRIC OSCILLATIONS ARE "NATURALLY" DEVELOPED IN THE ANTENNA CIRCUIT BY THE SUDDEN DISCHARGE OF ACCUMULATED ELECTRICAL FORCE THROUGH A SPARK GAP IN THAT CIRCUIT. SUCH OSCILLATIONS ARE "NECESSARILY OF A COMPLEX CHARACTER AND CONSIST OF A GREAT VARIETY OF SUPERIMPOSED SIMPLE HARMONIC VIBRATIONS OF DIFFERENT FREQUENCIES." "SIMILARLY THE VERTICAL CONDUCTOR AT THE RECEIVING STATION IS CAPABLE OF RECEIVING AND RESPONDING TO VIBRATIONS OF A GREAT VARIETY OF FREQUENCIES SO THAT THE ELECTRO-MAGNETIC WAVES WHICH EMANATE FROM ONE VERTICAL CONDUCTOR USED AS A TRANSMITTER ARE CAPABLE OF EXCITING VIBRATIONS IN ANY OTHER VERTICAL WIRE AS A RECEIVER ... AND THE MESSAGES FROM THE TRANSMITTING STATION WILL NOT BE SELECTIVELY RECEIVED BY THE PARTICULAR RECEIVING STATION WITH WHICH IT IS DESIROUS TO COMMUNICATE, AND WILL INTERFERE WITH THE OPERATION OF OTHER RECEIVING STATIONS WITHIN ITS SPHERE OF INFLUENCE." IN CONTRAST TO THE TWO-

CIRCUIT SYSTEM WHOSE INADEQUACIES HE HAD THUS DESCRIBED, STONE'S DRAWINGS AND SPECIFICATIONS DISCLOSE A FOUR-CIRCUIT SYSTEM FOR TRANSMITTING AND RECEIVING RADIO WAVES WHICH WAS VERY SIMILAR TO THAT LATER DISCLOSED BY MARCONI. THE TRANSMITTER INCLUDED A SOURCE OF LOW FREQUENCY OSCILLATING CURRENT AND A TELEGRAPH OR SIGNALLING KEY CONNECTED IN A CIRCUIT WHICH WAS INDUCTIVELY COUPLED WITH ANOTHER CLOSED CIRCUIT. THIS INCLUDED AN INDUCTION COIL, A CONDENSER, AND A SPARK GAP CAPABLE OF GENERATING HIGH FREQUENCY OSCILLATIONS. IT IN TURN WAS INDUCTIVELY COUPLED THROUGH A TRANSFORMER WITH AN OPEN ANTENNA CIRCUIT CONNECTED TO AN AERIAL CAPACITY AT ONE END AND THE EARTH AT THE OTHER. THE RECEIVER INCLUDED A SIMILAR ANTENNA CIRCUIT, INDUCTIVELY COUPLED WITH A CLOSED OSCILLATING CIRCUIT CONTAINING AN INDUCTION COIL, A CONDENSER, AND A COHERER OR OTHER DETECTOR OF RADIO WAVES. STONE THUS RECOGNIZED, ALTHOUGH HE USED DIFFERENT TERMINOLOGY, THE FACT, PREVIOUSLY OBSERVED BY LODGE, THAT AN OPEN ANTENNA CIRCUIT, SO CONSTRUCTED AS TO BE AN EFFICIENT RADIATOR, WAS NOT AN OSCILLATOR CAPABLE OF PRODUCING NATURAL WAVES OF A SINGLE WELL-DEFINED PERIODICITY, AND CONSEQUENTLY HAD A WIDE RANGE OF EXCITATION. HE ADOPTED THE SAME REMEDY FOR THIS DEFECT AS MARCONI LATER DID, NAMELY TO PRODUCE THE OSCILLATIONS IN A CLOSED CIRCUIT CAPABLE OF GENERATING PERSISTENT VIBRATIONS OF WELL-DEFINED PERIODICITY, AND THEN INDUCE THOSE OSCILLATIONS IN AN OPEN ANTENNA CIRCUIT CAPABLE OF RADIATING THEM EFFICIENTLY TO A DISTANT RESONANT RECEIVER. HE STATES THAT THE VIBRATIONS IN HIS CLOSED CIRCUIT "BEGIN WITH A MAXIMUM OF AMPLITUDE AND GRADUALLY DIE AWAY," A GOOD DESCRIPTION OF THE RESULTS OBTAINABLE BY A "PERSISTENT OSCILLATOR." FN13 SIMILARLY IN HIS RECEIVER STONE RECOGNIZED THAT AN OPEN ANTENNA CIRCUIT (LODGE'S "GOOD ABSORBER") WAS NOT A HIGHLY SENSITIVE RESPONDER TO WAVES OF A PARTICULAR FREQUENCY, AND ACCORDINGLY HE SOUGHT TO AUGMENT THE SELECTIVITY OF TUNING AT THE RECEIVER BY INTERPOSING BETWEEN THE ANTENNA CIRCUIT AND THE RESPONDING DEVICE A CLOSED CIRCUIT WHICH WOULD BE A MORE PERSISTENT VIBRATOR AND HENCE RENDER THE RECEIVING APPARATUS MORE SELECTIVELY RESPONSIVE TO WAVES OF A PARTICULAR FREQUENCY. IN SO DOING, HOWEVER, AS WILL PRESENTLY APPEAR, HE DID NOT DISREGARD THE FAVORABLE EFFECT ON SELECTIVITY OF TUNING AFFORDED BY MAKING THE ANTENNA CIRCUITS RESONANT TO THE TRANSMITTED FREQUENCY.

STONE'S APPLICATION RECOMMENDS THAT THE INDUCTANCE COILS IN THE CLOSED CIRCUITS AT TRANSMITTER AND RECEIVER "BE MADE ADJUSTABLE AND SERVE AS A MEANS WHEREBY THE OPERATORS MAY ADJUST THE APPARATUS TO THE PARTICULAR FREQUENCY WHICH IT IS INTENDED TO EMPLOY." HE THUS DISCLOSED A MEANS OF ADJUSTING THE TUNING OF THE CLOSED CIRCUITS BY VARIABLE INDUCTANCE. HIS ORIGINAL APPLICATION NOWHERE STATES IN SO MANY WORDS THAT THE ANTENNA CIRCUITS SHOULD BE TUNED, NOR DO ITS SPECIFICATIONS OR DRAWINGS EXPLICITLY DISCLOSE ANY MEANS FOR ADJUSTING THE TUNING OF THOSE CIRCUITS. BUT THERE IS NOTHING IN THEM TO SUGGEST THAT STONE DID NOT INTEND TO HAVE THE ANTENNA CIRCUITS TUNED, AND WE THINK THAT THE PRINCIPLES WHICH HE RECOGNIZED IN HIS APPLICATION, THE PURPOSE WHICH HE SOUGHT TO ACHIEVE, AND CERTAIN PASSAGES IN HIS SPECIFICATIONS, SHOW THAT HE RECOGNIZED, AS THEY PLAINLY SUGGEST TO THOSE SKILLED IN THE ART, THE DESIRABILITY OF TUNING THE ANTENNA CIRCUITS AS WELL. THE DISCLOSURES OF HIS APPLICATION WERE THUS AN ADEQUATE BASIS FOR THE SPECIFIC RECOMMENDATION, LATER ADDED BY AMENDMENT, AS TO THE DESIRABILITY OF CONSTRUCTING THE ANTENNA CIRCUITS SO AS TO BE RESONANT TO THE FREQUENCY PRODUCED IN THE CHARGING CIRCUIT OF THE TRANSMITTER.

THE MAJOR PURPOSE OF STONE'S SYSTEM WAS THE ACHIEVEMENT OF GREATER SELECTIVITY OF TUNING. HIS OBJECTIVE WAS TO TRANSMIT WAVES "OF BUT A SINGLE FREQUENCY" AND TO RECEIVE THEM AT A STATION WHICH "SHALL BE OPERATED ONLY BY

ELECTRIC WAVES OF A SINGLE FREQUENCY AND NO OTHERS." HE STATES: "BY MY INVENTION THE VERTICAL CONDUCTOR OF THE TRANSMITTING STATION IS MADE THE SOURCE OF ELECTRO-MAGNETIC WAVES OF BUT A SINGLE PERIODICITY, AND THE TRANSLATING APPARATUS AT THE RECEIVING STATION IS CAUSED TO BE SELECTIVELY RESPONSIVE TO WAVES OF BUT A SINGLE PERIODICITY SO THAT THE TRANSMITTING APPARATUS CORRESPONDS TO A TUNING FORK SENDING BUT A SINGLE SIMPLE MUSICAL TONE, AND THE RECEIVING APPARATUS CORRESPONDS TO AN ACOUSTIC RESONATOR CAPABLE OF ABSORBING THE ENERGY OF THAT SINGLE, SIMPLE MUSICAL TONE ONLY." HE SAYS THAT "WHEN THE APPARATUS AT A PARTICULAR (RECEIVING) STATION" IS PROPERLY TUNED TO A PARTICULAR TRANSMITTING STATION THE RECEIVER WILL SELECTIVELY RECEIVE MESSAGES FROM IT. HE ADDS: "MOREOVER, BY MY INVENTION THE OPERATOR AT THE TRANSMITTING OR RECEIVING STATION MAY AT WILL ADJUST THE APPARATUS AT HIS COMMAND IN SUCH A WAY AS TO PLACE HIMSELF IN COMMUNICATION WITH ANY ONE OF A NUMBER OF STATIONS ... BY BRINGING HIS APPARATUS INTO RESONANCE WITH THE PERIODICITY EMPLOYED." AND WITH RESPECT TO THE TRANSMITTER HE SAYS, "IT IS TO BE UNDERSTOOD THAT ANY SUITABLE DEVICE MAY BE EMPLOYED TO DEVELOP THE SIMPLE HARMONIC FORCE IMPRESSED UPON THE VERTICAL WIRE (ANTENNA). IT IS SUFFICIENT TO DEVELOP IN THE VERTICAL WIRE PRACTICALLY SIMPLE HARMONIC VIBRATIONS OF A FIXED AND HIGH FREQUENCY." THESE STATEMENTS SUFFICIENTLY INDICATE STONE'S BROAD PURPOSE OF PROVIDING A HIGH DEGREE OF TUNING AT SENDING AND RECEIVING STATIONS. IN SEEKING TO ACHIEVE THAT END HE NOT UNNATURALLY PLACED EMPHASIS ON THE TUNING OF THE CLOSED CIRCUITS, THE ASSOCIATION OF WHICH WITH THE ANTENNA CIRCUITS WAS AN IMPORTANT IMPROVEMENT WHICH HE WAS THE FIRST TO MAKE. BUT HE ALSO MADE IT PLAIN THAT IT WAS THE SENDING AND RECEIVING "APPARATUS" WHICH HE WISHED TO TUNE, SO THAT THE SENDING "APPARATUS" "WOULD CORRESPOND TO A TUNING FORK" AND THE RECEIVING "APPARATUS" TO "AN ACOUSTIC RESONATOR" CAPABLE OF ABSORBING THE ENERGY OF THE "SINGLE, SIMPLE MUSICAL TONE" TRANSMITTED. AND THIS HE SOUGHT TO ACHIEVE BY "ANY SUITABLE DEVICE." STONE THUS EMPHASIZED THE DESIRABILITY OF MAKING THE ENTIRE TRANSMITTING AND RECEIVING "APPARATUS" RESONANT TO A PARTICULAR FREQUENCY. AS NONE OF THE CIRCUITS ARE RESONANT TO A DESIRED FREQUENCY UNLESS THEY ARE TUNED TO THAT FREQUENCY, THIS REFERENCE TO THE TRANSMITTING AND RECEIVING APPARATUS AS BEING BROUGHT INTO RESONANCE WITH EACH OTHER CANNOT FAIRLY BE SAID TO MEAN THAT ONLY SOME OF THE CIRCUITS AT THE TRANSMITTER AND RECEIVER WERE TO BE TUNED. TO SAY THAT BY THIS REFERENCE TO THE TUNING OF SENDING AND RECEIVING APPARATUS HE MEANT TO CONFINE HIS INVENTION TO THE TUNING OF SOME ONLY OF THE CIRCUITS IN THAT APPARATUS IS TO READ INTO HIS SPECIFICATIONS A RESTRICTION WHICH IS PLAINLY NOT THERE AND WHICH CONTRADICTS EVERYTHING THEY SAY ABOUT THE DESIRABILITY OF RESONANCE OF THE APPARATUS. IT IS TO READ THE SPECIFICATIONS, WHICH TAKEN IN THEIR ENTIRETY ARE MERELY DESCRIPTIVE OR ILLUSTRATIVE OF HIS INVENTION, COMPARE CONTINENTAL PAPER BAG CO. V. EASTERN PAPER BAG CO., 210 U.S. 405, 418, 419-20, AS THOUGH THEY WERE CLAIMS WHOSE FUNCTION IS TO EXCLUDE FROM THE PATENT ALL THAT IS NOT SPECIFICALLY CLAIMED. MAHN V. HARWOOD, 112 U.S. 354, 361; MCCLAIN V. ORTMAYER, 141 U.S. 419, 423-5; MILCOR STEEL CO. V. FULLER CO., 316 U.S. 143, 146.

STONE HAD POINTED OUT THAT THE TUNING OF THE ANTENNA CIRCUITS SHOWN IN THE PRIOR ART DID NOT OF ITSELF AFFORD SUFFICIENT SELECTIVITY. IT WAS FOR THAT REASON THAT HE USED THE TUNED CLOSED CIRCUIT IN ASSOCIATION WITH THE ANTENNA CIRCUIT. BUT IN THE FACE OF HIS EMPHASIS ON THE DESIRABILITY OF TUNING THE TRANSMITTING AND RECEIVING APPARATUS, WE CANNOT IMPUTE TO HIM AN INTENTION TO EXCLUDE FROM HIS APPARATUS THE WELL KNOWN USE OF TUNING IN THE ANTENNA CIRCUITS AS AN AID TO THE SELECTIVITY WHICH IT WAS HIS PURPOSE TO ACHIEVE. THE INFERENCE TO BE DRAWN IS RATHER THAT HE INTENDED THE TUNED CLOSED CIRCUITS WHICH HE PROPOSED TO ADD TO

THE THEN KNOWN SYSTEMS OF RADIO COMMUNICATION, TO BE USED IN ASSOCIATION WITH ANY EXISTING TYPE OF VERTICAL WIRE ANTENNA CIRCUIT, INCLUDING ONE SO CONSTRUCTED AS TO BE EITHER RESONANT TO A PARTICULAR FREQUENCY, OR ADJUSTABLY RESONANT TO ANY DESIRED FREQUENCY, BOTH OF WHICH INVOLVED TUNING.

STONE'S FULL APPRECIATION OF THE VALUE OF MAKING ALL OF HIS CIRCUITS RESONANT TO THE SAME FREQUENCY IS SHOWN BY HIS SUGGESTION TO INSERT, BETWEEN THE CLOSED AND ANTENNA CIRCUITS AT THE TRANSMITTER AND RECEIVER, ONE OR MORE ADDITIONAL CLOSED CIRCUITS, SO CONSTRUCTED AS TO BE HIGHLY RESONANT TO THE PARTICULAR FREQUENCY EMPLOYED. HE SAYS THAT THE PURPOSE OF SUCH AN INTERMEDIATE CIRCUIT IS "TO WEED OUT AND THEREBY SCREEN" THE ANTENNA CIRCUIT AT THE TRANSMITTER AND THE DETECTING DEVICE AT THE RECEIVER FROM ANY HARMONICS OR OTHER IMPURITIES IN THE WAVE STRUCTURE.

HE STATES: "THIS SCREENING ACTION OF AN INTERPOSED RESONANT CIRCUIT IS DUE TO THE WELL KNOWN PROPERTY OF SUCH CIRCUITS BY WHICH A RESONANT CIRCUIT FAVORS THE DEVELOPMENT IN IT OF SIMPLE HARMONIC CURRENTS OF THE PERIOD TO WHICH IT IS ATTUNED AND STRONGLY OPPOSES THE DEVELOPMENT IN IT OF SIMPLE HARMONIC CURRENTS OF OTHER PERIODICITIES." HIS ORIGINAL APPLICATION THUS DISCLOSED THE ADVANTAGE, WHERE VIBRATIONS CREATED IN ONE CIRCUIT ARE TO BE IMPRESSED ON ANOTHER, OF MAKING THE LATTER CIRCUIT RESONANT TO THE SAME FREQUENCY AS THE FORMER, IN VIEW OF THE "WELL KNOWN PROPERTY" OF A RESONANT CIRCUIT TO FAVOR THE "DEVELOPMENT" IN IT OF FORCED VIBRATIONS OF THE SAME FREQUENCY AS ITS NATURAL PERIODICITY.

STONE'S APPLICATION SHOWS THAT THESE PRINCIPLES OF RESONANT CIRCUITS WERE NO LESS APPLICABLE TO THE ANTENNA CIRCUIT, AND SUGGESTS THE USE OF "ANY SUITABLE DEVICE" TO "DEVELOP" IN THE ANTENNA CIRCUIT THE "SIMPLE HARMONIC FORCE IMPRESSED" UPON IT. IT WAS THEN WELL KNOWN IN THE ART THAT EVERY ELECTRICAL CIRCUIT IS TO SOME DEGREE RESONANT TO A PARTICULAR FREQUENCY TO WHICH IT RESPONDS MORE READILY AND POWERFULLY THAN TO OTHERS. ALTHOUGH THE DEGREE OF RESONANCE ATTAINED BY A VERTICAL WIRE IS SMALL, ITS NATURAL RESONANCE IS NO DIFFERENT IN KIND FROM THAT OF A CLOSED CIRCUIT SUCH AS STONE'S SCREENING CIRCUIT. STONE RECOGNIZED THIS IN HIS APPLICATION. IN DESCRIBING THE COMPLEX NATURAL VIBRATIONS SET UP BY A SUDDEN DISCHARGE IN AN ANTENNA CIRCUIT, SUCH AS THAT COMMONLY USED AT THE TIME OF HIS APPLICATION, STONE SAID THAT "THE VIBRATIONS CONSIST OF A SIMPLE HARMONIC VIBRATION OF LOWER PERIOD THAN ALL THE OTHERS, KNOWN AS THE FUNDAMENTAL WITH A GREAT VARIETY OF SUPERIMPOSED SIMPLE HARMONICS OF HIGHER PERIODICITY SUPERIMPOSED THEREON." AND HE SAYS THAT THE OSCILLATIONS DEVELOPED IN THE CHARGING CIRCUIT OF HIS SYSTEM "INDUCE CORRESPONDING OSCILLATIONS IN THE VERTICAL WIRE," WHICH ARE "VIRTUALLY" FORCED VIBRATIONS, AND "PRACTICALLY INDEPENDENT, AS REGARDS THEIR FREQUENCY, OF THE CONSTANTS OF THE SECOND CIRCUIT IN WHICH THEY ARE INDUCED" - A PLAIN RECOGNITION THAT THE ANTENNA CIRCUIT HAS ELECTRO-MAGNETIC CONSTANTS WHICH AFFECT ITS NATURAL PERIODICITY, AND THAT THAT NATURAL PERIODICITY DOES HAVE SOME EFFECT ON THE FREQUENCY OF THE VIBRATIONS IMPRESSED UPON THE ANTENNA CIRCUIT. FN14 THUS STONE DID NOT, AS THE MARCONI COMPANY SUGGESTS, SAY THAT THE ANTENNA CIRCUIT HAD NO NATURAL PERIODICITY. HE RECOGNIZED THAT ITS NATURAL PERIODICITY WAS LESS STRONGLY MARKED THAN THAT OF HIS CLOSED CIRCUIT, AND HENCE THAT THE WAVE STRUCTURE COULD BE GREATLY IMPROVED BY CREATING THE OSCILLATIONS IN A CLOSED CIRCUIT SUCH AS HE DESCRIBED. BUT HE ALSO PLAINLY RECOGNIZED THAT THE ANTENNA CIRCUIT, LIKE HIS SCREENING CIRCUIT, WAS A CIRCUIT HAVING A NATURAL PERIOD OF VIBRATION WHICH WOULD THEREFORE BE MORE RESPONSIVE TO IMPRESSED OSCILLATIONS OF THAT SAME PERIODICITY. SINCE HE HAD PREVIOUSLY SAID THAT "ANY SUITABLE DEVICE MAY BE EMPLOYED TO DEVELOP THE SIMPLE HARMONIC FORCE IMPRESSED UPON THE VERTICAL WIRE," WE THINK THAT STONE'S

SPECIFICATIONS PLAINLY SUGGESTED TO THOSE SKILLED IN THE ART THAT THEY AVAIL THEMSELVES OF THIS MEANS OF DEVELOPING IN THE ANTENNA THIS SIMPLE HARMONIC FORCE, AND THAT THEY TUNE THE ANTENNA CIRCUIT IN ORDER TO IMPROVE THE STRENGTH AND QUALITY OF THE "FORCED" VIBRATIONS IMPRESSED UPON IT.

THE MARCONI COMPANY ARGUES THAT STONE'S THEORY OF "FORCED" OSCILLATIONS PRESUPPOSES THAT THE OPEN TRANSMITTER CIRCUIT BE UNTUNED. IT IS TRUE THAT STONE SAID THAT SUCH "FORCED" OSCILLATIONS HAVE A PERIOD OF VIBRATION WHICH IS "INDEPENDENT OF THE ELECTRICAL CONSTANTS OF THE CIRCUIT" ON WHICH THEY ARE IMPRESSED. BUT THE FACT THAT THE "FORCED" VIBRATION WILL RETAIN ITS NATURAL PERIOD WHATEVER THE FREQUENCY OF THE ANTENNA CIRCUIT MAY BE, DOES NOT PRECLUDE, AS STONE SHOWED, THE TUNING OF THAT CIRCUIT SO AS TO ACHIEVE MAXIMUM RESPONSIVENESS TO THE VIBRATIONS IMPRESSED UPON IT. STONE'S SPECIFICATIONS INDICATE THAT HE USED THE TERM "FORCED" MERELY AS MEANING THAT THE VIBRATIONS ARE DEVELOPED IN ANOTHER CIRCUIT AND THEN TRANSFERRED TO THE ANTENNA CIRCUIT BY INDUCTIVE COUPLING, AS DISTINGUISHED FROM "NATURAL" VIBRATIONS WHICH ORIGINATE IN THE ANTENNA OR RADIATING CIRCUIT - IN SHORT THAT "FORCED" IS MERELY USED AS A SYNONYM FOR "INDUCED." THUS HE STATES IN DESCRIBING THE OPERATION OF HIS TRANSMITTER, "THE HIGH FREQUENCY CURRENT ... PASSING THROUGH THE PRIMARY I1(OF THE ANTENNA TRANSFORMER) INDUCES A CORRESPONDING HIGH FREQUENCY ELECTROMOTIVE FORCE AND CURRENT IN THE SECONDARY I2 AND FORCED ELECTRIC VIBRATIONS RESULT IN THE VERTICAL CONDUCTOR V ... " FN15 HENCE THERE IS AMPLE SUPPORT FOR THE FINDING OF THE COURT BELOW THAT "BY FREE OSCILLATIONS IS MEANT THAT THEIR FREQUENCY WAS DETERMINED BY THE CONSTANTS OF THE CIRCUIT IN WHICH THEY WERE GENERATED. THE STONE APPLICATION AS FILED IMPRESSED THESE OSCILLATIONS UPON THE OPEN CIRCUIT, AND THEREFORE USED 'FORCED' OSCILLATIONS IN THE OPEN CIRCUIT OF THE TRANSMITTER, THAT IS, THE FREQUENCY OF THE OSCILLATIONS IN THE OPEN CIRCUIT WAS DETERMINED BY THE FREQUENCY OF THE OSCILLATIONS IN THE CLOSED CIRCUIT.

"THE EFFECT OF FORCING VIBRATIONS UPON A TUNED AND UNTUNED CIRCUIT MAY BE LIKENED UNTO THE EFFECT OF A TUNING FORK UPON A STRETCHED CORD IN A VISCOUS MEDIUM. WHEN THE CORD IS VIBRATED BY THE TUNING FORK IT HAS THE SAME PERIOD AS DOES THE FORK REGARDLESS OF WHETHER SUCH PERIOD BE THAT OF THE NATURAL PERIOD OF THE CORD, BUT WHEN THE FORK VIBRATIONS ARE IN TUNE WITH THE NATURAL PERIOD OR FUNDAMENTAL OF THE CORD, THEN THE AMPLITUDE OF VIBRATIONS IN THE CORD IS A MAXIMUM." THUS STONE'S APPLICATION, PRIOR TO MARCONI, SHOWED A FOUR-CIRCUIT SYSTEM, IN WHICH THE OSCILLATIONS WERE PRODUCED IN A CLOSED CHARGING CIRCUIT AND IMPRESSED ON AN OPEN ANTENNA CIRCUIT IN THE TRANSMITTER, AND WERE SIMILARLY RECEIVED IN AN OPEN ANTENNA CIRCUIT AND BY IT INDUCED IN A CLOSED CIRCUIT CONTAINING A DETECTOR. HE SHOWED THE EFFECT OF RESONANCE ON THE CIRCUITS RESULTING FROM THEIR TUNING TO A DESIRED FREQUENCY, AND EMPHASIZED THE IMPORTANCE OF MAKING THE TRANSMITTING AND RECEIVING APPARATUS RESONANT TO THAT FREQUENCY. STONE'S PATENT, FN16 GRANTED A YEAR AND A HALF BEFORE MARCONI - ALTHOUGH AFTER MARCONI'S APPLICATION WAS FILED - MAKES EXPLICIT, AS THE PATENT LAW PERMITS, WHAT WAS IMPLICIT IN STONE'S APPLICATION. BY AMENDMENTS TO HIS SPECIFICATIONS MADE APRIL 8, 1902, HE RECOMMENDED THAT THE FREQUENCY IMPRESSED UPON THE VERTICAL CONDUCTOR AT THE TRANSMITTER "MAY OR MAY NOT BE THE SAME AS THE NATURAL PERIOD OR FUNDAMENTAL OF SUCH CONDUCTOR" AND THAT THE ANTENNA CIRCUIT AT THE TRANSMITTER "MAY WITH ADVANTAGE BE SO CONSTRUCTED AS TO BE HIGHLY RESONANT TO A PARTICULAR FREQUENCY AND THE HARMONIC VIBRATIONS IMPRESSED THEREON MAY WITH ADVANTAGE BE OF THAT FREQUENCY." SINCE STONE USED A VARIABLE INDUCTANCE TO ALTER AT WILL THE FREQUENCY OF THE CHARGING CIRCUIT, THIS

DIRECTION PLAINLY INDICATED THAT THE FREQUENCY OF THE ANTENNA CIRCUIT MIGHT ALSO BE VARIABLE, AND SUGGESTED THE INCLUSION OF THE WELL-KNOWN LODGE VARIABLE INDUCTANCE IN THE CONSTRUCTION OF THE ANTENNA CIRCUIT TO ACHIEVE THAT RESULT. AND SINCE STONE HAD SPECIFIED THAT "BY MY INVENTION" THE OPERATOR AT THE RECEIVING STATION IS ABLE TO "ADJUST" THE RECEIVING APPARATUS SO AS TO PLACE IT IN RESONANCE WITH ANY PARTICULAR TRANSMITTING STATION, HIS PATENT EQUALLY PLAINLY SUGGESTED THE USE OF THE LODGE VARIABLE INDUCTANCE AS A MEANS OF ADJUSTING THE TUNING OF THE RECEIVING ANTENNA.

STONE'S 1902 AMENDMENTS ALSO SUGGESTED THAT AN "ELEVATED CONDUCTOR THAT IS APERIODIC MAY BE EMPLOYED" - I.E., ONE HAVING VERY WEAK NATURAL PERIODICITY AND CONSEQUENTLY "ADAPTED TO RECEIVE OR TRANSMIT ALL FREQUENCIES." BUT THIS SUGGESTION WAS ACCOMPANIED BY THE ALTERNATIVE RECOMMENDATION IN THE 1902 AMENDMENTS THAT THE ANTENNA CIRCUITS AT TRANSMITTER AND RECEIVER "MAY WITH ADVANTAGE BE MADE RESONANT TO A PARTICULAR FREQUENCY," I.E., BE PERIODIC. NO INFERENCE CAN BE DRAWN FROM THIS THAT ONLY AN APERIODIC ANTENNA WAS CONTEMPLATED EITHER BY THE APPLICATION OR THE AMENDMENTS. THE APPLICATION WAS SUFFICIENTLY BROAD TO COVER BOTH TYPES, SINCE BOTH WERE SUITABLE MEANS OF ACHIEVING UNDER DIFFERENT CONDITIONS THE RESULTS WHICH THE APPLICATION DESCRIBED AND SOUGHT TO ATTAIN. THE AMENDMENTS THUS MERELY CLARIFIED AND EXPLAINED IN FULLER DETAIL TWO ALTERNATIVE MEANS WHICH COULD BE EMPLOYED IN THE INVENTION DESCRIBED IN THE ORIGINAL APPLICATION, ONE OF THOSE MEANS BEING THE CONSTRUCTION OF THE ANTENNA SO AS TO BE HIGHLY RESONANT, I.E., TUNED, TO A PARTICULAR FREQUENCY. FN17 THE ONLY RESPECTS IN WHICH IT IS SERIOUSLY CONTENDED THAT MARCONI DISCLOSED INVENTION OVER STONE ARE THAT MARCONI EXPLICITLY CLAIMED FOUR CIRCUIT TUNING BEFORE STONE HAD MADE IT EXPLICIT BY HIS 1902 AMENDMENT, AND THAT MARCONI DISCLOSED MEANS OF ADJUSTING THE TUNING OF EACH OF HIS FOUR CIRCUITS WHEREAS STONE HAD EXPLICITLY SHOWN ADJUSTABLE TUNING ONLY IN THE TWO CLOSED CIRCUITS. BUT WE THINK THAT NEITHER MARCONI'S TUNING OF THE TWO ANTENNA CIRCUITS NOR HIS USE OF THE LODGE VARIABLE INDUCTANCE TO THAT END INVOLVED ANY INVENTION OVER STONE. TWO QUESTIONS ARE INVOLVED, FIRST, WHETHER THERE WAS ANY INVENTION OVER STONE IN TUNING THE ANTENNA CIRCUITS, AND, SECOND, WHETHER THERE WAS ANY INVENTION IN THE USE OF THE LODGE VARIABLE INDUCTANCE OR ANY OTHER KNOWN MEANS OF ADJUSTMENT IN ORDER TO MAKE THE TUNING OF THE ANTENNA CIRCUITS ADJUSTABLE.

FOR REASONS ALREADY INDICATED WE THINK IT CLEAR THAT STONE SHOWED TUNING OF THE ANTENNA CIRCUITS BEFORE MARCONI, AND IF THIS INVOLVED INVENTION STONE WAS THE FIRST INVENTOR. STONE'S APPLICATION EMPHASIZED THE DESIRABILITY OF TUNING, AND DISCLOSED MEANS OF ADJUSTING THE TUNING OF THE CLOSED CIRCUITS. HIS VERY EXPLICIT RECOGNITION OF THE INCREASED SELECTIVITY ATTAINED BY INDUCTIVE COUPLING OF SEVERAL RESONANT CIRCUITS PLAINLY SUGGESTED TO THOSE SKILLED IN THE ART THAT THE ANTENNA CIRCUIT COULD WITH ADVANTAGE BE A RESONANT CIRCUIT, THAT IS TO SAY A TUNED CIRCUIT, AND HENCE THAT IT WAS ONE OF THE CIRCUITS TO BE TUNED. HE STRESSED THE IMPORTANCE OF TUNING "BY ANY SUITABLE DEVICE" THE "APPARATUS" AT TRANSMITTER AND RECEIVER, WHICH INCLUDED AT BOTH AN ANTENNA CIRCUIT.

TUNING OF THE ANTENNA CIRCUIT WAS NOTHING NEW; LODGE HAD NOT ONLY TAUGHT THAT THE ANTENNA CIRCUITS AT TRANSMITTER AND RECEIVER SHOULD BE TUNED TO EACH OTHER BUT HAD SHOWN A MEANS OF ADJUSTING THE TUNING WHICH WAS THE PRECISE MEANS ADOPTED BY MARCONI, AND WHICH STONE HAD, PRIOR TO MARCONI, USED TO TUNE HIS CLOSED CIRCUIT - THE VARIABLE INDUCTANCE. TESLA, TOO, HAD SHOWN THE TUNING OF THE ANTENNA CIRCUIT AT THE TRANSMITTER TO THE FREQUENCY DEVELOPED BY THE CHARGING CIRCUIT, AND THE TUNING OF BOTH CIRCUITS AT THE RECEIVER TO THE FREQUENCY THUS TRANSMITTED. THUS MARCONI'S IMPROVEMENT IN TUNING THE ANTENNA CIRCUITS IS ONE

THE PRINCIPLES OF WHICH WERE WELL UNDERSTOOD AND STATED BY STONE HIMSELF BEFORE MARCONI, AND THE MECHANISM FOR ACHIEVING WHICH HAD PREVIOUSLY BEEN DISCLOSED BY LODGE AND STONE. FN18 SINCE NO INVENTION OVER STONE WAS INVOLVED IN TUNING THE ANTENNA CIRCUITS, NEITHER MARCONI NOR STONE MADE AN INVENTION BY PROVIDING ADJUSTABLE TUNING OF ANY OF THE CIRCUITS OR BY EMPLOYING LODGE'S VARIABLE INDUCTANCE AS A MEANS OF ADJUSTING THE TUNING OF THE RESONANT FOUR-CIRCUIT ARRANGEMENT EARLIER DISCLOSED BY STONE'S APPLICATION AND PATENTED BY HIM. NO INVENTION WAS INVOLVED IN EMPLOYING THE LODGE VARIABLE INDUCTANCE FOR TUNING EITHER THE CLOSED OR THE OPEN CIRCUITS IN LIEU OF OTHER STRUCTURAL MODES OF ADJUSTMENT FOR THAT PURPOSE. THE VARIABLE INDUCTANCE IMPARTED NO NEW FUNCTION TO THE CIRCUIT; AND MERELY MAKING A KNOWN ELEMENT OF A KNOWN COMBINATION ADJUSTABLE BY A MEANS OF ADJUSTMENT KNOWN TO THE ART, WHEN NO NEW OR UNEXPECTED RESULT IS OBTAINED, IS NOT INVENTION. PETERS V. HANSON, 129 U.S. 541, 550-51, 553; ELECTRIC CABLE CO. V. EDISON CO., 292 U.S. 69, 79, 80, AND CASES CITED; SMYTH MFG. CO. V. SHERIDAN, 149 F. 208, 211; CF. BASSICK MFG. CO. V. HOLLINGSHEAD CO., 298 U.S. 415, 424-5 AND CASES CITED.

STONE'S CONCEPTION OF HIS INVENTION AS DISCLOSED BY HIS PATENT ANTEDATED HIS APPLICATION. IT IS CARRIED BACK TO JUNE 30, 1899, SEVEN MONTHS BEFORE HIS APPLICATION, WHEN, IN A LETTER TO BAKER, HE DESCRIBED IN TEXT AND DRAWINGS HIS FOUR-CIRCUIT SYSTEM FOR WIRELESS TELEGRAPHY IN SUBSTANTIALLY THE SAME FORM AS THAT DISCLOSED BY THE APPLICATION. HIS LETTER IS EXPLICIT IN RECOMMENDING THE TUNING OF THE ANTENNA CIRCUITS. IN PART HE WROTE AS FOLLOWS: "INSTEAD OF UTILIZING THE VERTICAL WIRE (ANTENNA) ITSELF AT THE TRANSMITTING STATION AS THE OSCILLATOR, I PROPOSE TO IMPRESS UPON THIS VERTICAL WIRE, OSCILLATIONS FROM AN OSCILLATOR, WHICH OSCILLATIONS SHALL BE OF A FREQUENCY CORRESPONDING TO THE FUNDAMENTAL OF THE WIRE. SIMILARLY AT THE RECEIVING STATION, I SHALL DRAW FROM THE VERTICAL WIRE, ONLY THAT COMPONENT OF THE COMPLEX WAVE WHICH IS OF LOWEST FREQUENCY.

"IF NOW THE FUNDAMENTAL OF THE WIRE AT THE RECEIVING STATION BE THE SAME AS THAT OF THE WIRE AT THE TRANSMITTING STATION, THEN THE RECEIVING STATION MAY RECEIVE SIGNALS FROM THE TRANSMITTING STATION, BUT IF IT BE DIFFERENT FROM THAT OF THE TRANSMITTING STATION, IT MAY NOT RECEIVE THOSE SIGNALS.

. . . . .

"THE TUNING OF THESE CIRCUITS ONE TO ANOTHER AND ALL TO THE SAME FREQUENCY WILL PROBABLY BE BEST ACCOMPLISHED EMPIRICALLY, THOUGH THE BEST GENERAL PROPORTIONS MAY BE DETERMINED MATHEMATICALLY." ON JULY 18, 1899, STONE AGAIN WROTE TO BAKER, MATHEMATICALLY DEMONSTRATING HOW TO ACHIEVE THE SINGLE FREQUENCY BY MEANS OF FORCED VIBRATIONS. HE EXPRESSED AS A TRIGONOMETRIC FUNCTION THE FORM TAKEN BY THE FORCED WAVE "IF THE PERIOD OF THE IMPRESSED FORCE BE THE SAME AS THAT OF THE FUNDAMENTAL OF THE VERTICAL WIRE." HE ALSO POINTED OUT THAT THE TRANSMITTING CIRCUIT WHICH HE HAD DISCLOSED IN HIS EARLIER LETTER TO BAKER, "IS PRACTICALLY THE SAME AS THAT EMPLOYED BY TESLA," EXCEPT THAT STONE ADDED AN INDUCTANCE COIL IN THE CLOSED CIRCUIT "TO GIVE ADDITIONAL MEANS OF TUNING" AND TO "SWAMP" THE REACTIONS FROM THE COIL OF THE OSCILLATION TRANSFORMER AND THUS LOOSEN THE COUPLING BETWEEN THE OPEN AND CLOSED CIRCUIT OF THE TRANSMITTER N19. HIS RECOGNITION OF THE EFFECT UPON THE CURRENT IN THE ANTENNA IF IT IS OF THE SAME PERIOD AS THE CHARGING CIRCUIT; HIS STATEMENT THAT HIS TRANSMITTING SYSTEM WAS THE SAME AS THAT EMPLOYED BY TESLA; HIS RECOGNITION THAT THE FUNDAMENTAL OF THE RECEIVER SHOULD BE THE SAME AS THAT OF THE TRANSMITTER ANTENNA WHEN USED FOR THE TRANSMISSION OF A SINGLE FREQUENCY, AND FINALLY HIS STATEMENT THAT ALL FOUR CIRCUITS ARE TO BE TUNED, "ONE TO ANOTHER AND ALL TO THE SAME FREQUENCY," ALL INDICATE HIS UNDERSTANDING OF THE PRINCIPLES OF

RESONANCE AND OF THE SIGNIFICANCE OF TUNING THE ANTENNA CIRCUITS.

STONE DISCLOSED HIS INVENTION TO OTHERS, AND IN JANUARY, 1900, DESCRIBED IT TO HIS CLASS AT THE MASSACHUSETTS INSTITUTE OF TECHNOLOGY. BEFORE 1900 HE WAS DILIGENT IN OBTAINING CAPITAL TO PROMOTE HIS INVENTION. EARLY IN 1901 A SYNDICATE WAS ORGANIZED TO FINANCE LABORATORY EXPERIMENTS. THE STONE TELEGRAPH & TELEPHONE CO. WAS ORGANIZED IN DECEMBER, 1901. IT CONSTRUCTED SEVERAL EXPERIMENTAL STATIONS IN 1902 AND 1903; BEGINNING IN 1904 OR 1905 IT BUILT WIRELESS STATIONS AND SOLD APPARATUS, EQUIPPED A NAVY COLLIER AND SOME BATTLESHIPS, AND IT APPLIED FOR A LARGE NUMBER OF PATENTS. THE APPARATUS USED IN THE STATIONS IS DESCRIBED BY STONE'S TESTIMONY IN THIS SUIT AS HAVING RESONANT OPEN AND CLOSED CIRCUITS LOOSELY COUPLED INDUCTIVELY TO EACH OTHER, AT BOTH THE TRANSMITTER AND RECEIVER, AND ALL TUNED TO THE SAME WAVE LENGTH, AS DESCRIBED IN HIS LETTERS TO BAKER AND HIS PATENT.

WE THINK THAT STONE'S ORIGINAL APPLICATION SUFFICIENTLY DISCLOSED THE DESIRABILITY THAT THE ANTENNA CIRCUITS IN TRANSMITTER AND RECEIVER BE RESONANT TO THE SAME FREQUENCY AS THE CLOSED CIRCUITS, AS HE EXPRESSLY RECOMMENDED IN HIS PATENT. BUT IN ANY EVENT IT IS PLAIN THAT NO DEPARTURE FROM OR IMPROPER ADDITION TO THE SPECIFICATIONS WAS INVOLVED IN THE 1902 AMENDMENTS, WHICH MERELY MADE EXPLICIT WHAT WAS ALREADY IMPLICIT. HOBBS V. BEACH, 180 U.S. 383, 395-7. WE WOULD ORDINARILY BE SLOW TO RECOGNIZE AMENDMENTS MADE AFTER THE FILING OF MARCONI'S APPLICATION AND DISCLOSING FEATURES SHOWN IN THAT APPLICATION. CF. SCHRIBER-SCHROTH CO. V. CLEVELAND TRUST CO., 305 U.S. 47, 57; POWERS KENNEDY CORPORATION V. CONCRETE CO., 282 U.S. 175, 185-6; MACKAY RADIO CO. V. RADIO CORPORATION, 306 U.S. 86. BUT HERE STONE'S LETTERS TO BAKER, WHOSE AUTHENTICITY HAS NOT BEEN QUESTIONED IN THIS CASE, AFFORD CONVINCING PROOF THAT STONE HAD CONCEIVED OF THE IDEA OF TUNING ALL FOUR CIRCUITS PRIOR TO THE DATE OF MARCONI'S INVENTION. CF. BICKELL V. SMITH-HAMBURY-SCOTT WELDING CO., 53 F.2D 356, 358.

IT IS WELL ESTABLISHED THAT AS BETWEEN TWO INVENTORS PRIORITY OF INVENTION WILL BE AWARDED TO THE ONE WHO BY SATISFYING PROOF CAN SHOW THAT HE FIRST CONCEIVED OF THE INVENTION. PHILADELPHIA & TRENTON R. CO. V. STIMPSON, 14 PET. 448, 462; LOOM CO. V. HIGGINS, 105 U.S. 580, 593; RADIO CORPORATION V. RADIO LABORATORIES, 293 U.S. 1, 11-13; CHRISTIE V. SEYBOLD, 55 F. 69, 76; AUTOMATIC WEIGHING MACH. CO. V. PNEUMATIC SCALE CORP., 158 F. 415, 417-22; HARPER V. ZIMMERMANN, 41 F.2D 261, 265; SACHS V. HARTFORD ELECTRIC SUPPLY CO., 47 F.2D 743, 748.

COMMERCIAL SUCCESS ACHIEVED BY THE LATER INVENTOR AND PATENTEE CANNOT SAVE HIS PATENT FROM THE DEFENSE OF ANTICIPATION BY A PRIOR INVENTOR. FN20 COMPARE SMITH V. HALL, 301 U.S. 216 WITH SMITH V. SNOW, 294 U.S. 1. TO OBTAIN THE BENEFIT OF HIS PRIOR CONCEPTION, THE INVENTOR MUST NOT ABANDON HIS INVENTION, GAYLER V. WILDER, 10 HOW. 477, 481, BUT MUST PROCEED WITH DILIGENCE TO REDUCE IT TO PRACTICE. WE THINK STONE HAS SHOWN THE NECESSARY DILIGENCE. COMPARE RADIO CORPORATION V. RADIO LABORATORIES, SUPRA, 13, 14. THE DELAY UNTIL 1902 IN INCLUDING IN HIS PATENT SPECIFICATIONS THE SENTENCES ALREADY REFERRED TO, WHICH EXPLICITLY PROVIDE FOR TUNING OF THE ANTENNA CIRCUITS, DOES NOT IN THE CIRCUMSTANCES OF THIS CASE SHOW ANY ABANDONMENT OF THAT FEATURE OF STONE'S INVENTION SINCE, AS WE HAVE SEEN, THE IDEA OF SUCH TUNING WAS AT LEAST IMPLICIT IN HIS ORIGINAL APPLICATION, AND THE 1902 AMENDMENTS MERELY CLARIFIED THAT APPLICATION'S EFFECT AND PURPORT.

MARCONI'S PATENT NO. 763,772 WAS SUSTAINED BY A UNITED STATES DISTRICT COURT IN MARCONI WIRELESS TELEGRAPH CO. V. NATIONAL SIGNALLING CO., 213 F. 815, AND HIS INVENTION AS SPECIFIED IN HIS CORRESPONDING BRITISH PATENT NO. 7777 OF 1900, WAS UPHELD IN MARCONI V. BRITISH RADIO & TELEGRAPH CO., 27 T.L.R. 274, 28 R.P.C. 18. THE FRENCH COURT LIKEWISE SUSTAINED HIS FRENCH PATENT, CIVIL TRIBUNAL OF THE SEINE,

DEC. 24, 1912. NONE OF THESE COURTS CONSIDERED THE STONE PATENT OR HIS LETTERS. ALL REST THEIR FINDINGS OF INVENTION ON MARCONI'S DISCLOSURE OF A FOUR-CIRCUIT SYSTEM AND ON HIS TUNING OF THE FOUR CIRCUITS, IN THE SENSE OF RENDERING THEM RESONANT TO THE SAME FREQUENCY, IN BOTH OF WHICH RESPECTS STONE ANTICIPATED MARCONI, AS WE HAVE SEEN. NONE OF THESE OPINIONS SUGGESTS THAT IF THE COURTS HAD KNOWN OF STONE'S ANTICIPATION, THEY WOULD HAVE HELD THAT MARCONI SHOWED INVENTION OVER STONE BY MAKING THE TUNING OF HIS ANTENNA CIRCUIT ADJUSTABLE, OR BY USING LODGE'S VARIABLE INDUCTANCE FOR THAT PURPOSE. IN MARCONI WIRELESS TELEGRAPH CO. V. KILBOURNE & CLARK MFG. CO., 239 F. 328, AFFIRMED 265 F. 644, THE DISTRICT COURT HELD THAT THE ACCUSED DEVICE DID NOT INFRINGE. WHILE IT ENTERED FORMAL FINDINGS OF VALIDITY WHICH THE CIRCUIT COURT OF APPEALS APPROVED, NEITHER COURT'S OPINION DISCUSSED THE QUESTION OF VALIDITY AND THAT QUESTION WAS NOT ARGUED IN THE CIRCUIT COURT OF APPEALS. FN21 MARCONI'S REPUTATION AS THE MAN WHO FIRST ACHIEVED SUCCESSFUL RADIO TRANSMISSION RESTS ON HIS ORIGINAL PATENT, WHICH BECAME REISSUE NO. 11,913, AND WHICH IS NOT HERE IN QUESTION. THAT REPUTATION, HOWEVER WELL-DESERVED, DOES NOT ENTITLE HIM TO A PATENT FOR EVERY LATER IMPROVEMENT WHICH HE CLAIMS IN THE RADIO FIELD. PATENT CASES, LIKE OTHERS, MUST BE DECIDED NOT BY WEIGHING THE REPUTATIONS OF THE LITIGANTS, BUT BY CAREFUL STUDY OF THE MERITS OF THEIR RESPECTIVE CONTENTIONS AND PROOFS. AS THE RESULT OF SUCH A STUDY WE ARE FORCED TO CONCLUDE, WITHOUT UNDERTAKING TO DETERMINE WHETHER STONE'S PATENT INVOLVED INVENTION, THAT THE COURT OF CLAIMS WAS RIGHT IN DECIDING THAT STONE ANTICIPATED MARCONI, AND THAT MARCONI'S PATENT DID NOT DISCLOSE INVENTION OVER STONE. HENCE THE JUDGMENT BELOW HOLDING INVALID THE BROAD CLAIMS OF THE MARCONI PATENT MUST BE AFFIRMED. IN VIEW OF OUR INTERPRETATION OF THE STONE APPLICATION AND PATENT WE NEED NOT CONSIDER THE CORRECTNESS OF THE COURT'S CONCLUSION THAT EVEN IF STONE'S DISCLOSURES SHOULD BE READ AS FAILING TO DIRECT THAT THE ANTENNA CIRCUITS BE MADE RESONANT TO A PARTICULAR FREQUENCY, MARCONI'S PATENT INVOLVED NO INVENTION OVER LODGE, TESLA, AND STONE.

CLAIM 16 OF MARCONI PATENT NO. 763,772.

THE GOVERNMENT ASKS US TO REVIEW SO MUCH OF THE DECISION OF THE COURT OF CLAIMS AS HELD VALID AND INFRINGED CLAIM 16 OF MARCONI'S PATENT NO. 763,772. THAT CLAIM IS FOR AN ANTENNA CIRCUIT AT THE RECEIVER CONNECTED AT ONE END TO "AN OSCILLATION-RECEIVING CONDUCTOR" AND AT THE OTHER TO A CAPACITY (WHICH COULD BE THE EARTH), CONTAINING THE PRIMARY WINDING OF A TRANSFORMER, "MEANS FOR ADJUSTING THE TWO TRANSFORMER CIRCUITS IN ELECTRICAL RESONANCE WITH EACH OTHER," AND "AN ADJUSTABLE CONDENSER IN A SHUNT CONNECTED WITH THE OPEN CIRCUIT, AND AROUND SAID TRANSFORMER-COIL." MARCONI THUS DISCLOSES AND CLAIMS THE ADDITION TO THE RECEIVER ANTENNA OF AN ADJUSTABLE CONDENSER CONNECTED IN A SHUNT AROUND THE PRIMARY OF THE TRANSFORMER. THE SPECIFICATIONS DESCRIBE THE CONDENSER AS "PREFERABLY ONE PROVIDED WITH TWO TELESCOPING METALLIC TUBES SEPARATED BY A DIELECTRIC AND ARRANGED TO READILY VARY THE CAPACITY BY BEING SLID UPON EACH OTHER." MARCONI, HOWEVER, MAKES NO CLAIM FOR THE PARTICULAR CONSTRUCTION OF THE CONDENSER.

ALTHOUGH THE CLAIM BROADLY PROVIDES FOR "MEANS OF ADJUSTING THE TWO TRANSFORMER-CIRCUITS IN ELECTRICAL RESONANCE," MARCONI'S DRAWINGS DISCLOSE THE USE OF A VARIABLE INDUCTANCE CONNECTED BETWEEN THE AERIAL CONDUCTOR AND THE TRANSFORMER-COIL IN SUCH A MANNER THAT THE VARIABLE INDUCTANCE IS NOT INCLUDED IN THAT PART OF THE ANTENNA CIRCUIT WHICH IS BRIDGED BY THE CONDENSER. THE CONDENSER IS THUS ARRANGED IN PARALLEL WITH THE TRANSFORMER COIL AND IN SERIES WITH THE VARIABLE INDUCTANCE. IN HIS SPECIFICATIONS MARCONI ENUMERATES A NUMBER OF PREFERRED ADJUSTMENTS FOR TUNING THE TRANSMITTING AND RECEIVING STATIONS,

SHOWING THE PRECISE EQUIPMENT TO BE USED TO ACHIEVE TUNING TO THE DESIRED WAVE LENGTH. THE TWO TUNINGS WHICH SHOW THE USE OF THE ADJUSTABLE CONDENSER IN THE RECEIVER ANTENNA ALSO MAKE USE OF THE VARIABLE INDUCTANCE. AND HIS SPECIFICATIONS STATE: "IN A SHUNT AROUND SAID PRIMARY J1(THE PRIMARY OF THE TRANSFORMER) I USUALLY PLACE A CONDENSER H ... AN INDUCTANCE COIL G1 OF VARIABLE INDUCTANCE IS INTERPOSED IN THE PRIMARY CIRCUIT OF THE TRANSFORMER, BEING PREFERABLY LOCATED BETWEEN THE CYLINDER F1(THE AERIAL CAPACITY) AND THE COIL J1." IN THIS RESPECT THE DEVICES WHICH THE COURT BELOW FOUND TO INFRINGE CLAIM 16 EXHIBIT SOMEWHAT DIFFERENT ARRANGEMENTS. APPARATUS MANUFACTURED BY THE KILBOURNE AND CLARK COMPANY, AND USED BY THE GOVERNMENT, HAD A RECEIVER ANTENNA CIRCUIT CONTAINING A VARIABLE INDUCTANCE IN ADDITION TO THE TRANSFORMER COIL, AND HAVING AN ADJUSTABLE CONDENSER SO CONSTRUCTED THAT IT COULD BE CONNECTED EITHER IN SERIES WITH THE TWO INDUCTANCES, OR IN A SHUNT BRIDGING BOTH OF THEM. APPARATUS MANUFACTURED BY THE TELEFUNKEN COMPANY SHOWED A SIMILAR ANTENNA CIRCUIT HAVING NO VARIABLE INDUCTANCE, BUT HAVING AN ADJUSTABLE CONDENSER SO ARRANGED THAT IT COULD BE CONNECTED EITHER IN SERIES WITH THE TRANSFORMER COIL, OR IN PARALLEL WITH IT BY PLACING THE CONDENSER IN A SHUNT CIRCUIT WHICH WOULD THUS BRIDGE ALL THE INDUCTANCE IN THE ANTENNA CIRCUIT.

THE MARCONI PATENT DOES NOT DISCLOSE THE FUNCTION WHICH IS SERVED BY THE ADJUSTABLE CONDENSER DISCLOSED BY CLAIM 16, EXCEPT IN SO FAR AS MARCONI IN HIS SPECIFICATIONS, IN DESCRIBING THE MEANS OF TUNING THE RECEIVER CIRCUITS TO A PARTICULAR DESIRED FREQUENCY, PRESCRIBES SPECIFIC VALUES FOR BOTH THE VARIABLE INDUCTANCE AND THE ADJUSTABLE CONDENSER IN THE RECEIVER ANTENNA CIRCUIT. THE COURT OF CLAIMS FOUND THAT THIS INDICATED "THAT THE PURPOSE OF THE CONDENSER CONNECTED IN SHUNT WITH THE PRIMARY WINDING OF THE TRANSFORMER OF THE RECEIVER, IS TO ENABLE THE ELECTRICAL PERIODICITY OR TUNING OF THE OPEN CIRCUIT OF THE RECEIVER TO BE ALTERED." THE COURT THUS BASED ITS HOLDING THAT CLAIM 16 DISCLOSED PATENTABLE INVENTION ON ITS FINDING THAT MARCONI, BY THE USE OF AN ADJUSTABLE CONDENSER IN THE ANTENNA CIRCUIT, DISCLOSED A NEW AND USEFUL METHOD OF TUNING THAT CIRCUIT. THE GOVERNMENT CONTENDS THAT THE ARRANGEMENT OF THE ANTENNA CIRCUIT DISCLOSED BY MARCONI'S SPECIFICATIONS - WITH THE CONDENSER SHUNTED AROUND THE TRANSFORMER COIL BUT NOT AROUND THE VARIABLE INDUCTANCE - IS SUCH THAT THE CONDENSER CANNOT INCREASE THE WAVE-LENGTH OVER WHAT IT WOULD BE WITHOUT SUCH A CONDENSER, AND THAT IT CAN DECREASE THAT WAVE-LENGTH ONLY WHEN ADJUSTED TO HAVE A VERY SMALL CAPACITY. THE GOVERNMENT CONTENDS THEREFORE THAT ITS PRINCIPAL FUNCTION IS NOT THAT OF TUNING BUT OF PROVIDING "LOOSE COUPLING." FN22 THE GOVERNMENT DOES NOT DENY THAT THIS PRECISE ARRANGEMENT IS NOVEL AND USEFUL, BUT IT CONTENDS THAT ITS DEVICES DO NOT INFRINGE THAT PRECISE ARRANGEMENT, AND THAT CLAIM 16, IF MORE BROADLY CONSTRUED SO AS TO COVER ITS APPARATUS, IS INVALID BECAUSE ANTICIPATED BY THE PRIOR ART, PARTICULARLY THE PATENTS OF PUPIN AND FESSENDEN.

AS WE HAVE SEEN FROM OUR DISCUSSION OF THE OTHER CLAIMS OF THE MARCONI PATENT, THE IDEA OF TUNING THE ANTENNA CIRCUITS INVOLVED NO PATENTABLE INVENTION. IT WAS WELL KNOWN THAT TUNING WAS ACHIEVED BY THE PROPER ADJUSTMENT OF EITHER THE INDUCTANCE OR THE CAPACITY IN A CIRCUIT, OR BOTH. LODGE AND STONE HAD ACHIEVED TUNING BY THE USE OF AN ADJUSTABLE INDUCTION COIL, SO ARRANGED THAT ITS EFFECTIVE INDUCTANCE COULD READILY BE VARIED.

BUT CAPACITY WAS NO LESS IMPORTANT IN TUNING. DE TUNZELMANN'S DESCRIPTIONS OF HERTZ'S EXPERIMENTS SHOW THAT HERTZ, IN ORDER TO MAKE HIS RECEIVING APPARATUS RESONANT TO THE PARTICULAR FREQUENCY RADIATED BY THE TRANSMITTER, CAREFULLY DETERMINED THE CAPACITY OF BOTH, AND INDEED DISCLOSED A MEANS OF ADJUSTING THE

CAPACITY OF THE RECEIVER BY ATTACHING TO IT WIRES WHOSE LENGTH COULD READILY BE VARIED. MARCONI IN HIS PRIOR PATENT NO. 586,193, GRANTED JULY 13, 1897, WHICH BECAME REISSUE NO. 11,913, HAD DISCLOSED A TWO-CIRCUIT SYSTEM FOR THE TRANSMISSION OF RADIO WAVES IN WHICH BOTH TRANSMITTER AND RECEIVER HAD LARGE METAL PLATES SERVING AS CAPACITY AREAS. HIS SPECIFICATIONS DESCRIBE THE CONSTRUCTION OF TRANSMITTING AND RECEIVING STATIONS SO AS TO BE RESONANT TO THE SAME FREQUENCY BY CALCULATION OF THE LENGTH OF THESE METAL PLATES, THEREBY DETERMINING THE CAPACITY OF THE ANTENNÁ CIRCUITS OF TRANSMITTER AND RECEIVER RESPECTIVELY. HE STATES THAT THE PLATES ARE "PREFERABLY OF SUCH A LENGTH AS TO BE ELECTRICALLY TUNED WITH THE ELECTRIC OSCILLATIONS TRANSMITTED," AND DESCRIBES MEANS OF ACHIEVING THIS RESULT SO AS TO DETERMINE "THE LENGTH MOST APPROPRIATE TO THE LENGTH OF WAVE EMITTED BY THE OSCILLATOR." CLAIM 24 OF HIS PATENT CLAIMS "THE COMBINATION OF A TRANSMITTER CAPABLE OF PRODUCING ELECTRICAL OSCILLATIONS OR RAYS OF DEFINITE CHARACTER AT THE WILL OF THE OPERATOR, AND A RECEIVER LOCATED AT A DISTANCE AND HAVING A CONDUCTOR TUNED TO RESPOND TO SUCH OSCILLATIONS ... " THE ONLY MEANS OF ACHIEVING THIS TUNING DISCLOSED BY THE SPECIFICATIONS IS THE DETERMINATION OF THE CAPACITY OF THE ANTENNA OF TRANSMITTER AND RECEIVER IN THE MANNER DESCRIBED.

MOREOVER THE USE OF AN ADJUSTABLE CONDENSER AS A MEANS OF TUNING WAS KNOWN TO THE PRIOR ART. PUPIN IN PATENT NO. 640,516, APPLIED FOR MAY 28, 1895, AND GRANTED JANUARY 2, 1900, BEFORE MARCONI, DISCLOSED THE USE OF AN ADJUSTABLE CONDENSER AS A MEANS OF TUNING A RECEIVING CIRCUIT IN A SYSTEM OF WIRED TELEGRAPHY. PUPIN'S PATENT WAS DESIGNED TO PERMIT THE SIMULTANEOUS TRANSMISSION OVER A WIRE OF SEVERAL MESSAGES AT DIFFERENT FREQUENCIES, AND THE SELECTIVE RECEPTION AT A GIVEN RECEIVING STATION OF THE PARTICULAR MESSAGE DESIRED, BY TUNING THE RECEIVING CIRCUIT TO THE FREQUENCY AT WHICH THAT MESSAGE WAS TRANSMITTED. HIS SPECIFICATIONS AND DRAWINGS DISCLOSE AT THE RECEIVER A TELEGRAPH KEY OR OTHER SUITABLE DETECTING INSTRUMENT LOCATED IN A SHUNT FROM THE WIRE ALONG WHICH THE MESSAGES WERE PASSED. THE SHUNT CIRCUIT INCLUDED A CONDENSER "OF ADJUSTABLE CAPACITY," AN ADJUSTABLE INDUCTION COIL, AND A DETECTING INSTRUMENT. HIS SPECIFICATIONS STATE THAT "THE CAPACITY OF THE CONDENSER H AND THE SELF-INDUCTION OF THE (INDUCTION) COIL I BEING SUCH THAT THE NATURAL PERIOD OR FREQUENCY OF THE SHUNT OR RESONANCE CIRCUIT HI IS THE SAME AS THE PERIOD OF ONE OF THE ELECTROMOTIVE FORCES WHICH PRODUCE THE CURRENT COMING OVER THE LINE ... THIS CIRCUIT HI WILL BE IN RESONANCE WITH THE CURRENT AND THEREFORE WILL ACT SELECTIVELY WITH RESPECT TO IT." HE DISCLOSED AN ALTERNATIVE SYSTEM IN WHICH A SIMILAR SHUNT CIRCUIT CONTAINING A CONDENSER, ALREADY DESCRIBED AS OF ADJUSTABLE CAPACITY, AND THE PRIMARY OF A TRANSFORMER, WAS INDUCTIVELY COUPLED WITH ANOTHER CIRCUIT CONTAINING THE SECONDARY OF THE TRANSFORMER, AN INDUCTION COIL, AN ADJUSTABLE CONDENSER, AND A RECEIVING DEVICE. HE THUS IN EFFECT DISCLOSED AN OPEN RECEIVING CIRCUIT WITH EARTH CONNECTION INCLUDING THE PRIMARY OF AN OSCILLATION TRANSFORMER - THE SECONDARY OF WHICH IS CONNECTED IN A CIRCUIT WITH A TELEGRAPH KEY OR OTHER SUITABLE DETECTING INSTRUMENT - AND AN ADJUSTABLE CONDENSER IN A SHUNT BRIDGING THE PRIMARY OF THE TRANSFORMER AND THUS CONNECTED IN PARALLEL WITH IT.

THUS PUPIN SHOWED THE USE OF AN ADJUSTABLE CONDENSER AS A MEANS OF TUNING AN ELECTRICAL CIRCUIT SO AS TO BE SELECTIVELY RECEPTIVE TO IMPULSES OF A PARTICULAR FREQUENCY. IT IS TRUE THAT HIS PATENT RELATED NOT TO THE RADIO ART BUT TO THE ART OF WIRED TELEGRAPHY, AN ART WHICH EMPLOYED MUCH LOWER FREQUENCIES. BUT SO FAR AS WE ARE INFORMED THE PRINCIPLES OF RESONANCE, AND THE METHODS OF ACHIEVING IT, APPLICABLE TO THE LOW FREQUENCIES USED BY PUPIN ARE THE SAME AS THOSE APPLICABLE

TO HIGH FREQUENCY RADIO TRANSMISSION AND RECEPTION.

FESSENDEN, IN PATENT NO. 706,735, APPLIED FOR DEC. 15, 1899, BEFORE MARCONI, AND GRANTED AUG. 12, 1902, DISCLOSED, IN THE ANTENNA CIRCUIT OF A RADIO RECEIVER, A CONDENSER IN A SHUNT AROUND A COIL. THE COIL WAS USED IN EFFECT AS A TRANSFORMER; BY THE MAGNETIC LINES OF FORCE SET UP WHEN A CURRENT PASSED THROUGH IT AN INDICATOR WAS CAUSED TO MOVE, THEREBY EITHER CLOSING AN ELECTRICAL CONNECTION OR GIVING A VISIBLE SIGNAL. FESSENDEN'S SPECIFICATIONS DO NOT CLEARLY DISCLOSE THE PURPOSE OF HIS CONDENSER, BUT THEY SPECIFY THAT IT MUST BE "OF THE PROPER SIZE." HE ALSO DISCLOSES A CONDENSER IN A SHUNT CIRCUIT AROUND THE TERMINALS OF A SPARK GAP IN THE ANTENNA CIRCUIT OF THE TRANSMITTER, AND HIS SPECIFICATIONS PRESCRIBE THAT "THIS SHUNT-CIRCUIT MUST BE TUNED TO THE RECEIVING-CONDUCTOR; OTHERWISE THE OSCILLATIONS PRODUCED BY IT WILL HAVE NO ACTION UPON THE WAVE-RESPONSIVE DEVICE AT THE RECEIVING STATION." WE HAVE REFERRED TO THE PUPIN AND FESSENDEN PATENTS, NOT FOR THE PURPOSE OF DETERMINING WHETHER THEY ANTICIPATE CLAIM 16 OF MARCONI, AS THE GOVERNMENT INSISTS, BUT TO INDICATE THE IMPORTANCE OF CONSIDERING THEM IN THAT ASPECT, TOGETHER WITH THE RELEVANT TESTIMONY, WHICH THE COURT BELOW DID NOT DO. IN THE PRESENT STATE OF THE RECORD WE DO NOT UNDERTAKE TO DETERMINE WHETHER AND TO WHAT EXTENT THESE DISCLOSURES EITHER ANTICIPATE CLAIM 16 OF THE MARCONI PATENT OR REQUIRE THAT CLAIM TO BE SO NARROWLY CONSTRUED THAT DEFENDANTS' ACCUSED DEVICES OR SOME OF THEM DO NOT INFRINGE MARCONI.

ALTHOUGH THE PUPIN AND FESSENDEN PATENTS WERE IN THE RECORD BEFORE THE COURT OF CLAIMS WHEN IT ENTERED ITS DECISION FINDING CLAIM 16 VALID AND INFRINGED, THEY WERE NOT REFERRED TO IN CONNECTION WITH CLAIM 16 EITHER IN THE COURT'S OPINION OR IN ITS FINDINGS, EVIDENTLY BECAUSE NOT URGED UPON THAT COURT BY THE GOVERNMENT AS ANTICIPATING CLAIM 16. BUT THIS COURT, IN THE EXERCISE OF ITS APPELLATE POWER, IS NOT PRECLUDED FROM LOOKING AT ANY EVIDENCE OF RECORD WHICH, WHETHER OR NOT CALLED TO THE ATTENTION OF THE COURT BELOW, IS RELEVANT TO AND MAY AFFECT THE CORRECTNESS OF ITS DECISION SUSTAINING OR DENYING ANY CONTENTION WHICH A PARTY HAS MADE BEFORE IT. MUNCIE GEAR CO. V. OUTBOARD MOTOR CO., 315 U.S. 759, 766-8; ACT OF MAY 22, 1939, 28 U.S.C. SEC. 288; CF. HORMEL V. HELVERING, 312 U.S. 552, 556.

IN ORDER TO DETERMINE WHETHER THIS COURT SHOULD CONSIDER THE EVIDENCE WHICH THE GOVERNMENT NOW PRESSES UPON IT, AND SHOULD ON THE BASIS OF THAT EVIDENCE EITHER DECIDE FOR ITSELF WHETHER CLAIM 16 IS VALID AND INFRINGED OR REMAND THAT QUESTION TO THE COURT OF CLAIMS FOR FURTHER CONSIDERATION, IT IS NECESSARY TO SET OUT IN SOME DETAIL THE RELEVANT PROCEEDINGS BELOW. THE CASE WAS REFERRED TO A SPECIAL COMMISSIONER FOR THE TAKING OF TESTIMONY UNDER A STIPULATION THAT THE ISSUE OF REASONABLE COMPENSATION FOR DAMAGES AND PROFITS BE POSTPONED UNTIL THE DETERMINATION OF THE ISSUES OF VALIDITY AND INFRINGEMENT. ON JUNE 26, 1933, THE COMMISSIONER FILED A REPORT IN WHICH HE MADE THE FOLLOWING FINDINGS WITH REGARD TO CLAIM 16, WHICH THE COURT OF CLAIMS LATER ADOPTED IN SUBSTANCE: "LXII. CLAIM 16 OF MARCONI #763772 IS DIRECTED TO SUBJECT MATTER WHICH IS NEW AND USEFUL ...

"LXV. THE RECEIVING APPARATUS OF THE KILBOURNE & CLARK COMPANY, SHOWN IN EXHIBIT 95, AND THE RECEIVER MADE BY THE TELEFUNKEN COMPANY, ILLUSTRATED IN EXHIBIT 79, EACH HAS APPARATUS COMING WITHIN THE TERMINOLOGY OF CLAIM 16." BOTH PARTIES FILED EXCEPTIONS TO THE COMMISSIONER'S REPORT. THE MARCONI COMPANY EXCEPTED TO PART OF FINDING LXII, AND TOOK SEVERAL EXCEPTIONS WHICH WERE FORMALLY ADDRESSED TO FINDING LXV. THE GOVERNMENT, IN A MEMORANDUM, OPPOSED THE SUGGESTED AMENDMENTS TO THESE FINDINGS. BUT THE GOVERNMENT FILED NO EXCEPTIONS TO THESE TWO FINDINGS, NOR DID IT, IN ITS EXTENSIVE BRIEF BEFORE THE COURT OF CLAIMS, MAKE ANY CONTENTION THAT CLAIM 16 EITHER IS INVALID OR WAS NOT

INFRINGED.

AFTER THE COURT HAD RENDERED ITS INTERLOCUTORY DECISION HOLDING CLAIM 16 VALID AND INFRINGED, THE CASE WAS SENT BACK TO THE COMMISSIONER TO TAKE EVIDENCE ON THE ACCOUNTING. MUCH EVIDENCE WAS TAKEN BEARING ON THE FUNCTION SERVED BY THE CONDENSER IN THE ARRANGEMENT DESCRIBED IN CLAIM 16 AND IN THE GOVERNMENT'S RECEIVERS, AND IN THAT CONNECTION THE PUPIN AND FESSENDEN PATENTS WERE AGAIN INTRODUCED IN EVIDENCE BY THE GOVERNMENT. WHEN THE PUPIN PATENT WAS OFFERED THE COMMISSIONER STATED: "OBVIOUSLY, AS I UNDERSTAND THE OFFER OF THIS PATENT OF PUPIN, IT DOES NOT IN ANY WAY ATTACK THE VALIDITY OF CLAIM 16 OF THE MARCONI PATENT IN SUIT. AS YOU STATE MR. BLACKMAR, THAT HAS BEEN DECIDED BY THE COURT, AND I DO NOT RECALL JUST NOW WHAT PROCEDURE WAS FOLLOWED AFTER THE DECISION AND PRIOR TO THIS ACCOUNTING PROCEEDING; BUT THE DEFENDANT HAD AT THAT TIME OPPORTUNITY FOR A MOTION FOR A NEW TRIAL AND PRESENTATION OF NEWLY-DISCOVERED EVIDENCE AND ALL THOSE MATTERS." ACCORDINGLY, THE COMMISSIONER STATED THAT HE RECEIVED THE PATENT IN EVIDENCE "FOR THE SOLE PURPOSE OF AIDING THE WITNESS AND THE COMMISSIONER AND THE COURT IN AN UNDERSTANDING OF HOW THE CONDENSER IN THE MARCONI PATENT OPERATES." AND IN OFFERING THE FESSENDEN PATENT COUNSEL FOR THE GOVERNMENT SIMILARLY STATED THAT IT WAS OFFERED "NOT TO SHOW INVALIDITY BUT AS SHOWING JUSTIFICATION FOR THE DEFENDANT'S USE." IN ITS EXCEPTIONS TO THE COMMISSIONER'S REPORT ON THE ACCOUNTING THE GOVERNMENT ASKED THE COURT OF CLAIMS TO MAKE CERTAIN SPECIFIC FINDINGS AS TO THE MODE OF OPERATION OF THE ARRANGEMENTS DISCLOSED IN THE PUPIN AND FESSENDEN PATENTS, AND ALSO TO FIND THAT "THE MODE OF CONNECTING THE PRIMARY CONDENSER IN PARALLEL WITH THE ANTENNA-TO-EARTH CAPACITY USED BY THE DEFENDANT FOLLOWED THE DISCLOSURE OF PUPIN 640,516 AND THE FESSENDEN PATENT 706,735 ... AND HENCE DOES NOT INFRINGE THE MARCONI CLAIM 16 WHICH IS BASED UPON A DIFFERENT ARRANGEMENT, OPERATING IN A DIFFERENT MANNER TO OBTAIN A DIFFERENT RESULT." THE GOVERNMENT CONTENDED THAT THERE WAS NO FINDING OF FACT THAT CLAIM 16 HAD BEEN INFRINGED, AND THAT THE COURT, IN THE COURSE OF THE ACCOUNTING PROCEEDING HAD BY AN ORDER OF OCTOBER 22, 1937, REOPENED THE ENTIRE SUBJECT OF INFRINGEMENT. WE AGREE WITH THE COURT THAT THE COMMISSIONER'S FINDING LXV, WHICH THE COURT ADOPTED AS FINDING LXIII, WAS A FINDING OF INFRINGEMENT, AND WE SEE NO REASON TO QUESTION THE COURT'S CONCLUSION THAT ITS ORDER HAD NOT REOPENED THE SUBJECT OF INFRINGEMENT.

IN VIEW, HOWEVER, OF THE GOVERNMENT'S APPARENT MISUNDERSTANDING OF THE SCOPE OF THE ISSUES LEFT OPEN ON THE ACCOUNTING WE THINK THAT ITS REQUEST FOR A FINDING OF NON-INFRINGEMENT SPECIFICALLY ADDRESSED TO THE PUPIN AND FESSENDEN PATENTS WAS A SUFFICIENT REQUEST TO THE COURT TO RECONSIDER ITS PREVIOUS DECISION OF INFRINGEMENT. AND WHILE MOST OF THE ARGUMENT ON THE GOVERNMENT'S EXCEPTIONS TO THE COMMISSIONER'S REPORT WAS BASED ON EVIDENCE TAKEN UPON THE ACCOUNTING, THE GOVERNMENT'S BRIEFS SUFFICIENTLY DISCLOSED TO THE COURT THAT THE PUPIN AND FESSENDEN PATENTS, AT LEAST, HAD BEEN IN THE RECORD PRIOR TO THE INTERLOCUTORY DECISION.

THE COURT, IN REJECTING THE GOVERNMENT'S REQUEST FOR A FINDING OF NON INFRINGEMENT, STATED: "THE QUESTION OF INFRINGEMENT OF MARCONI CLAIM 16 ... IS NOT BEFORE US IN THE PRESENT ACCOUNTING." "THE SOLE PURPOSE AND FUNCTION OF AN ACCOUNTING IN A PATENT INFRINGEMENT CASE IS TO ASCERTAIN THE AMOUNT OF COMPENSATION DUE, AND NO OTHER ISSUE CAN BE BROUGHT INTO THE ACCOUNTING TO CHANGE OR ALTER THE COURT'S PRIOR DECISION." WE CANNOT SAY WITH CERTAINTY WHETHER IN REJECTING THE GOVERNMENT'S REQUEST THE COURT THOUGHT THAT IT LACKED POWER TO RECONSIDER ITS PRIOR DECISION, OR WHETHER IT HELD MERELY THAT IN THE EXERCISE OF ITS DISCRETION IT SHOULD NOT DO SO. NOR DOES IT APPEAR THAT, ASSUMING

IT CONSIDERED THE QUESTION TO BE ONE OF DISCRETION, IT RECOGNIZED THAT IN PART AT LEAST THE GOVERNMENT'S REQUEST WAS BASED ON EVIDENCE, HAVING AN IMPORTANT BEARING ON THE VALIDITY AND CONSTRUCTION OF CLAIM 16, WHICH HAD BEEN BEFORE THE COURT BUT HAD NOT BEEN CONSIDERED BY IT WHEN IT HELD CLAIM 16 VALID AND INFRINGED.

ALTHOUGH THE INTERLOCUTORY DECISION OF THE COURT OF CLAIMS ON THE QUESTION OF VALIDITY AND INFRINGEMENT WAS APPEALABLE, UNITED STATES V. ESNAULT-PELTERIE, 299 U.S. 201, 303 U.S. 26; 28 U.S.C. SEC. 288(B), AS ARE INTERLOCUTORY ORDERS OF DISTRICT COURTS IN SUITS TO ENJOIN INFRINGMENT, 28 U.S.C. SEC. 227(A); SIMMONS CO. V. GRIER BROS. CO., 258 U.S. 82, 89, THE DECISION WAS NOT FINAL UNTIL THE CONCLUSION OF THE ACCOUNTING. BARNARD V. GIBSON, 7 HOW. 649; HUMISTON V. STAINTHORP, 2 WALL. 106; SIMMONS CO. V. GRIER BROS. CO., SUPRA, 89. HENCE THE COURT DID NOT LACK POWER AT ANY TIME PRIOR TO ENTRY OF ITS FINAL JUDGMENT AT THE CLOSE OF THE ACCOUNTING TO RECONSIDER ANY PORTION OF ITS DECISION AND REOPEN ANY PART OF THE CASE. PERKINS V. FOURNIQUET, 6 HOW. 206, 208; MCGOURKEY V. TOLEDO ,& OHIO CENTRAL RY. CO., 146 U.S. 536, 544; SIMMONS CO. V. GRIER BROS. CO., SUPRA, 90-91. IT WAS FREE IN ITS DISCRETION TO GRANT A REARGUMENT BASED EITHER ON ALL THE EVIDENCE THEN OF RECORD OR ONLY THE EVIDENCE BEFORE THE COURT WHEN IT RENDERED ITS INTERLOCUTORY DECISION, OR TO REOPEN THE CASE FOR FURTHER EVIDENCE.

WHETHER IT SHOULD HAVE TAKEN ANY OF THESE COURSES WAS A MATTER PRIMARILY FOR ITS DISCRETION, TO BE EXERCISED IN THE LIGHT OF VARIOUS CONSIDERATIONS WHICH THIS COURT CANNOT PROPERLY APPRAISE WITHOUT MORE INTIMATE KNOWLEDGE THAN IT HAS OF THE PROCEEDINGS IN A LONG AND COMPLEX TRIAL. AMONG THOSE CONSIDERATIONS ARE THE QUESTIONS WHETHER, AS APPEARS TO BE THE CASE FROM SUCH PORTIONS OF THE RECORD AS HAVE BEEN FILED IN THIS COURT OR CITED TO US BY COUNSEL, THE GOVERNMENT FAILED TO MAKE ANY CONTENTION AS TO THE VALIDITY OR CONSTRUCTION OF CLAIM 16 IN THE PROCEEDINGS LEADING TO THE INTERLOCUTORY DECISION; WHETHER THE SHOWING OF NON-INFRINGEMENT WHICH IT NOW MAKES IS SUFFICIENTLY STRONG, AND THE PUBLIC INTEREST THAT AN INVALID PATENT BE NOT SUSTAINED IS SUFFICIENTLY GREAT, TO JUSTIFY RECONSIDERING THE DECISION AS TO CLAIM 16 DESPITE THE FAILURE OF GOVERNMENT COUNSEL TO PRESS ITS CONTENTION AT THE PROPER TIME; WHETHER ADEQUATE CONSIDERATION OF THE QUESTION OF NON INFRINGEMENT CAN BE HAD ON THE EXISTING RECORD, OR WHETHER ADDITIONAL TESTIMONY SHOULD BE RECEIVED; AND WHETHER, BALANCING THE STRENGTH OR WEAKNESS OF THE GOVERNMENT'S PRESENT SHOWING OF NON-INFRINGEMENT AGAINST THE UNDESIRABILITY OF FURTHER PROLONGING THIS ALREADY EXTENDED LITIGATION, THE CASE IS ONE WHICH JUSTIFIES RECONSIDERATION.

THESE ARE ALL MATTERS REQUIRING CAREFUL CONSIDERATION BY THE TRIAL COURT. IN ORDER THAT THE CASE MAY RECEIVE THAT CONSIDERATION, WE VACATE THE JUDGMENT AS TO CLAIM 16 AND REMAND THE CAUSE TO THE COURT OF CLAIMS FOR FURTHER PROCEEDINGS IN CONFORMITY TO THIS OPINION.

IF ON THE REMAND THE COURT SHOULD EITHER DECLINE TO RECONSIDER ITS DECISION OF INFRINGEMENT, OR SHOULD UPON RECONSIDERATION ADHERE TO THAT DECISION, IT SHOULD PASS UPON THE CONTENTION OF THE GOVERNMENT, URGED HERE AND BELOW, AS TO THE MEASURE OF DAMAGES, WITH RESPECT TO WHICH THE COURT MADE NO FINDINGS. THE GOVERNMENT'S CONTENTION IS THAT THE VARIABLE CAPACITY SHUNT OF THE ACCUSED DEVICES BRIDGED ALL THE INDUCTANCE IN THE RECEIVING ANTENNA CIRCUIT, AND THAT EVEN THOUGH THOSE DEVICES INFRINGED THEY NEVERTHELESS EMBODY AN IMPROVEMENT OVER MARCONI'S CLAIM 16, IN WHICH ONLY THE TRANSFORMER COIL WAS BRIDGED. IN COMPUTING THE DAMAGES THE COURT MEASURED THEM BY 65% OF THE COST TO THE GOVERNMENT OF THE INDUCTION COILS WHICH WOULD BE REQUIRED TO REPLACE IN THE ACCUSED DEVICES THE ADJUSTABLE CONDENSERS AS A MEANS OF TUNING, TAKING INTO ACCOUNT THE GREATER CONVENIENCE AND EFFICIENCY OF CONDENSER TUNING. THE

ALLOWANCE OF ONLY 65% WAS ON THE THEORY THAT IF THE PARTIES HAD NEGOTIATED FOR THE USE OF THE INVENTION THE PRICE WOULD HAVE BEEN LESS THAN THE COST TO THE GOVERNMENT OF THE AVAILABLE ALTERNATIVE MEANS OF TUNING.

IN COMPUTING THE DAMAGES THE COURT APPARENTLY DID NOT TAKE INTO ACCOUNT OR ATTEMPT TO APPRAISE ANY CONTRIBUTION WHICH MAY HAVE BEEN MADE BY THE IMPROVEMENT OVER MARCONI WHICH THE GOVERNMENT ASSERTS WAS INCLUDED IN THE ACCUSED DEVICES. THE COURT FOUND THAT WHERE THE CONDENSER IS CONNECTED IN SERIES WITH THE INDUCTANCE COILS IN THE ANTENNA IT "CAN BE USED TO SHORTEN THE NATURAL RESONANT WAVE LENGTH OF THE ANTENNA CIRCUIT BUT CANNOT LENGTHEN IT BEYOND WHAT WOULD BE THE RESONANT WAVE LENGTH IF THE CONDENSER WERE NOT PRESENT." ON THE OTHER HAND, IT FOUND THAT WHEN THE CONDENSER IS CONNECTED IN PARALLEL IT ENABLES THE PERIODICITY OF THE ANTENNA TO BE LOWERED, PERMITTING THE RECEPTION OF LONGER WAVE-LENGTHS.

THE COMPUTATION OF DAMAGES WAS BASED ON THE PREMISE THAT THE ADVANTAGE TO THE GOVERNMENT RESULTING FROM THE INFRINGEMENT WAS DERIVED FROM THE ABILITY WHICH THE ACCUSED DEVICES HAD THUS ACQUIRED TO RECEIVE LONGER WAVE-LENGTHS. BUT THERE WAS SUBSTANTIAL TESTIMONY THAT THE ARRANGEMENT DISCLOSED BY MARCONI'S SPECIFICATIONS WAS IN EFFECT A CONNECTION IN SERIES WHICH DID NOT MAKE POSSIBLE RECEPTION OF LONGER WAVE-LENGTHS, AS DID THE ARRANGEMENT IN THE ACCUSED DEVICES. AND THE COURT NOWHERE FOUND THAT THE ARRANGEMENT COVERED BY MARCONI'S CLAIM 16 DID MAKE POSSIBLE SUCH RECEPTION. THE APPROPRIATE EFFECT TO BE GIVEN TO THIS TESTIMONY IS IMPORTANT IN THE LIGHT OF THE RECOGNIZED DOCTRINE THAT IF A DEFENDANT HAS ADDED "NON-INFRINGING AND VALUABLE IMPROVEMENTS WHICH HAD CONTRIBUTED TO THE MAKING OF THE PROFITS," IT IS NOT LIABLE FOR BENEFITS RESULTING FROM SUCH IMPROVEMENTS. WESTINGHOUSE ELECTRIC CO. V. WAGNER MFG. CO., 225 U.S. 604, 614-15, 616-17; SHELDON V. METRO GOLDWYN CORP., 309 U.S. 390, 402-406, AND CASES CITED. FINDING LXIII THAT THE GOVERNMENT WAS USING "APPARATUS COMING WITHIN THE TERMINOLOGY OF CLAIM 16," AND FINDING 23 ON THE ACCOUNTING THAT THE ACCUSED DEVICES "INFRINGE CLAIM 16 OF THE MARCONI PATENT," GIVE NO AID IN SOLVING THIS PROBLEM FOR THEY ARE NOT ADDRESSED TO THE QUESTION WHETHER, ASSUMING INFRINGEMENT, THE GOVERNMENT HAS MADE IMPROVEMENTS WHICH OF THEMSELVES ARE NON-INFRINGING. THAT CAN ONLY BE AFFORDED BY FINDINGS WHICH APPRAISE THE EVIDENCE, ESTABLISH THE SCOPE OF MARCONI'S CLAIM AND THE NATURE AND EXTENT OF THE DIFFERENCE IN FUNCTION, IF ANY, BETWEEN THE DEVICE CLAIMED BY MARCONI AND THOSE USED BY THE GOVERNMENT, AND DETERMINE WHETHER ANY DIFFERENCES SHOWN TO EXIST CONSTITUTE A "NON INFRINGING IMPROVEMENT" FOR WHICH MARCONI DESERVES NO CREDIT.

THE JUDGMENT AS TO CLAIM 16 WILL BE VACATED AND THE CAUSE REMANDED FOR FURTHER PROCEEDINGS.

THE FLEMING PATENT NO. 803,684.

THE FLEMING PATENT, ENTITLED: "INSTRUMENT FOR CONVERTING ALTERNATING ELECTRIC CURRENTS INTO CONTINUOUS CURRENTS" WAS APPLIED FOR APRIL 19, 1905, AND GRANTED ON NOVEMBER 7, 1905 TO THE MARCONI COMPANY, AS ASSIGNEE OF FLEMING. ITS SPECIFICATIONS STATE THAT "THIS INVENTION RELATES TO CERTAIN NEW AND USEFUL DEVICES FOR CONVERTING ALTERNATING ELECTRIC CURRENTS, AND ESPECIALLY HIGH-FREQUENCY ALTERNATING ELECTRIC CURRENTS OR ELECTRIC OSCILLATIONS, INTO CONTINUOUS ELECTRIC CURRENTS FOR THE PURPOSE OF MAKING THEM DETECTABLE BY AND MEASURABLE WITH ORDINARY DIRECTCURRENT INSTRUMENTS, SUCH AS A 'MIRROR-GALVANOMETER' OF THE USUAL TYPE OR ANY ORDINARY DIRECT-CURRENT AMMETER." FLEMING'S DRAWINGS AND SPECIFICATIONS SHOW A COMBINATION APPARATUS BY WHICH ALTERNATING CURRENT IMPULSES RECEIVED THROUGH AN ANTENNA CIRCUIT CONTAINING

THE PRIMARY OF A TRANSFORMER ARE INDUCED IN THE SECONDARY OF THE TRANSFORMER. TO ONE END OF THE SECONDARY COIL IS CONNECTED A CARBON FILAMENT LIKE THAT OF AN INCANDESCENT ELECTRIC LAMP, WHICH IS HEATED BY A BATTERY. SURROUNDING, BUT NOT TOUCHING THE FILAMENT, IS A CYLINDER OF ALUMINUM OPEN AT THE TOP AND BOTTOM, WHICH IS CONNECTED WITH THE OTHER END OF THE SECONDARY. THE CYLINDER AND FILAMENT ARE ENCLOSED IN AN EVACUATED VESSEL SUCH AS AN ORDINARY ELECTRIC LAMP BULB. AN INDICATING INSTRUMENT OR GALVANOMETER IS SO LOCATED IN THIS CIRCUIT AS TO RESPOND TO THE FLOW OF CURRENT IN IT. THE SPECIFICATIONS EXPLAIN THE OPERATION OF THIS DEVICE: "THIS ARRANGEMENT DESCRIBED ABOVE OPERATES AS AN ELECTRIC VALVE AND PERMITS NEGATIVE ELECTRICITY TO FLOW FROM THE HOT CARBON B TO THE METAL CYLINDER C, BUT NOT IN THE REVERSE DIRECTION, SO THAT THE ALTERNATIONS INDUCED IN THE COIL K BY THE HERTZIAN WAVES RECEIVED BY THE AERIAL WIRE N ARE RECTIFIED OR TRANSFORMED INTO A MORE OR LESS CONTINUOUS CURRENT CAPABLE OF ACTUATING THE GALVANOMETER L BY WHICH THE SIGNALS CAN BE READ." THE SPECIFICATIONS FURTHER STATE: " ... THE AERIAL WIRE N MAY BE REPLACED BY ANY CIRCUIT IN WHICH THERE IS AN ALTERNATING ELECTROMOTIVE FORCE, WHETHER OF LOW FREQUENCY OR OF HIGH FREQUENCY ... " "HENCE THE DEVICE MAY BE USED FOR RECTIFYING EITHER HIGH-FREQUENCY OR LOW-FREQUENCY ALTERNATING CURRENTS OF ELECTRICAL OSCILLATIONS ... " ONLY CLAIMS 1 AND 37 OF THE PATENT ARE IN SUIT. THEY READ AS FOLLOWS: "1. THE COMBINATION OF A VACUOUS VESSEL, TWO CONDUCTORS ADJACENT TO BUT NOT TOUCHING EACH OTHER IN THE VESSEL, MEANS FOR HEATING ONE OF THE CONDUCTORS, AND A CIRCUIT OUTSIDE THE VESSEL CONNECTING THE TWO CONDUCTORS.

"37. AT A RECEIVING-STATION IN A SYSTEM OF WIRELESS TELEGRAPHY EMPLOYING ELECTRICAL OSCILLATIONS OF HIGH FREQUENCY A DETECTOR COMPRISING A VACUOUS VESSEL, TWO CONDUCTORS ADJACENT TO BUT NOT TOUCHING EACH OTHER IN THE VESSEL, MEANS FOR HEATING ONE OF THE CONDUCTORS, A CIRCUIT OUTSIDE OF THE VESSEL CONNECTING THE TWO CONDUCTORS, MEANS FOR DETECTING A CONTINUOUS CURRENT IN THE CIRCUIT, AND MEANS FOR IMPRESSING UPON THE CIRCUIT THE RECEIVED OSCILLATIONS." THE CURRENT APPLIED TO THE FILAMENT OR CATHODE BY THE BATTERY SETS UP A FLOW OF ELECTRONS (NEGATIVE ELECTRIC CHARGES) FROM THE HEATED CATHODE, WHICH ARE ATTRACTED TO THE COLD PLATE OR ANODE WHEN THE LATTER IS POSITIVELY CHARGED. WHEN AN ALTERNATING CURRENT IS SET UP IN THE CIRCUIT CONTAINING THE CATHODE, ANODE, AND SECONDARY OF THE TRANSFORMER, THE ELECTRONIC DISCHARGE FROM THE CATHODE CLOSES THE CIRCUIT AND PERMITS A CONTINUOUS FLOW OF ELECTRICITY THROUGH IT WHEN THE PHASE OF THE CURRENT IS SUCH THAT THE ANODE IS POSITIVELY CHARGED, WHILE PREVENTING ANY FLOW OF CURRENT THROUGH THE TUBE WHEN THE ANODE IS NEGATIVELY CHARGED. THE ALTERNATING CURRENT IS THUS RECTIFIED SO AS TO PRODUCE A CURRENT FLOWING ONLY IN ONE DIRECTION. SEE DEFOREST RADIO CO. V. GENERAL ELECTRIC CO., 283 U.S. 664; RADIO CORPORATION V. RADIO LABORATORIES, 293 U.S. 1; DETROLA RADIO CORP. V. HAZELTINE CORPORATION, 313 U.S. 259.

CLAIMS 1 AND 37 OF THE FLEMING PATENT ARE IDENTICAL IN THEIR STRUCTURAL ELEMENTS. BOTH CLAIM THE VACUUM TUBE, AND THE TWO ELECTRODES CONNECTED BY A CIRCUIT OUTSIDE THE TUBE, ONE ELEMENT BEING HEATED. THE CLAIMS DIFFER ONLY IN THAT CLAIM 37 INCLUDES "MEANS FOR DETECTING" THE CONTINUOUS OR DIRECT CURRENT IN THE ANODE-CATHODE CIRCUIT, AND "MEANS FOR IMPRESSING UPON THE CIRCUIT THE RECEIVED OSCILLATIONS" FROM THE TRANSFORMER COIL OF THE ANTENNA CIRCUIT.

IN THE PATENT AS ORIGINALLY ISSUED THERE HAD BEEN ANOTHER DIFFERENCE BETWEEN THE TWO CLAIMS. CLAIM 37 DESCRIBES THE TUBE AS BEING USED "IN A SYSTEM OF WIRELESS TELEGRAPHY EMPLOYING ELECTRICAL OSCILLATIONS OF HIGH FREQUENCY." NO SUCH LIMITATION WAS PLACED ON CLAIM 1 AS ORIGINALLY CLAIMED, AND THE SPECIFICATIONS ALREADY QUOTED PLAINLY CONTEMPLATED THE USE OF THE CLAIMED DEVICE WITH LOW AS

WELL AS HIGH FREQUENCY CURRENTS. THIS DISTINCTION WAS ELIMINATED BY A DISCLAIMER FILED BY THE MARCONI COMPANY NOVEMBER 17, 1915, RESTRICTING THE COMBINATION OF THE ELEMENTS OF CLAIM 1 TO A USE "IN CONNECTION WITH HIGH FREQUENCY ALTERNATING ELECTRIC CURRENTS OR ELECTRIC OSCILLATIONS OF THE ORDER EMPLOYED IN HERTZIAN WAVE TRANSMISSION," AND DELETING CERTAIN REFERENCES TO LOW FREQUENCIES IN THE SPECIFICATIONS. THE RESULT OF THE DISCLAIMER WAS TO LIMIT BOTH CLAIMS TO THE USE OF THE PATENTED DEVICE FOR RECTIFYING HIGH FREQUENCY ALTERNATING WAVES OR CURRENTS SUCH AS WERE EMPLOYED IN WIRELESS TELEGRAPHY.

THE EARLIEST DATE ASSERTED FOR FLEMING'S INVENTION, AS LIMITED BY THE DISCLAIMER, IS NOVEMBER 16, 1904. TWENTY YEARS BEFORE, ON OCTOBER 21, 1884, EDISON HAD SECURED UNITED STATES PATENT NO. 307,031. IN HIS SPECIFICATIONS HE STATED: "I HAVE DISCOVERED THAT IF A CONDUCTING SUBSTANCE IS INTERPOSED ANYWHERE IN THE VACUOUS SPACE WITHIN THE GLOBE OF AN INCANDESCENT ELECTRIC LAMP, AND SAID CONDUCTING SUBSTANCE IS CONNECTED OUTSIDE OF THE LAMP WITH ONE TERMINAL, PREFERABLY THE POSITIVE ONE, OF THE INCANDESCENT CONDUCTOR, A PORTION OF THE CURRENT WILL, WHEN THE LAMP IS IN OPERATION, PASS THROUGH THE SHUNT-CIRCUIT THUS FORMED, WHICH SHUNT INCLUDES A PORTION OF THE VACUOUS SPACE WITHIN THE LAMP. THIS CURRENT I HAVE FOUND TO BE PROPORTIONAL TO THE DEGREE OF INCANDESCENCE OF THE CONDUCTOR OR CANDLEPOWER OF THE LAMP." EDISON PROPOSED TO USE THIS DISCOVERY AS A MEANS OF "INDICATING, VARIATIONS IN THE ELECTRO-MOTIVE FORCE IN AN ELECTRIC CIRCUIT," BY CONNECTING A LAMP THUS EQUIPPED AT A POINT WHERE THE CURRENT WAS TO BE MEASURED. THE DRAWINGS OF HIS PATENT SHOW AN ELECTRIC CIRCUIT, INCLUDING A FILAMENT (CATHODE) AND A PLATE (ANODE) BOTH "IN THE VACUOUS SPACE WITHIN THE GLOBE" - AN ELECTRIC LIGHT BULB. THE SHUNT-CIRCUIT EXTENDS FROM THE PLATE THROUGH A GALVANOMETER TO THE FILAMENT. HIS SPECIFICATIONS DISCLOSE THAT THE VACUOUS SPACE WITHIN THE GLOBE IS A CONDUCTOR OF CURRENT BETWEEN THE PLATE ANODE AND THE FILAMENT; THAT THE STRENGTH OF THE CURRENT IN THE FILAMENT-TO-PLATE CIRCUIT THROUGH THE VACUUM DEPENDS UPON THE DEGREE OF INCANDESCENCE AT THE FILAMENT; AND THAT THE PLATE ANODE IS PREFERABLY CONNECTED TO THE POSITIVE SIDE OF THE CURRENT SUPPLY. THE CLAIMS OF THE PATENT ARE FOR THE COMBINATION OF THE FILAMENT, PLATE AND INTERCONNECTING CIRCUIT, INCLUDING THE GALVANOMETER. CLAIM 5, A TYPICAL CLAIM, READS AS FOLLOWS: "THE COMBINATION, WITH AN INCANDESCENT ELECTRIC LAMP, OF A CIRCUIT HAVING ONE TERMINAL IN THE VACUOUS SPACE WITHIN THE GLOBE OF SAID LAMP, AND THE OTHER CONNECTED WITH ONE SIDE OF THE LAMP-CIRCUIT, AND ELECTRICALLY CONTROLLED OR OPERATED APPARATUS IN SAID CIRCUIT, SUBSTANTIALLY AS SET FORTH." THE STRUCTURE DISCLOSED IN FLEMING'S CLAIMS 1 AND 37 THUS DIFFERED IN NO MATERIAL RESPECT FROM THAT DISCLOSED BY EDISON. SINCE FLEMING'S ORIGINAL CLAIM 1 IS MERELY FOR THE STRUCTURE, IT READS DIRECTLY ON EDISON'S CLAIM 5 AND COULD NOT BE TAKEN AS INVENTION OVER IT.

FLEMING USED THIS STRUCTURE FOR A DIFFERENT PURPOSE THAN EDISON. EDISON DISCLOSED THAT HIS DEVICE OPERATED TO PASS A CURRENT ACROSS THE VACUOUS SPACE WITHIN THE TUBE BETWEEN FILAMENT AND PLATE. HE USED THIS CURRENT AS A MEANS OF MEASURING THE CURRENT PASSING THROUGH THE FILAMENT CIRCUIT. FLEMING, IN HIS SPECIFICATIONS, DISCLOSED THE USE OF HIS TUBE AS A RECTIFIER OF ALTERNATING CURRENTS, AND IN CLAIM 37 HE CLAIMED THE USE OF THAT APPARATUS AS A MEANS OF RECTIFYING ALTERNATING CURRENTS OF RADIO FREQUENCY. BUT IN THIS USE OF THE TUBE TO CONVERT ALTERNATING INTO DIRECT CURRENTS THERE WAS NO NOVELTY FOR IT HAD BEEN DISCLOSED BY OTHERS AND BY FLEMING HIMSELF LONG BEFORE FLEMING'S INVENTION DATE.

ON JANUARY 9, 1890, TEN YEARS BEFORE FLEMING FILED HIS APPLICATION, HE STATED IN A PAPER READ BEFORE THE ROYAL SOCIETY OF LONDON: "IT HAS BEEN KNOWN FOR SOME TIME

THAT IF A PLATINUM PLATE OR WIRE IS SEALED THROUGH THE GLASS BULB OF AN ORDINARY CARBON FILAMENT INCANDESCENT LAMP, THIS METALLIC PLATE BEING QUITE OUT OF CONTACT WITH THE CARBON CONDUCTOR, A SENSITIVE GALVANOMETER CONNECTED BETWEEN THIS INSULATED METAL PLATE ENCLOSED IN THE VACUUM AND THE EXTERNAL POSITIVE ELECTRODE OF THE LAMP INDICATES A CURRENT OF SOME MILLIAMPERES PASSING THROUGH IT WHEN THE LAMP IS SET IN ACTION, BUT THE SAME INSTRUMENT WHEN CONNECTED BETWEEN THE NEGATIVE ELECTRODE OF THE LAMP AND THE INSULATED METAL PLATE INDICATES NO SENSIBLE CURRENT. THIS PHENOMENON IN CARBON INCANDESCENCE LAMPS WAS FIRST OBSERVED BY MR. EDISON, IN 1884, AND FURTHER EXAMINED BY MR. W. H. PREECE, IN 1885." PROCEEDINGS OF THE ROYAL SOCIETY OF LONDON, VOL. 47, PP. 118-9.

FLEMING'S 1890 PAPER FURTHER POINTED OUT THAT THE VACUOUS SPACE "POSSESSES A CURIOUS UNILATERAL CONDUCTIVITY"; THAT IS, IT PERMITS CURRENT TO "FLOW ACROSS THE VACUOUS SPACE FROM THE HOT CARBON (CATHODE) TO THE COOLER METAL PLATE (ANODE), BUT NOT IN THE REVERSE DIRECTION." ID. 122. HE NOTED THE ABILITY OF THE TUBE TO ACT AS A RECTIFIER OF ALTERNATING CURRENT, SAYING: "WHEN THE LAMP IS ACTUATED BY AN ALTERNATING CURRENT A CONTINUOUS CURRENT IS FOUND FLOWING THROUGH A GALVANOMETER, CONNECTED BETWEEN THE INSULATED PLATE AND EITHER TERMINAL OF THE LAMP. THE DIRECTION OF THE CURRENT THROUGH THE GALVANOMETER IS SUCH AS TO SHOW THAT NEGATIVE ELECTRICITY IS FLOWING FROM THE PLATE THROUGH THE GALVANOMETER TO THE LAMP TERMINAL." ID. 120.

FLEMING'S PAPER THUS NOTED, CONTRARY TO THE THEN POPULAR CONCEPTION, THAT IT IS NEGATIVE ELECTRICITY WHICH FLOWS FROM CATHODE TO ANODE, BUT HE EMPHASIZED THAT EVEN THIS HAD BEEN A PART OF GENERAL SCIENTIFIC KNOWLEDGE, AS FOLLOWS: "THE EFFECT OF HEATING THE NEGATIVE ELECTRODE IN FACILITATING DISCHARGE THROUGH VACUOUS SPACES HAS PREVIOUSLY BEEN DESCRIBED BY W. HITTORF ('ANNALEN DER PHYSIK UND CHEMIE,' VOL. 21, 1884, P. 90-139), AND IT IS ABUNDANTLY CONFIRMED BY THE ABOVE EXPERIMENTS. WE MAY SAY THAT A VACUOUS SPACE BOUNDED BY TWO ELECTRODES - ONE INCANDESCENT, AND THE OTHER COLD - POSSESSES A UNILATERAL CONDUCTIVITY FOR ELECTRIC DISCHARGE WHEN THESE ELECTRODES ARE WITHIN A DISTANCE OF THE MEAN FREE PATH OF PROJECTION OF THE MOLECULES WHICH THE IMPRESSED ELECTROMOTIVE FORCE CAN DETACH AND SEND OFF FROM THE HOT NEGATIVE ELECTRODE.

"THIS UNILATERAL CONDUCTIVITY OF VACUOUS SPACES HAVING UNEQUALLY HEATED ELECTRODES HAS BEEN EXAMINED BY MM. ELSTER AND GEITEL (SEE 'WIEDEMANN'S ANNALEN,' VOL. 38, 1889, P. 40), AND ALSO BY GOLDSTEIN ('WIED. ANN.,' VOL. 24, 1885, P. 83), WHO IN EXPERIMENTS OF VARIOUS KINDS HAVE DEMONSTRATED THAT WHEN AN ELECTRIC DISCHARGE ACROSS A VACUOUS SPACE TAKES PLACE FROM A CARBON CONDUCTOR TO ANOTHER ELECTRODE, THE DISCHARGE TAKES PLACE AT LOWER ELECTROMOTIVE FORCE WHEN THE CARBON CONDUCTOR IS THE NEGATIVE ELECTRODE AND IS RENDERED INCANDESCENT." ID. 125-6.

FLEMING'S REFERENCE IN THIS PUBLICATION TO THE UNILATERAL CONDUCTIVITY OF THE VACUOUS SPACE BETWEEN CATHODE AND ANODE, AND THE CONSEQUENT ABILITY OF THE TWO TO DERIVE A CONTINUOUS UNIDIRECTIONAL CURRENT FROM AN ALTERNATING CURRENT WAS A RECOGNITION THAT THE EDISON TUBE EMBODYING THE STRUCTURE DESCRIBED COULD BE USED AS A RECTIFIER OF ALTERNATING CURRENT. THIS KNOWLEDGE, DISCLOSED BY PUBLICATION MORE THAN TWO YEARS BEFORE FLEMING'S APPLICATION, WAS A BAR TO ANY CLAIM FOR A PATENT FOR AN INVENTION EMBODYING THE PUBLISHED DISCLOSURE. R.S. SECS. 4886, 4920; 35 U.S.C. SECS. 31, 69. WAGNER V. MECCANO LTD., 246 F. 603, 607; CF. MUNCIE GEAR CO. V. OUTBOARD CO., SUPRA, 766.

IT IS UNNECESSARY TO DECIDE WHETHER FLEMING'S USE OF THE EDISON DEVICE FOR THE PURPOSE OF RECTIFYING HIGH FREQUENCY HERTZIAN WAVES, AS DISTINGUISHED FROM LOW FREQUENCY WAVES, INVOLVED INVENTION OVER THE PRIOR ART, OR WHETHER THE COURT

BELOW RIGHTLY HELD THAT THE DEVICES USED BY THE GOVERNMENT DID NOT INFRINGE THE CLAIMS SUED UPON, FOR WE ARE OF THE OPINION THAT THE COURT WAS RIGHT IN HOLDING THAT FLEMING'S PATENT WAS RENDERED INVALID BY AN IMPROPER DISCLAIMER. IT IS PLAIN THAT FLEMING'S ORIGINAL CLAIM 1, SO FAR AS APPLICABLE TO USE WITH LOW FREQUENCY ALTERNATING CURRENTS, INVOLVED NOTHING NEW, AS FLEMING HIMSELF MUST HAVE KNOWN IN VIEW OF HIS 1890 PAPER, AND AS HE RECOGNIZED BY HIS DISCLAIMER IN 1915, MADE TWENTY-FIVE YEARS AFTER HIS PAPER WAS PUBLISHED AND TEN YEARS AFTER HIS PATENT HAD BEEN ALLOWED. ITS INVALIDITY WOULD DEFEAT THE ENTIRE PATENT UNLESS THE INVALID PORTION HAD BEEN CLAIMED "THROUGH INADVERTENCE, ACCIDENT, OR MISTAKE, AND WITHOUT ANY FRAUDULENT OR DECEPTIVE INTENTION," AND WAS ALSO DISCLAIMED WITHOUT "UNREASONABLE" NEGLECT OR DELAY. R.S. SECS. 4917, 4922; 35 U.S.C. SECS. 65, 71; ENSTEN V. SIMON ASCHER & CO., 282 U.S. 445, 452; ALTOONA THEATRES V. TRI-ERGON CORP., 294 U.S. 477, 493; MAYTAG CO. V. HURLEY CO., 307 U.S. 243.

WE NEED NOT STOP TO INQUIRE WHETHER, AS THE GOVERNMENT CONTENDS, THE SUBJECT MATTER OF THE DISCLAIMER WAS IMPROPER AS IN EFFECT ADDING A NEW ELEMENT TO THE CLAIM. SEE MILCOR STEEL CO. V. FULLER CO., 316 U.S. 143, 147-8. FOR WE THINK THAT THE COURT BELOW WAS CORRECT IN HOLDING THAT THE FLEMING PATENT WAS INVALID BECAUSE FLEMING'S CLAIM FOR "MORE THAN HE HAD INVENTED" WAS NOT INADVERTENT, AND HIS DELAY IN MAKING THE DISCLAIMER WAS "UNREASONABLE." BOTH OF THESE ARE QUESTIONS OF FACT, BUT SINCE THE COURT IN ITS OPINION PLAINLY STATES ITS CONCLUSIONS AS TO THEM, AND THOSE CONCLUSIONS ARE SUPPORTED BY SUBSTANTIAL EVIDENCE, ITS OMISSION TO MAKE FORMAL FINDINGS OF FACT IS IMMATERIAL. ACT OF MAY 22, 1939, 53 STAT. 752, 28 U.S.C. SEC. 288(B); CF. AMERICAN PROPELLER CO. V. UNITED STATES, 300 U.S. 475, 479-80; GREAT LAKES DREDGE & DOCK CO. V. HUFFMAN, 319 U.S. 293.

THE PURPOSE OF THE RULE THAT A PATENT IS INVALID IN ITS ENTIRETY IF ANY PART OF IT BE INVALID IS THE PROTECTION OF THE PUBLIC FROM THE THREAT OF AN INVALID PATENT, AND THE PURPOSE OF THE DISCLAIMER STATUTE IS TO ENABLE THE PATENTEE TO RELIEVE HIMSELF FROM THE CONSEQUENCES OF MAKING AN INVALID CLAIM IF HE IS ABLE TO SHOW BOTH THAT THE INVALID CLAIM WAS INADVERTENT AND THAT THE DISCLAIMER WAS MADE WITHOUT UNREASONABLE NEGLECT OR DELAY. ENSTEN V. SIMON ASCHER & CO., SUPRA. HERE THE PATENTEE HAS SUSTAINED NEITHER BURDEN.

FLEMING'S PAPER OF 1890 SHOWED HIS OWN RECOGNITION THAT HIS CLAIM OF USE OF HIS PATENT FOR LOW FREQUENCY CURRENTS WAS ANTICIPATED BY EDISON AND OTHERS. IT TAXES CREDULITY TO SUPPOSE, IN THE FACE OF THIS PUBLICATION, THAT FLEMING'S CLAIM FOR USE OF THE EDISON TUBE WITH LOW FREQUENCY CURRENTS WAS MADE "THROUGH INADVERTENCE, ACCIDENT OR MISTAKE," WHICH IS PREREQUISITE TO A LAWFUL DISCLAIMER. NO EXPLANATION OR EXCUSE IS FORTHCOMING FOR HIS CLAIM OF INVENTION OF A DEVICE WHICH HE HAD SO OFTEN DEMONSTRATED TO BE OLD IN THE ART, AND WHICH HE HAD SPECIFICALLY AND CONSISTENTLY ATTRIBUTED TO EDISON. NOR IS ANY EXPLANATION OFFERED FOR THE DELAY OF THE PATENTEE - THE MARCONI COMPANY - IN WAITING TEN YEARS TO DISCLAIM THE USE OF THE DEVICE WITH LOW FREQUENCY CURRENTS AND TO RESTRICT IT TO A USE WITH HIGH FREQUENCY HERTZIAN WAVES WHICH EDISON HAD PLAINLY FORESHADOWED BUT NOT CLAIMED. FOR TEN YEARS THE FLEMING PATENT WAS HELD OUT TO THE PUBLIC AS A MONOPOLY OF ALL ITS CLAIMED FEATURES. THAT WAS TOO LONG IN THE ABSENCE OF ANY EXPLANATION OR EXCUSE FOR THE DELAY, AND HENCE IN THIS CASE WAS LONG ENOUGH TO INVALIDATE THE PATENT. THE CONCLUSION OF THE COURT OF CLAIMS NOT ONLY HAS SUPPORT IN THE EVIDENCE, BUT WE CAN HARDLY SEE HOW ON THIS RECORD ANY OTHER COULD HAVE BEEN REACHED. THE MARCONI COMPANY'S CONTENTION THAT IT NOWHERE APPEARS THAT FLEMING WAS NOT THE FIRST INVENTOR OF THE USE OF THE PATENTED DEVICE TO RECTIFY HIGH FREQUENCY ALTERNATING CURRENTS IS IRRELEVANT TO THE QUESTION OF THE SUFFICIENCY OF THE DISCLAIMER. THE DISCLAIMER ITSELF IS AN

ASSERTION THAT THE CLAIMED USE OF THE INVENTION WITH LOW FREQUENCIES WAS NOT THE INVENTION OF THE PATENTEE, WHOSE RIGHTS WERE DERIVED WHOLLY FROM FLEMING. THIS IMPROPER CLAIM FOR SOMETHING NOT THE INVENTION OF THE PATENTEE RENDERED THE WHOLE PATENT INVALID UNLESS SAVED BY A TIMELY DISCLAIMER WHICH WAS NOT MADE.

THE MARCONI COMPANY ALSO ASSERTS THAT, AS IT IS SUING AS ASSIGNEE OF THE PATENTEE, IT IS UNAFFECTED BY THE PROVISIONS OF THE DISCLAIMER STATUTES, WHICH IT CONSTRUES AS RESTRICTING TO THE "PATENTEE" THE CONSEQUENCES OF UNREASONABLE DELAY IN MAKING THE DISCLAIMER AND AS EXEMPTING THE ASSIGNEE FROM THOSE CONSEQUENCES BY THE SENTENCE "BUT NO PATENTEE SHALL BE ENTITLED TO THE BENEFITS OF THIS SECTION IF HE HAS UNREASONABLY NEGLECTED OR DELAYED TO ENTER A DISCLAIMER." 35 U.S.C. 71. AS THE COURT BELOW FOUND, THE MARCONI COMPANY WAS ITSELF THE PATENTEE TO WHOM THE PATENT WAS ISSUED ON THE ASSIGNMENT OF FLEMING'S APPLICATION IN CONFORMITY TO 35 U.S.C. SEC. 44. THE RIGHT GIVEN BY SEC. 71 TO THE PATENTEE OR HIS ASSIGNEES TO SUE FOR INFRINGEMENT UPON A PROPER DISCLAIMER OBVIOUSLY DOES NOT RELIEVE THE PATENTEE FROM THE CONSEQUENCES OF HIS FAILURE TO COMPLY WITH THE STATUTE BECAUSE HE ACQUIRED HIS PATENT UNDER AN ASSIGNMENT OF THE APPLICATION. ALTOONA THEATRES V. TRI-ERGON CORP., SUPRA; MAYTAG CO. V. HURLEY CO., SUPRA; FRANCE MFG. CO. V. JEFFERSON ELECTRIC CO., 106 F.2D 605, 610. SUCH A CONTENTION IS NOT SUPPORTED BY THE WORDS OF THE STATUTE AND IF ALLOWED WOULD PERMIT THE NULLIFICATION OF THE DISCLAIMER STATUTE BY THE EXPEDIENT OF AN ASSIGNMENT OF THE APPLICATION. WE NEED NOT CONSIDER WHETHER ONE WHO HAS TAKEN AN ASSIGNMENT OF A PATENT AFTER ITS ISSUANCE WOULD HAVE ANY GREATER RIGHTS THAN HIS ASSIGNOR IN THE EVENT OF THE LATTER'S UNDUE DELAY IN FILING A DISCLAIMER. COMPARE APEX ELECTRICAL MFG. CO. V. MAYTAG CO., 122 F.2D 182, 189.

THE JUDGMENT IN NO. 373 IS VACATED AND THE CAUSE REMANDED TO THE COURT OF CLAIMS FOR FURTHER PROCEEDINGS NOT INCONSISTENT WITH THIS OPINION.

THE JUDGMENT IN NO. 369 IS AFFIRMED. SO ORDERED.

FN1 ON NOVEMBER 20, 1919, THE MARCONI COMPANY ASSIGNED TO THE RADIO CORPORATION OF AMERICA ALL OF ITS ASSETS, INCLUDING THE PATENTS HERE IN SUIT, BUT RESERVED, AND AGREED TO PROSECUTE, THE PRESENT CLAIMS AGAINST THE UNITED STATES, ON WHICH IT HAD INSTITUTED SUIT ON JULY 29, 1916.

FN2 SEE MARCONI WIRELESS TEL. CO. V. NATIONAL ELECTRIC SIGNALLING CO., 213 F. 815, 825, 829-31; ENCYCLOPEDIA BRITANNICA (14TH ED.) VOL. 14, P. 869; DUNLAP, MARCONI, THE MAN AND HIS WIRELESS; JACOT AND COLLIER, MARCONI - MASTER OF SPACE; VYVYAN, WIRELESS OVER THIRTY YEARS; FLEMING, ELECTRIC WAVE TELEGRAPHY, 426-443.

MARCONI WAS GRANTED EIGHT OTHER UNITED STATES PATENTS FOR WIRELESS APPARATUS ON APPLICATIONS FILED BETWEEN THE FILING DATES OF NOS. 586,193 AND 763,772. THEY ARE NOS. 624,516, 627,650, 647,007, 647,008, 647,009, 650,109, 650,110, 668,315.

FN3 CAPACITY IS THE PROPERTY OF AN ELECTRICAL CIRCUIT WHICH ENABLES IT TO RECEIVE AND STORE AN ELECTRICAL CHARGE WHEN A VOLTAGE IS APPLIED TO IT, AND TO RELEASE THAT CHARGE AS THE APPLIED VOLTAGE IS WITHDRAWN, THEREBY CAUSING A CURRENT TO FLOW IN THE CIRCUIT. ALTHOUGH ANY CONDUCTOR OF ELECTRICITY HAS CAPACITY TO SOME DEGREE, THAT PROPERTY IS SUBSTANTIALLY ENHANCED IN A CIRCUIT BY THE USE OF A CONDENSER, CONSISTING OF TWO OR MORE METAL PLATES SEPARATED BY A NON-CONDUCTOR, SUCH THAT WHEN A VOLTAGE IS APPLIED TO THE CIRCUIT ONE PLATE WILL BECOME POSITIVELY AND THE OTHER NEGATIVELY CHARGED.

SELF-INDUCTANCE IS THE PROPERTY OF A CIRCUIT BY WHICH, WHEN THE AMOUNT OR DIRECTION OF THE CURRENT PASSING THROUGH IT IS CHANGED, THE MAGNETIC STRESSES CREATED INDUCE A VOLTAGE OPPOSED TO THE CHANGE. ALTHOUGH ANY CONDUCTOR HAS SELF-INDUCTANCE TO SOME DEGREE, THAT PROPERTY IS MOST MARKED IN A COIL.

SEE GENERALLY ALBERT, ELECTRICAL FUNDAMENTALS OF COMMUNICATION, CHS. V, VI,

VII, AND IX; TERMAN, RADIO ENGINEERING, CHS. II AND III; MORECROFT, PRINCIPLES OF RADIO COMMUNICATION, CHS. I, II, III; LAUER AND BROWN, RADIO ENGINEERING PRINCIPLES, CHS. I AND II.

FN4 A COHERER WAS A DEVICE DISCLOSED BY BRANLY AS EARLY AS 1891. IT WAS USED BY LODGE IN EXPERIMENTS DESCRIBED IN THE LONDON ELECTRICIAN FOR JUNE 15, 1894, P. 189, AND WAS IN COMMON USE THEREAFTER AS A DETECTOR OF RADIO WAVES UNTIL REPLACED BY THE CRYSTAL AND THE CATHODEANODE TUBE. THE MOST COMMON FORM CONSISTED OF A TUBE CONTAINING METAL FILINGS WHICH, IN THEIR NORMAL STATE, WERE A NON-CONDUCTOR. WHEN PLACED IN A CIRCUIT THROUGH WHICH HIGH FREQUENCY OSCILLATIONS PASSED, THE FILINGS ALIGNED THEMSELVES IN A CONTINUOUS STREAM THROUGH WHICH THE LOW FREQUENCY ELECTRICAL CURRENT OPERATING A KEY OR OTHER SIGNALLING DEVICE COULD PASS. BY MEANS OF A DEVICE WHICH TAPPED THE SIDES OF THE TUBE, THE STREAM OF FILINGS WAS BROKEN WHEN THE HIGH-FREQUENCY OSCILLATIONS CEASED. THUS THE COHERER WAS A SENSITIVE DEVICE BY WHICH WEAK, HIGH-FREQUENCY SIGNALS COULD BE MADE TO ACTUATE A LOW-FREQUENCY CURRENT OF SUFFICIENT POWER TO OPERATE A TELEGRAPHIC KEY OR OTHER DEVICE PRODUCING A VISIBLE OR AUDIBLE SIGNAL.

FN5 OF THE CLAIMS IN SUIT IN NO. 369, CLAIMS 10 AND 20 COVER THE FOUR-CIRCUIT SYSTEM, WHILE CLAIMS 1, 3, 6, 8, 11 AND 12 COVER THE TWO TRANSMITTER CIRCUITS AND CLAIMS 2, 13, 14, 17, 18 AND 19 COVER THE TWO RECEIVER CIRCUITS. CLAIM 10 MERELY PROVIDES THAT THE FOUR CIRCUITS BE IN RESONANCE WITH EACH OTHER AND HENCE DOES NOT PRESCRIBE MEANS OF ADJUSTING THE TUNING. CLAIM 11 LIKEWISE PRESCRIBES NO MEANS OF ADJUSTMENT. THE OTHER CLAIMS PROVIDE MEANS OF ADJUSTMENT, EITHER A "VARIABLE INDUCTANCE" (CLAIMS 1, 2, 3, 8, 12, 13, 18, AND 19) OR MORE GENERALLY "MEANS" FOR ADJUSTING THE PERIOD OF THE CIRCUITS (CLAIMS 3, 6, 14 AND 17). SOME OF THE CLAIMS MERELY PROVIDE MEANS OF ADJUSTING THE TUNING OF THE ANTENNA CIRCUIT (CLAIMS 1, 2, 8, 12, AND 13) AND HENCE DO NOT REQUIRE THAT THE CLOSED CIRCUITS BE TUNED. OTHERS EITHER SPECIFICALLY PRESCRIBE THE ADJUSTABLE TUNING OF BOTH CIRCUITS AT TRANSMITTER (CLAIMS 3, 6) OR RECEIVER (CLAIMS 18 AND 19) OR BOTH (CLAIM 20) OR ELSE PRESCRIBE "MEANS FOR ADJUSTING THE TWO TRANSFORMER-CIRCUITS IN ELECTRICAL RESONANCE WITH EACH OTHER, SUBSTANTIALLY AS DESCRIBED" (CLAIMS 14 AND 17).

FN6 A DYNAMICAL THEORY OF THE ELECTROMAGNETIC FIELD (1864), 155 PHILOSOPHICAL TRANSACTIONS OF THE ROYAL SOCIETY 459; 1 SCIENTIFIC PAPERS OF JAMES CLERK MAXWELL 526.

FN7 SEE THE LONDON ELECTRICIAN FOR SEPTEMBER 21, 1888, P. 628.

EBERT, IN THE LONDON ELECTRICIAN FOR JULY 6, 1894, P. 333, LIKEWISE POINTED OUT THAT HERTZ'S RECEIVERS ARE "SO ARRANGED THAT THEY SHOW THE MAXIMUM RESONANT EFFECT WITH A GIVEN EXCITER; THEY ARE 'ELECTRICALLY TUNED.'" FN8 DE TUNZELMANN SHOWS THAT HERTZ CLEARLY UNDERSTOOD THE PRINCIPLES OF ELECTRICAL RESONANCE. SOME OF HIS EARLY EXPERIMENTS WERE DESIGNED TO DETERMINE WHETHER PRINCIPLES OF RESONANCE WERE APPLICABLE TO HIGH FREQUENCY ELECTRICAL CIRCUITS. FROM THEM HERTZ CONCLUDED THAT "AN OSCILLATORY CURRENT OF DEFINITE PERIOD WOULD, OTHER CONDITIONS BEING THE SAME, EXERT A MUCH GREATER INDUCTIVE EFFECT UPON ONE OF EQUAL PERIOD THAN UPON ONE DIFFERING EVEN SLIGHTLY FROM IT." ID. P. 626. HERTZ KNEW THAT THE FREQUENCY TO WHICH A CIRCUIT WAS RESONANT WAS A FUNCTION OF THE SQUARE ROOT OF THE PRODUCT OF THE SELF-INDUCTANCE AND CAPACITY IN THE CIRCUIT AND BY A FORMULA SIMILAR TO THAT NOW USED HE CALCULATED THE APPROXIMATE FREQUENCY OF THE OSCILLATIONS PRODUCED BY HIS TRANSMITTER. ID., SEPTEMBER 28, 1888, 664-5.

FN9 FORTNIGHTLY REVIEW, NO. 101, FEBRUARY, 1892, 173, 174-5.

FN10 MARTIN, INVENTIONS, RESEARCHES AND WRITINGS OF NIKOLA TESLA, PP. 346-8.

FN11 TESLA'S SPECIFICATIONS STATE THAT THE CURRENT SHOULD PREFERABLY BE "OF VERY CONSIDERABLE FREQUENCY." IN DESCRIBING APPARATUS USED EXPERIMENTALLY BY

HIM, THE SPECIFICATIONS STATE THAT THE OSCILLATIONS ARE GENERATED IN THE CHARGING CIRCUIT BY THE PERIODIC DISCHARGE OF A CONDENSER BY MEANS OF "A MECHANICALLY OPERATED BREAK," A MEANS WHOSE EFFECTS ARE SIMILAR TO THOSE OF THE SPARK GAP GENERALLY USED AT THIS PERIOD IN THE RADIO ART. HE FURTHER STATES THAT THE INDUCTANCE OF THE CHARGING CIRCUIT IS SO CALCULATED THAT THE "PRIMARY CIRCUIT VIBRATES GENERALLY ACCORDING TO ADJUSTMENT, FROM TWO HUNDRED AND THIRTY THOUSAND TO TWO HUNDRED AND FIFTY THOUSAND TIMES PER SECOND." THE RANGE OF RADIO FREQUENCIES IN USE IN 1917 WAS SAID BY A WITNESS FOR THE PLAINTIFF TO EXTEND FROM 30,000 TO 1,500,000 CYCLES PER SECOND. THE RANGE OF FREQUENCIES ALLOCATED FOR RADIO USE BY THE INTERNATIONAL TELECOMMUNICATION CONVENTION, PROCLAIMED JUNE 27, 1934, 49 STAT. 2391, 2459, IS FROM 10 TO 60,000 KILOCYCLES (10,000 TO 60,000,000 CYCLES) PER SECOND, AND THE SPECTRUM OF WAVES OVER WHICH THE FEDERAL COMMUNICATIONS COMMISSION CURRENTLY EXERCISES JURISDICTION EXTENDS FROM 10 TO 500,000 KILOCYCLES. CODE OF FEDERAL REGULATIONS, TITLE 47, CH. I, SEC. 2.71. THUS TESLA'S APPARATUS WAS INTENDED TO OPERATE AT RADIO FREQUENCIES.

FN12 MARCONI'S PATENT NO. 627,650, OF JUNE 27, 1899, SIMILARLY SHOWED A TWO-CIRCUIT RECEIVING SYSTEM, IN WHICH THE COHERER WAS PLACED IN A CLOSED CIRCUIT WHICH WAS INDUCTIVELY COUPLED WITH A TUNED ANTENNA CIRCUIT. THE COURT OF CLAIMS FOUND, HOWEVER, THAT THIS PATENT DID NOT CLEARLY DISCLOSE THE DESIRABILITY OF TUNING BOTH CIRCUITS.

FN13 THAT THE CLOSED CIRCUIT WAS INTENDED TO BE A "PERSISTENT OSCILLATOR" IS ALSO BROUGHT OUT BY STONE'S EMPHASIS ON "LOOSE COUPLING." STONE'S APPLICATION EXPLAINED IN DETAIL THE FACT THAT WHEN TWO CIRCUITS ARE INDUCTIVELY COUPLED TOGETHER THERE NORMALLY RESULT "TWO DEGREES OF FREEDOM," THAT IS TO SAY, THE SUPERPOSITION OF TWO FREQUENCIES IN THE SAME CIRCUIT BECAUSE OF THE EFFECT ON EACH OF THE MAGNETIC LINES OF FORCE SET UP BY THE OTHER. HE DISCUSSED IN DETAIL METHODS OF ELIMINATING THIS SUPERPOSITION, WHICH INTERFERED WITH ACCURATE SELECTIVITY OF TUNING, BY SO CONSTRUCTING HIS CIRCUITS AS TO BE "LOOSELY COUPLED." THIS HE ACHIEVED BY INCLUDING IN THE CLOSED CIRCUITS A LARGE INDUCTANCE COIL, WHICH HAD THE EFFECT OF "SWAMPING" THE UNDESIRABLE EFFECT OF THE LINES OF FORCE SET UP IN THE PRIMARY OF THE TRANSFORMER BY THE CURRENT INDUCED IN THE SECONDARY. SINCE THE TURNS OF WIRE IN THE PRIMARY OF THE TRANSFORMER CONSTITUTED A RELATIVELY SMALL PART OF THE TOTAL INDUCTANCE IN THE CLOSED CIRCUIT THE EFFECT OF THOSE TURNS ON THE FREQUENCY OF THE CIRCUIT WAS MINIMIZED.

BUT THE TESTIMONY AT THE TRIAL WAS IN SUBSTANTIAL AGREEMENT THAT THE LOOSER THE COUPLING THE SLOWER IS THE TRANSFER OF ENERGY FROM THE CLOSED CHARGING CIRCUIT TO THE OPEN ANTENNA CIRCUIT. HENCE THE USE OF LOOSE COUPLING PRESUPPOSES A CHARGING CIRCUIT THAT WILL STORE ITS ENERGY FOR A CONSIDERABLE PERIOD, I.E., THAT WILL MAINTAIN PERSISTENT OSCILLATIONS.

FN14 STONE'S RECOGNITION OF THE SIMILARITY BETWEEN HIS ANTENNA CIRCUIT AND HIS SCREENING CIRCUIT IS FURTHER SHOWN BY HIS DIRECTION THAT THE COUPLING BETWEEN THE SCREENING CIRCUIT AND THE CHARGING CIRCUIT, LIKE THAT BETWEEN THE ANTENNA AND CHARGING CIRCUITS WHERE NO SCREENING CIRCUIT IS USED, BE LOOSE. SEE NOTE 12, SUPRA.

FN15 STONE'S LANGUAGE HERE MAKES IT PLAIN THAT THROUGHOUT HIS ALLUSIONS TO A FREQUENCY DEVELOPED IN ONE CIRCUIT AS BEING "IMPRESSED" OR "FORCED" ON ANOTHER CIRCUIT WHEN THE TWO CIRCUITS ARE COUPLED THROUGH A TRANSFORMER, ARE USED FIGURATIVELY OR METAPHORICALLY ONLY AS SYNONYMOUS WITH "INDUCED." SCIENTIFICALLY THE OSCILLATIONS IN THE CHARGING CIRCUIT ARE NOT IMPRESSED OR FORCED ON THE OTHER. THE STRESS IN THE MAGNETIC FIELD OF THE FIRST CIRCUIT SETS UP OR INDUCES CORRESPONDING STRESSES IN THE MAGNETIC FIELD OF THE OTHER CIRCUIT. THE RESULTING FREQUENCY IN THE SECOND CIRCUIT IS AFFECTED BOTH BY THE FREQUENCY OF

THE OSCILLATIONS IN THE CHARGING CIRCUIT AND THE INDUCTANCE AND CAPACITY IN THE SECOND CIRCUIT. THE RESULT MAY BE THE SUPERPOSITION OF TWO FREQUENCIES IN THE SECOND CIRCUIT. THIS MAY BE AVOIDED AND A SINGLE FREQUENCY DEVELOPED, AS STONE SHOWED, BY TUNING THE SECOND CIRCUIT SO AS TO BE RESONANT TO THE FREQUENCIES CREATED IN THE FIRST.

FN16 AT THE INSISTENCE OF THE PATENT OFFICE STONE DIVIDED HIS ORIGINAL APPLICATION, AND WAS GRANTED TWO PATENTS, NO. 714,756 FOR A METHOD AND NO. 714,831 FOR APPARATUS. THE FORMER IS THE ONE PARTICULARLY RELIED ON HERE.

FN17 THIS IS BORNE OUT BY THE SUBSEQUENT LETTER FROM STONE TO THE COMMISSIONER OF PATENTS DATED JUNE 7, 1902. STONE THERE REFERS TO A LETTER BY THE PATENT OFFICE SAYING THAT THE STATEMENT THAT A SIMPLE HARMONIC WAVE DEVELOPED IN THE CLOSED CIRCUIT "CAN BE TRANSFERRED TO THE ELEVATED CONDUCTOR AND FROM THE LATTER TO THE ETHER WITHOUT CHANGE OF FORM" IS "AN ARGUMENT THE SOUNDNESS OF WHICH THE OFFICE HAS NO MEANS OF TESTING." STONE REPLIED WITH ARGUMENTS TO SHOW THAT THE VIBRATIONS RADIATED BY THE ANTENNA CIRCUIT WOULD BE SUFFICIENTLY PURE FOR PRACTICAL PURPOSES EITHER IF THE ANTENNA CIRCUIT WERE APERIODIC, OR IF IT HAD A FUNDAMENTAL WHICH WAS OF THE SAME FREQUENCY AS THAT OF THE FORCED VIBRATIONS IMPRESSED UPON IT, ALTHOUGH THEY WOULD NOT BE PURE IF THE ANTENNA CIRCUIT HAD A MARKED NATURAL PERIODICITY AND WAS UNTUNED. THIS LETTER, WHILE SOMEWHAT LATER IN DATE THAN THE AMENDMENTS, REINFORCES THE CONCLUSION THAT THE PURPOSE OF THOSE AMENDMENTS WAS TO EXPLAIN MORE FULLY THE DETAILS OF THEORY AND PRACTICE NECESSARY TO THE SUCCESS OF THE IDEA UNDERLYING STONE'S ORIGINAL INVENTION. FN18 IT IS NOT WITHOUT SIGNIFICANCE THAT MARCONI'S APPLICATION WAS AT ONE TIME REJECTED BY THE PATENT OFFICE BECAUSE ANTICIPATED BY STONE, AND WAS ULTIMATELY ALLOWED, ON RENEWAL OF HIS APPLICATION, ON THE SOLE GROUND THAT MARCONI SHOWED THE USE OF A VARIABLE INDUCTANCE AS A MEANS OF TUNING THE ANTENNA CIRCUITS, WHEREAS STONE, IN THE OPINION OF THE EXAMINER, TUNED HIS ANTENNA CIRCUITS BY ADJUSTING THE LENGTH OF THE AERIAL CONDUCTOR. ALL OF MARCONI'S CLAIMS WHICH INCLUDED THAT ELEMENT WERE ALLOWED, AND THE EXAMINER STATED THAT THE REMAINING CLAIMS WOULD BE ALLOWED IF AMENDED TO INCLUDE A VARIABLE INDUCTANCE. APPARENTLY THROUGH OVERSIGHT, CLAIMS 10 AND 11, WHICH FAILED TO INCLUDE THAT ELEMENT, WERE INCLUDED IN THE PATENT AS GRANTED. IN ALLOWING THESE CLAIMS THE EXAMINER MADE NO REFERENCE TO LODGE'S PRIOR DISCLOSURE OF A VARIABLE INDUCTANCE IN THE ANTENNA CIRCUIT.

FN19 SEE FOOTNOTE 13, SUPRA.

FN20 EVEN IF THE LACK OF INVENTION IN MARCONI'S IMPROVEMENT OVER STONE - MAKING ADJUSTABLE THE TUNING OF THE ANTENNA CIRCUITS WHICH STONE HAD SAID SHOULD BE TUNED - COULD BE SAID TO BE IN SUFFICIENT DOUBT SO THAT COMMERCIAL SUCCESS COULD AID IN RESOLVING THE DOUBT, THROPP'S SONS CO. V. SEIBERLING, 264 U.S. 320, 330; DEFOREST RADIO CO. V. GENERAL ELECTRIC CO., 283 U.S. 664, 685; ALTOONA THEATRES V. TRI ERGON CORP., 294 U.S. 477, 488, IT HAS NOT BEEN ESTABLISHED THAT THE ALLEGED IMPROVEMENT CONTRIBUTED IN ANY MATERIAL DEGREE TO THAT SUCCESS. COMPARE ALTOONA THEATRES V. TRI-ERGON CORP., SUPRA. MARCONI'S SPECIFICATIONS DISCLOSE A LARGE NUMBER OF DETAILS OF CONSTRUCTION, NONE OF WHICH IS CLAIMED AS INVENTION IN THIS PATENT, IN WHICH HIS APPARATUS DIFFERED FROM, AND MAY HAVE BEEN GREATLY SUPERIOR TO, STONE'S. MANY OF THESE FORMED THE SUBJECT OF PRIOR PATENTS. AFTER HIS APPLICATION FOR HIS PATENT, AS WELL AS BEFORE, MARCONI MADE OR ADOPTED A GREAT NUMBER OF IMPROVEMENTS IN HIS SYSTEM OF WIRELESS TELEGRAPHY. TWO OF HIS ENGINEERS HAVE WRITTEN THAT A MAJOR FACTOR IN HIS SUCCESSFUL TRANSMISSION ACROSS THE ATLANTIC IN DECEMBER, 1901, WAS THE USE OF MUCH GREATER POWER AND HIGHER ANTENNAE THAN HAD PREVIOUSLY BEEN ATTEMPTED, AN IMPROVEMENT IN NO WAY

SUGGESTED BY THE PATENT HERE IN SUIT. FLEMING, ELECTRIC WAVE TELEGRAPHY, 449-53; VYVYAN, WIRELESS OVER THIRTY YEARS, 22-33. INDEED BOTH ARE AGREED THAT IN THE ACTUAL TRANSMISSION ACROSS THE ATLANTIC TUNING PLAYED NO PART; THE RECEIVER ANTENNA CONSISTED OF A WIRE SUSPENDED BY A KITE WHICH ROSE AND FELL WITH THE WIND, VARYING THE CAPACITY SO MUCH AS TO MAKE TUNING IMPOSSIBLE. IBID.

BY 1913, WHEN HE TESTIFIED IN THE NATIONAL ELECTRIC SIGNALLING CO. CASE, THAT "DUE TO THE UTILIZATION OF THE INVENTION" OF THIS PATENT HE HAD SUCCESSFULLY TRANSMITTED MESSAGES 6,600 MILES, HE HAD, AFTER ALMOST CONTINUOUS EXPERIMENTATION, FURTHER INCREASED THE POWER USED, DEVELOPED NEW APPARATUS CAPABLE OF USE WITH HEAVY POWER, ENLARGED HIS ANTENNAE AND ADOPTED THE USE OF HORIZONTAL, "DIRECTIONAL" ANTENNAE, AND MADE USE OF IMPROVED TYPES OF SPARK GAPS AND DETECTING APPARATUS, INCLUDING THE FLEMING CATHODE-ANODE TUBE, THE CRYSTAL DETECTOR, AND SOUND RECORDING OF THE SIGNALS - TO MENTION BUT A FEW OF THE IMPROVEMENTS MADE. HE HAD ALSO DISCOVERED THAT MUCH GREATER DISTANCES COULD BE ATTAINED AT NIGHT. SEE VYVYAN, SUPRA, 34-47, 55-60. THE SUCCESS ATTAINED BY THE APPARATUS DEVELOPED BY MARCONI AND HIS FELLOW ENGINEERS BY CONTINUOUS EXPERIMENTATION OVER A PERIOD OF YEARS - HOWEVER RELEVANT IT MIGHT BE IN RESOLVING DOUBTS WHETHER THE BASIC FOUR-CIRCUIT, TUNED SYSTEM DISCLOSED BY MARCONI, AND BEFORE HIM BY STONE, INVOLVED INVENTION - CANNOT, WITHOUT FURTHER PROOF, BE ATTRIBUTED IN SIGNIFICANT DEGREE TO ANY PARTICULAR ONE OF THE MANY IMPROVEMENTS MADE BY MARCONI OVER STONE DURING A PERIOD OF YEARS. THE FACT THAT MARCONI'S APPARATUS AS A WHOLE WAS SUCCESSFUL DOES NOT ENTITLE HIM TO RECEIVE A PATENT FOR EVERY FEATURE OF ITS STRUCTURE.

FN21 A PRELIMINARY INJUNCTION RESTRAINING INFRINGEMENT WAS ENTERED IN MARCONI WIRELESS TEL. CO. V. DEFOREST CO., 225 F. 65, AFFIRMED, 225 F. 373, BOTH COURTS, WITHOUT INDEPENDENT DISCUSSION OF THE VALIDITY OF THE PATENT, DETERMINING THAT THE DECISION IN THE NATIONAL SIGNALLING CO. CASE JUSTIFIED THE GRANT OF PRELIMINARY RELIEF. A PRELIMINARY INJUNCTION WAS ALSO GRANTED IN MARCONI WIRELESS TEL. CO. V. ATLANTIC COMMUNICATIONS CO., AN ACTION BROUGHT IN THE EASTERN DISTRICT OF NEW YORK.

STONE'S LETTERS WERE INTRODUCED IN EVIDENCE IN THE ATLANTIC COMMUNICATIONS COMPANY CASE AND THE KILBOURNE & CLARK CASE. HIS DEPOSITION IN THE LATTER CASE, TAKEN FEBRUARY 28 AND 29, 1916, WAS INCORPORATED IN THE RECORD IN THIS CASE. HE THERE TESTIFIED THAT HE HAD REFRAINED FROM PRODUCING PROOFS OF THE PRIORITY OF HIS INVENTION WHEN CALLED UPON TO TESTIFY IN PRIOR LITIGATION IN 1911 AND 1914 BECAUSE HE WISHED THE PRIORITY OF HIS INVENTION TO BE ESTABLISHED BY THE OWNERS OF THE PATENT - THE STONE TELEGRAPH CO. AND ITS BONDHOLDERS IN ORDER TO BE SURE THAT A BONA FIDE DEFENSE WOULD BE MADE. HE SAID THAT BY MAY 1915, WHEN HE TESTIFIED IN THE ATLANTIC COMMUNICATIONS CO. CASE, HE HAD CONCLUDED THAT THE OWNERS OF THE PATENT WERE NOT IN A FINANCIAL POSITION TO LITIGATE, AND THAT THE ATLANTIC CO. "WOULD MAKE A BONA FIDE STONE DEFENSE." FN22 SEE NOTE 13, SUPRA. MOST OF THE CURRENT IN THE ANTENNA CIRCUIT IS SAID TO PASS THROUGH THE CONDENSER SHUNT AND NOT THROUGH THE TRANSFORMER COIL, THUS MINIMIZING THE EFFECT UPON THE FREQUENCY OF VIBRATIONS IN THE ANTENNA CIRCUIT OF THE MAGNETIC STRESSES SET UP IN THE PRIMARY OF THE TRANSFORMER BY THE CURRENT INDUCED IN THE SECONDARY.

MR. JUSTICE MURPHY TOOK NO PART IN THE CONSIDERATION OR DECISION OF THIS CASE.

MR. JUSTICE FRANKFURTER, DISSENTING IN PART: I REGRET TO FIND MYSELF UNABLE TO AGREE TO THE COURT'S CONCLUSION REGARDING THE INVALIDITY OF THE BROAD CLAIMS OF MARCONI'S PATENT. SINCE BROAD CONSIDERATIONS CONTROL THE SIGNIFICANCE AND ASSESSMENT OF THE DETAILS ON WHICH JUDGMENT IN THE CIRCUMSTANCES OF A CASE LIKE THIS IS BASED, I SHALL INDICATE THE GENERAL DIRECTION OF MY VIEWS.

IT IS AN OLD OBSERVATION THAT THE TRAINING OF ANGLO-AMERICAN JUDGES ILL FITS THEM TO DISCHARGE THE DUTIES CAST UPON THEM BY PATENT LEGISLATION. FN1 THE SCIENTIFIC ATTAINMENTS OF A LORD MOULTON ARE PERHAPS UNIQUE IN THE ANNALS OF THE ENGLISH-SPEAKING JUDICIARY. HOWEVER, SO LONG AS THE CONGRESS, FOR THE PURPOSES OF PATENTABILITY, MAKES THE DETERMINATION OF ORIGINALITY A JUDICIAL FUNCTION, JUDGES MUST OVERCOME THEIR SCIENTIFIC INCOMPETENCE AS BEST THEY CAN. BUT CONSCIOUSNESS OF THEIR LIMITATIONS SHOULD MAKE THEM VIGILANT AGAINST IMPORTING THEIR OWN NOTIONS OF THE NATURE OF THE CREATIVE PROCESS INTO CONGRESSIONAL LEGISLATION, WHEREBY CONGRESS "TO PROMOTE THE PROGRESS OF SCIENCE AND USEFUL ARTS" HAS SECURED "FOR LIMITED TIMES TO ... INVENTORS THE EXCLUSIVE RIGHT TO THEIR ... DISCOVERIES." ABOVE ALL, JUDGES MUST AVOID THE SUBTLE TEMPTATION OF TAKING SCIENTIFIC PHENOMENA OUT OF THEIR CONTEMPORANEOUS SETTING AND READING THEM WITH A RETROSPECTIVE EYE.

THE DISCOVERIES OF SCIENCE ARE THE DISCOVERIES OF THE LAWS OF NATURE, AND LIKE NATURE DO NOT GO BY LEAPS. EVEN NEWTON AND EINSTEIN, HARVEY AND DARWIN, BUILT ON THE PAST AND ON THEIR PREDECESSORS. SELDOM INDEED HAS A GREAT DISCOVERER OR INVENTOR WANDERED LONELY AS A CLOUD. GREAT INVENTIONS HAVE ALWAYS BEEN PARTS OF AN EVOLUTION, THE CULMINATION AT A PARTICULAR MOMENT OF AN ANTECEDENT PROCESS. SO TRUE IS THIS THAT THE HISTORY OF THOUGHT RECORDS STRIKING COINCIDENTAL DISCOVERIES - SHOWING THAT THE NEW INSIGHT FIRST DECLARED TO THE WORLD BY A PARTICULAR INDIVIDUAL WAS "IN THE AIR" AND RIPE FOR DISCOVERY AND DISCLOSURE.

THE REAL QUESTION IS HOW SIGNIFICANT A JUMP IS THE NEW DISCLOSURE FROM THE OLD KNOWLEDGE. RECONSTRUCTION BY HINDSIGHT, MAKING OBVIOUS SOMETHING THAT WAS NOT AT ALL OBVIOUS TO SUPERIOR MINDS UNTIL SOMEONE POINTED IT OUT, - THIS IS TOO OFTEN A TEMPTING EXERCISE FOR ASTUTE MINDS. THE RESULT IS TO REMOVE THE OPPORTUNITY OF OBTAINING WHAT CONGRESS HAS SEEN FIT TO MAKE AVAILABLE.

THE INESCAPABLE FACT IS THAT MARCONI IN HIS BASIC PATENT HIT UPON SOMETHING THAT HAD ELUDED THE BEST BRAINS OF THE TIME WORKING ON THE PROBLEM OF WIRELESS COMMUNICATION - CLERK MAXWELL AND SIR OLIVER LODGE AND NIKOLA TESLA. GENIUS IS A WORD THAT OUGHT TO BE RESERVED FOR THE RAREST OF GIFTS. I AM NOT QUALIFIED TO SAY WHETHER MARCONI WAS A GENIUS. CERTAINLY THE GREAT EMINENCE OF CLERK MAXWELL AND SIR OLIVER LODGE AND NIKOLA TESLA IN THE FIELD IN WHICH MARCONI WAS WORKING IS NOT QUESTIONED. THEY WERE, I SUPPOSE, MEN OF GENIUS. THE FACT IS THAT THEY DID NOT HAVE THE "FLASH" (A CURRENT TERM IN PATENT OPINIONS HAPPILY NOT USED IN THIS DECISION) THAT BEGOT THE IDEA IN MARCONI WHICH HE GAVE TO THE WORLD THROUGH THE INVENTION EMBODYING THE IDEA. BUT IT IS NOW HELD THAT IN THE IMPORTANT ADVANCE UPON HIS BASIC PATENT MARCONI DID NOTHING THAT HAD NOT ALREADY BEEN SEEN AND DISCLOSED.

TO FIND IN 1943 THAT WHAT MARCONI DID REALLY DID NOT PROMOTE THE PROGRESS OF SCIENCE BECAUSE IT HAD BEEN ANTICIPATED IS MORE THAN A MIRAGE OF HINDSIGHT. WIRELESS IS SO UNCONSCIOUS A PART OF US, LIKE THE AUTOMOBILE TO THE MODERN CHILD, THAT IT IS ALMOST IMPOSSIBLE TO IMAGINE OURSELVES BACK INTO THE TIME WHEN MARCONI GAVE TO THE WORLD WHAT FOR US IS PART OF THE ORDER OF OUR UNIVERSE. AND YET, BECAUSE A JUDGE OF UNUSUAL CAPACITY FOR UNDERSTANDING SCIENTIFIC MATTERS IS ABLE TO DEMONSTRATE BY A PROCESS OF INTRICATE RATIOCINATION THAT ANYONE COULD HAVE DRAWN PRECISELY THE INFERENCES THAT MARCONI DREW AND THAT STONE HINTED AT ON PAPER, THE COURT FINDS THAT MARCONI'S PATENT WAS INVALID ALTHOUGH NOBODY EXCEPT MARCONI DID IN FACT DRAW THE RIGHT INFERENCES THAT WERE EMBODIED INTO A WORKABLE BOON FOR MANKIND. FOR ME IT SPEAKS VOLUMES THAT IT SHOULD HAVE TAKEN FORTY YEARS TO REVEAL THE FATAL BEARING OF STONE'S RELATION TO MARCONI'S

ACHIEVEMENT BY A RETROSPECTIVE READING OF HIS APPLICATION TO MEAN THIS RATHER THAN THAT. THIS IS FOR ME, AND I SAY IT WITH MUCH DIFFIDENCE, TOO EASY A TRANSITION FROM WHAT WAS NOT TO WHAT BECAME.

I HAVE LITTLE DOUBT, IN SO FAR AS I AM ENTITLED TO EXPRESS AN OPINION, THAT THE VAST TRANSFORMING FORCES OF TECHNOLOGY HAVE RENDERED OBSOLETE MUCH IN OUR PATENT LAW. FOR ALL I KNOW THE BASIC ASSUMPTION OF OUR PATENT LAW MAY BE FALSE, AND INVENTORS AND THEIR FINANCIAL BACKERS DO NOT NEED THE INCENTIVE OF A LIMITED MONOPOLY TO STIMULATE INVENTION. BUT WHATEVER REVAMPING OUR PATENT LAWS MAY NEED, IT IS THE BUSINESS OF CONGRESS TO DO THE REVAMPING. WE HAVE NEITHER CONSTITUTIONAL AUTHORITY NOR SCIENTIFIC COMPETENCE FOR THE TASK.

FN1 "CONSIDERING THE EXCLUSIVE RIGHT TO INVENTION AS GIVEN NOT OF NATURAL RIGHT, BUT FOR THE BENEFIT OF SOCIETY, I KNOW WELL THE DIFFICULTY OF DRAWING A LINE BETWEEN THE THINGS WHICH ARE WORTH TO THE PUBLIC THE EMBARRASSMENT OF AN EXCLUSIVE PATENT, AND THOSE WHICH ARE NOT. AS A MEMBER OF THE PATENT BOARD FOR SEVERAL YEARS, WHILE THE LAW AUTHORIZED A BOARD TO GRANT OR REFUSE PATENTS, I SAW WITH WHAT SLOW PROGRESS A SYSTEM OF GENERAL RULES COULD BE MATURED. ... INSTEAD OF REFUSING A PATENT IN THE FIRST INSTANCE, AS THE BOARD WAS AUTHORIZED TO DO, THE PATENT NOW ISSUES OF COURSE, SUBJECT TO BE DECLARED VOID ON SUCH PRINCIPLES AS SHOULD BE ESTABLISHED BY THE COURTS OF LAW. THIS BUSINESS, HOWEVER, IS BUT LITTLE ANALOGOUS TO THEIR COURSE OF READING, SINCE WE MIGHT IN VAIN TURN OVER ALL THE LUBBERLY VOLUMES OF THE LAW TO FIND A SINGLE RAY WHICH WOULD LIGHTEN THE PATH OF THE MECHANIC OR THE MATHEMATICIAN. IT IS MORE WITHIN THE INFORMATION OF A BOARD OF ACADEMICAL PROFESSORS, AND A PREVIOUS REFUSAL OF PATENT WOULD BETTER GUARD OUR CITIZENS AGAINST HARASSMENT BY LAW-SUITS. BUT ENGLAND HAD GIVEN IT TO HER JUDGES, AND THE USUAL PREDOMINANCY OF HER EXAMPLES CARRIED IT TO OURS." THOMAS JEFFERSON TO MR. ISAAC M'PHERSON, AUGUST 13, 1813, WORKS OF THOMAS JEFFERSON, WASH. ED., VOL. VI, PP. 181-82.

"I CANNOT STOP WITHOUT CALLING ATTENTION TO THE EXTRAORDINARY CONDITION OF THE LAW WHICH MAKES IT POSSIBLE FOR A MAN WITHOUT ANY KNOWLEDGE OF EVEN THE RUDIMENTS OF CHEMISTRY TO PASS UPON SUCH QUESTIONS AS THESE. THE INORDINATE EXPENSE OF TIME IS THE LEAST OF THE RESULTING EVILS, FOR ONLY A TRAINED CHEMIST IS REALLY CAPABLE OF PASSING UPON SUCH FACTS, E.G., IN THIS CASE THE CHEMICAL CHARACTER OF VON FURTH'S SOCALLED 'ZINC COMPOUND,' OR THE PRESENCE OF INACTIVE ORGANIC SUBSTANCES. ... HOW LONG WE SHALL CONTINUE TO BLUNDER ALONG WITHOUT THE AID OF UNPARTISAN AND AUTHORITATIVE SCIENTIFIC ASSISTANCE IN THE ADMINISTRATION OF JUSTICE, NO ONE KNOWS; BUT ALL FAIR PERSONS NOT CONVENTIONALIZED BY PROVINCIAL LEGAL HABITS OF MIND OUGHT, I SHOULD THINK, UNITE TO EFFECT SOME SUCH ADVANCE." JUDGE LEARNED HAND IN PARKE DAVIS & CO. V. MULFORD CO., 189 F. 95, 115(1911).

MR. JUSTICE ROBERTS JOINS IN THIS OPINION.

MR. JUSTICE RUTLEDGE, DISSENTING IN PART: UNTIL NOW LAW FN1 HAS UNITED WITH ALMOST UNIVERSAL REPUTE N2 IN ACKNOWLEDGING MARCONI AS THE FIRST TO ESTABLISH WIRELESS TELEGRAPHY ON A COMMERCIAL BASIS. BEFORE HIS INVENTION, NOW IN ISSUE, FN3 ETHER BORNE COMMUNICATION TRAVELED SOME EIGHTY MILES. HE LENGTHENED THE ARC TO 6,000. WHETHER OR NOT THIS WAS "INVENTIVE" LEGALLY, IT WAS A GREAT AND BENEFICIAL ACHIEVEMENT. FN4 TODAY, FORTY YEARS AFTER THE EVENT, THE COURT'S DECISION REDUCES IT TO AN ELECTRICAL MECHANIC'S APPLICATION OF MERE SKILL IN THE ART.

BY PRESENT KNOWLEDGE, IT WOULD BE NO MORE. SCHOOL BOYS AND MECHANICS NOW COULD PERFORM WHAT MARCONI DID IN 1900. BUT BEFORE THEN WIZARDS HAD TRIED AND FAILED. THE SEARCH WAS AT THE PINNACLE OF ELECTRICAL KNOWLEDGE. THERE, SEEKING,

AMONG OTHERS, WERE TESLA, LODGE AND STONE, OLD HANDS AND GREAT ONES. WITH THEM WAS MARCONI, STILL YOUNG AS THE COMPANY WENT FN5 OBSESSED WITH YOUTH'S ZEAL FOR THE HUNT. AT SUCH AN ALTITUDE, TO WORK AT ALL WITH SUCCESS IS TO QUALIFY FOR GENIUS, IF THAT IS IMPORTANT. AND A SHORT STEP FORWARD GIVES EVIDENCE OF INVENTIVE POWER. FOR AT THAT HEIGHT A MERELY SLIGHT ADVANCE COMES THROUGH INSIGHT ONLY A FIRST-RATE MIND CAN PRODUCE. THIS IS SO, WHETHER IT COMES BY YEARS OF HARD WORK TRACKING DOWN THE SOUGHT SECRET OR BY INTUITION FLASHED FROM SUBCONSCIOUSNESS MADE FERTILE BY LONG EXPERIENCE OR SHORTER INTENSIVE CONCENTRATION. AT THIS LEVEL AND IN THIS COMPANY MARCONI WORKED AND WON. HE WON BY THE TEST OF RESULTS. NO ONE DISPUTES THIS. HIS INVENTION HAD IMMEDIATE AND VAST SUCCESS, WHERE ALL THAT HAD BEEN DONE BEFORE, INCLUDING HIS OWN WORK, GAVE BUT NARROWLY LIMITED UTILITY. TO MAKE USEFUL IMPROVEMENT AT THIS PLANE, BY SUCH A LEAP, ITSELF SHOWS HIGH CAPACITY. AND THAT IS TRUE, ALTHOUGH IT WAS INHERENT IN THE SITUATION THAT MARCONI'S SUCCESS SHOULD COME BY ONLY A SMALL MARGIN OF DIFFERENCE IN CONCEPTION. THERE WAS NOT ROOM FOR ANY GREAT LEAP OF THOUGHT, BEYOND WHAT HE AND OTHERS HAD DONE, TO BRING TO BIRTH THE PRACTICAL AND USEFUL RESULT. THE MOST EMINENT MEN OF THE TIME WERE CONSCIOUS OF THE PROBLEM, WERE INTERESTED IN IT, HAD SOUGHT FOR YEARS THE EXACTLY RIGHT ARRANGEMENT, ALWAYS APPROACHING MORE NEARLY BUT NEVER QUITE REACHING THE STAGE OF PRACTICAL SUCCESS. THE INVENTION WAS, SO TO SPEAK, HOVERING IN THE GENERAL CLIMATE OF SCIENCE, MOMENTARILY AWAITING BIRTH. BUT JUST THE RIGHT RELEASING TOUCH HAD NOT BEEN FOUND. MARCONI ADDED IT.

WHEN TO ALTITUDE OF THE PLANE OF CONCEPTION AND RESULTS SO IMMEDIATE AND USEFUL IS ADDED WELL-NIGH UNANIMOUS CONTEMPORARY JUDGMENT, ONE WHO LONG AFTERWARD WOULD OVERTURN THE INVENTION ASSUMES A DOUBLE BURDEN. HE UNDERTAKES TO OVERCOME WHAT WOULD OFFER STRONG RESISTANCE FRESH IN ITS ORIGINAL SETTING. HE SEEKS ALSO TO OVERTHROW THE VERDICT OF TIME. LONG-RANGE RETROACTIVE DIAGNOSIS, HOWEVER COMPETENT THE PHYSICIAN, BECOMES HAZARDOUS BY PROGRESSION AS THE PASSING YEARS ADD DISTORTIONS OF THE PAST AND DESTROY ITS PERSPECTIVE. NO LIGHT TASK IS ACCEPTED THEREFORE IN UNDERTAKING TO OVERTHROW A VERDICT SETTLED SO LONG AND SO WELL, AND ESPECIALLY ONE SO FOREIGN TO THE ART OF JUDGES.

IN LAWYERS' TERMS THIS MEANS A BURDEN OF PROOF, NOT INSURMOUNTABLE, BUT INHOSPITABLE TO IMPLICATIONS AND INFERENCES WHICH IN LESS SETTLED SITUATIONS WOULD BE PERMISSIBLE TO SWING THE BALANCE OF JUDGMENT AGAINST THE CLAIMED INVENTION. THAT MARCONI RECEIVED PATENTS ELSEWHERE WHICH, ONCE ESTABLISHED, HAVE STOOD THE TEST OF TIME AS WELL AS OF CONTEMPORARY JUDGMENT, AND SECURED HIS AMERICAN PATENT ONLY AFTER YEARS WERE REQUIRED TO CONVINCE OUR OFFICE HE HAD FOUND WHAT SO MANY OTHERS SOUGHT, BUT EMPHASIZES THE WEIGHT AND CLARITY OF PROOF REQUIRED TO OVERCOME HIS CLAIM.

MARCONI RECEIVED PATENTS HERE, IN ENGLAND, AND IN FRANCE. FN6 THE AMERICAN PATENT WAS NOT ISSUED PERFUNCTORILY. IT CAME FORTH ONLY AFTER A LONG STRUGGLE HAD BROUGHT ABOUT REVERSAL OF THE PATENT OFFICE'S ORIGINAL AND LATER REJECTIONS. THE APPLICATION WAS FILED IN NOVEMBER, 1900. IN DECEMBER IT WAS REJECTED ON LODGE, FN7 AND AN EARLIER PATENT TO MARCONI. FN8 IT WAS AMENDED AND AGAIN REJECTED. FURTHER AMENDMENTS FOLLOWED AND OPERATION OF THE SYSTEM WAS EXPLAINED. AGAIN REJECTION TOOK PLACE, THIS TIME ON LODGE, THE EARLIER MARCONI, BRAUN AND OTHER PATENTS. AFTER FURTHER PROCEEDINGS, THE CLAIMS WERE REJECTED ON TESLA. FN9 A YEAR ELAPSED, BUT IN MARCH, 1904, RECONSIDERATION WAS GRANTED. SOME CLAIMS THEN WERE REJECTED ON STONE, FN10 OTHERS WERE AMENDED, STILL OTHERS WERE CANCELLED, AND FINALLY ON JUNE 28, 1904, THE PATENT ISSUED. FRENCH AND BRITISH PATENTS HAD BEEN

GRANTED IN 1900.

LITIGATION FOLLOWED AT ONCE. AMONG MARCONI'S AMERICAN VICTORIES WERE THE DECISIONS CITED ABOVE. FN11 ABROAD THE RESULTS WERE SIMILAR. FN12 UNTIL 1935, WHEN THE COURT OF CLAIMS HELD IT INVALID IN THIS CASE, 81 CT. CL. 671, NO COURT HAD FOUND MARCONI'S PATENT WANTING IN INVENTION. IT STOOD WITHOUT ADVERSE JUDICIAL DECISION FOR OVER THIRTY YEARS. IN THE FACE OF THE BURDEN THIS HISTORY CREATES, WE TURN TO THE REFERENCES, CHIEFLY TESLA, LODGE AND STONE. THE COURT RELIES PRINCIPALLY ON STONE, BUT WITHOUT DECIDING WHETHER THIS WAS INVENTIVE.

IT IS IMPORTANT, IN CONSIDERING THE REFERENCES, TO STATE THE PARTIES' CONTENTIONS CONCISELY. THE GOVERNMENT'S STATEMENT IS THAT THEY DIFFER OVER WHETHER MARCONI WAS FIRST TO CONCEIVE FOUR-CIRCUIT "TUNING" FOR TRANSMISSION OF SOUND BY HERTZIAN WAVES. IT SAYS THIS WAS TAUGHT PREVIOUSLY BY TESLA, LODGE AND STONE. PETITIONER HOWEVER SAYS NONE OF THEM TAUGHT WHAT MARCONI DID. IT CONTENDS THAT MARCONI WAS THE FIRST TO ACCOMPLISH THE KIND OF TUNING HE ACHIEVED, AND IN EFFECT URGES THIS WAS PATENTABLY DIFFERENT FROM OTHER FORMS FOUND EARLIER.

SPECIFICALLY PETITIONER URGES THAT TESLA HAD NOTHING TO DO WITH EITHER HERTZIAN WAVES OR TUNING, BUT IN FACT HIS TRANSMITTING AND RECEIVING WIRES COULD NOT BE TUNED. FN13 LODGE, IT CLAIMS, DISCLOSED A TUNED ANTENNA, FOR EITHER TRANSMITTER OR RECEIVER OR BOTH, BUT THE CLOSED CIRCUITS ASSOCIATED WITH THE ANTENNA ONES WERE NOT TUNED. FINALLY IT IS SAID STONE DOES NOT DESCRIBE TUNING THE ANTENNA, BUT DOES SHOW TUNING OF THE ASSOCIATED CLOSED CIRCUIT. AND MARCONI TUNED BOTH.

PETITIONER DOES NOT CLAIM THE GENERAL PRINCIPLES OF TUNING. IT ADMITS THEY HAD LONG BEEN FAMILIAR TO PHYSICISTS AND THAT LODGE AND OTHERS FULLY UNDERSTOOD THEM. BUT IT ASSERTS LODGE DID NOT KNOW WHAT CIRCUITS SHOULD BE TUNED, TO ACCOMPLISH WHAT MARCONI ACHIEVED, AND THAT, TO SECURE THIS, "KNOWLEDGE THAT TUNING IS POSSIBLE IS NOT ENOUGH THERE IS ALSO REQUIRED THE KNOWLEDGE OF WHETHER OR NOT TO TUNE AND HOW MUCH." LIKEWISE, PETITIONER DOES NOT DENY THAT STONE KNEW AND UTILIZED THE PRINCIPLES OF TUNING; BUT URGES, WITH RESPECT TO THE CLAIM HE APPLIED THEM TO ALL OF THE FOUR CIRCUITS, THAT THE ONLY ONES TUNED, IN HIS ORIGINAL APPLICATION, WERE THE CLOSED CIRCUITS AND THEREFORE THAT THE ANTENNA CIRCUITS WERE NOT TUNED; ALTHOUGH IT IS NOT DENIED THAT THE EFFECTS OF TUNING THE CLOSED CIRCUITS WERE REFLECTED IN THE OPEN ONES BY WHAT STONE DESCRIBES AS "PRODUCING FORCED SIMPLE HARMONIC ELECTRIC VIBRATIONS OF THE SAME PERIODICITY IN AN ELEVATED CONDUCTOR." THE STONE AMENDMENTS OF 1902, MADE MORE THAN A YEAR AFTER MARCONI'S FILING DATE, ADMITTEDLY DISCLOSE TUNING OF BOTH THE CLOSED AND THE OPEN CIRCUITS, AND WERE MADE FOR THE PURPOSE OF STATING EXPRESSLY THE LATTER EFFECT, CLAIMED TO BE IMPLICIT IN THE ORIGINAL APPLICATION. PETITIONER DENIES THIS WAS IMPLICIT AND ARGUES, IN EFFECT, THAT WHAT STONE ORIGINALLY MEANT BY "PRODUCING FORCED ... VIBRATIONS" WAS CREATING THE DESIRED EFFECTS IN THE ANTENNA BY FORCE, NOT BY TUNING; AND THEREFORE THAT THE TWO METHODS WERE PATENTABLY DIFFERENT.

IT SEEMS CLEAR THAT THE PARTIES USE THE WORD "TUNING" TO MEAN DIFFERENT THINGS AND THE AMBIGUITY, IF THERE IS ONE, MUST BE RESOLVED BEFORE THE CRUCIAL QUESTIONS CAN BE STATED WITH MEANING. IT WILL AID, IN DECIDING WHETHER THERE IS AMBIGUITY OR ONLY CONFUSION, TO CONSIDER THE TERM AND THE POSSIBLE CONCEPTIONS IT MAY CONVEY IN THE LIGHT OF THE PROBLEMS MARCONI AND STONE, AS WELL AS OTHER REFERENCES, WERE SEEKING TO SOLVE.

MARCONI HAD IN MIND FIRST A SPECIFIC DIFFICULTY, AS DID THE PRINCIPAL REFERENCES. IT AROSE FROM WHAT, TO THE TIME OF HIS INVENTION, HAD BEEN A BAFFLING PROBLEM IN

THE ART. SHORTLY AND SIMPLY, IT WAS THAT AN ELECTRICAL CIRCUIT WHICH IS A GOOD CONSERVER OF ENERGY IS A BAD RADIATOR AND, CONVERSELY, A GOOD RADIATOR IS A BAD CONSERVER OF ENERGY. EFFECTIVE USE OF HERTZIAN WAVES OVER LONG DISTANCES REQUIRED BOTH EFFECTS. TO STATE THE MATTER DIFFERENTLY, LODGE HAD EXPLAINED IN 1894 THE DIFFICULTIES OF FULLY UTILIZING THE PRINCIPLE OF SYMPATHETIC RESONANCE IN DETECTING ETHER WAVES. TO SECURE THIS, IT WAS NECESSARY, ON THE ONE HAND, TO DISCHARGE A LONG SERIES OF WAVES OF EQUAL OR APPROXIMATELY EQUAL LENGTH. SUCH A SERIES CAN BE PRODUCED ONLY BY A CIRCUIT WHICH CONSERVES ITS ENERGY WELL, WHAT MARCONI CALLS A PERSISTENT OSCILLATOR. ON THE OTHER HAND, FOR DISTANT DETECTION, THE WAVES MUST BE OF SUBSTANTIAL AMPLITUDE, AND ONLY A CIRCUIT WHICH LOSES ITS ENERGY RAPIDLY CAN TRANSMIT SUCH WAVES WITH MAXIMUM EFFICIENCY. OBVIOUSLY IN A SINGLE CIRCUIT THE TWO DESIRED EFFECTS TEND TO CANCEL EACH OTHER, AND THEREFORE TO LIMIT THE DISTANCE OF DETECTION. SIMILAR DIFFICULTY CHARACTERIZED THE RECEIVER, FOR A GOOD RADIATOR IS A GOOD ABSORBER, AND THAT VERY QUALITY DISABLES IT TO STORE UP AND HOLD THE EFFECT OF A TRAIN OF WAVES, UNTIL ENOUGH IS ACCUMULATED TO BREAK DOWN THE COHERER, AS DETECTION REQUIRES.

SINCE THE DIFFICULTY WAS INHERENT IN A SINGLE CIRCUIT, WHETHER AT ONE END OR THE OTHER, MARCONI USED TWO IN BOTH TRANSMITTER AND RECEIVER, FOUR IN ALL. IN EACH STATION HE USED ONE CIRCUIT TO OBTAIN ONE OF THE NECESSARY ADVANTAGES AND THE OTHER CIRCUIT TO SECURE THE OTHER ADVANTAGE. THE ANTENNA (OR OPEN) CIRCUITS HE MADE "GOOD RADIATORS" (OR ABSORBERS). THE CLOSED CIRCUITS HE CONSTRUCTED AS "GOOD CONSERVERS." BY COUPLING THE TWO AT EACH END LOOSELY HE SECURED FROM THEIR COMBINATION THE DUAL ADVANTAGES HE SOUGHT. AT THE TRANSMITTER, THE CLOSED CIRCUIT, BY VIRTUE OF ITS CAPACITY FOR CONSERVING ENERGY, GAVE PERSISTENT OSCILLATION, WHICH PASSED SUBSTANTIALLY UNDIMINISHED THROUGH THE COUPLING TRANSFORMER TO THE "GOOD RADIATOR" OPEN CIRCUIT AND FROM IT WAS DISCHARGED WITH LITTLE LOSS OF ENERGY INTO THE ETHER. THENCE IT WAS PICKED UP BY THE "GOOD ABSORBER" OPEN CIRCUIT AND PASSED, WITHOUT SERIOUS LOSS OF ENERGY, THROUGH THE COUPLING TRANSFORMER, INTO THE CLOSED "GOOD CONSERVING" CIRCUIT, WHERE IT ACCUMULATED TO BREAK THE COHERER AND GIVE DETECTION.

MOREOVER, AND FOR PRESENT PURPOSES THIS IS THE IMPORTANT THING, MARCONI BROUGHT THE CLOSED AND OPEN CIRCUITS INTO ALMOST COMPLETE HARMONY BY PLACING VARIABLE INDUCTANCE IN EACH. THROUGH THIS THE PERIODICITY OF THE OPEN CIRCUIT WAS ADJUSTED AUTOMATICALLY TO THAT OF THE CLOSED ONE; AND, SINCE THE CIRCUITS OF THE RECEIVING STATION WERE SIMILARLY ADJUSTABLE, THE MAXIMUM RESONANCE WAS SECURED THROUGHOUT THE SYSTEM. MARCONI THUS NOT ONLY SOLVED THE DILEMMA OF A SINGLE CIRCUIT ARRANGEMENT; HE ATTAINED THE MAXIMUM OF RESONANCE AND SELECTIVITY BY PROVIDING IN EACH CIRCUIT INDEPENDENT MEANS OF TUNING.

IN 1911 THIS SOLUTION WAS HELD INVENTIVE, AS AGAINST LODGE, MARCONI'S PRIOR PATENTS, BRAUN AND OTHER REFERENCES, IN MARCONI V. BRITISH RADIO TEL. & TEL. CO., 27 T.L. R. 274. MR. JUSTICE PARKER CAREFULLY REVIEWED THE PRIOR ART, STATED THE PROBLEM, MARCONI'S SOLUTION, AND IN DISPOSING OF BRAUN'S SPECIFICATION CONCLUDED IT "DID NOT CONTAIN EVEN THE REMOTEST SUGGESTION OF THE PROBLEM ... , MUCH LESS ANY SUGGESTION BEARING ON ITS SOLUTION. ... " AS TO LODGE, MR. JUSTICE PARKER OBSERVED, REFERRING FIRST TO MARCONI: " ... IT IS IMPORTANT TO NOTICE THAT IN THE RECEIVER THE MERE INTRODUCTION OF TWO CIRCUITS INSTEAD OF ONE WAS NO NOVELTY. A FIGURE IN LODGE'S 1897 PATENT SHOWS THE OPEN CIRCUIT OF HIS RECEIVING AERIAL LINKED THROUGH A TRANSFORMER WITH A CLOSED CIRCUIT CONTAINING THE COHERER, HIS IDEA BEING, AS HE STATES, TO LEAVE HIS RECEIVING AERIAL FREER TO VIBRATE ELECTRICALLY WITHOUT DISTURBANCE FROM ATTACHED WIRES. THIS SECONDARY CIRCUIT, AS SHOWN, IS NOT TUNED TO, NOR CAN IT BE TUNED TO, THE CIRCUIT OF THE AERIAL. THIS, IN MY OPINION,

IS EXCEEDINGLY STRONG EVIDENCE THAT MARCONI'S 1900 INVENTION WAS NOT SO OBVIOUS AS TO DEPRIVE IT OF SUBJECT MATTER. IN THE LITERATURE QUOTED THERE IS NO TRACE OF THE IDEA UNDERLYING MR. MARCONI'S INVENTION, NOR, SO FAR AS I CAN SEE, A SINGLE SUGGESTION FROM WHICH A COMPETENT ENGINEER COULD ARRIVE AT THIS IDEA." IT WAS THEREFORE CLEARLY MR. JUSTICE PARKER'S VIEW, IN HIS CLOSER PERSPECTIVE TO THE ORIGIN OF THE INVENTION AND THE REFERENCES HE CONSIDERED, THAT IN NONE OF THEM, AND PARTICULARLY NOT IN LODGE OR BRAUN, WAS THERE ANTICIPATION OF MARCONI'S SOLUTION.

HE DID NOT MEAN THAT THE REFERENCES DID NOT APPLY "THE PRINCIPLE OF RESONANCE AS BETWEEN TRANSMITTER AND RECEIVER" OR UTILIZE "THE PRINCIPLE OF SYMPATHETIC RESONANCE FOR THE PURPOSE OF DETECTION OF ETHER WAVES." FOR HE EXPRESSLY ATTRIBUTED TO LODGE, IN HIS 1894 LECTURES, EXPLANATION "WITH GREAT EXACTNESS (OF) THE VARIOUS DIFFICULTIES ATTENDING THE FULL UTILIZATION" OF THAT PRINCIPLE. AND IN REFERRING TO MARCONI'S FIRST PATENT, OF 1896, THE OPINION STATES THAT MARCONI "FOR WHAT IT WAS WORTH ... TUNED THE TWO CIRCUITS (I.E., THE SENDING AND RECEIVING ONES) TOGETHER AS HERTZ HAD DONE." FROM THESE AND OTHER STATEMENTS IN THE OPINION IT IS OBVIOUS THAT MR. JUSTICE PARKER FOUND MARCONI'S INVENTION IN SOMETHING MORE THAN MERELY THE APPLICATION OF THE "PRINCIPLE OF RESONANCE," OR "SYMPATHETIC RESONANCE," OR ITS USE TO "TUNE" TOGETHER THE TRANSMITTING AND RECEIVING CIRCUITS. FOR MARCONI IN HIS OWN PRIOR INVENTIONS, LODGE AND THE OTHER REFERENCES, IN FACT ALL WHO HAD CONSTRUCTED ANY SYSTEM USING HERTZIAN WAVES CAPABLE OF TRANSMITTING AND DETECTING SOUND, NECESSARILY HAD MADE USE, IN SOME MANNER AND TO SOME EXTENT, OF "THE PRINCIPLE OF RESONANCE" OR "SYMPATHETIC RESONANCE." THAT PRINCIPLE IS INHERENT IN THE IDEA OF WIRELESS COMMUNICATION BY HERTZIAN WAVES. SO THAT, NECESSARILY, ALL THE PRIOR CONCEPTIONS INCLUDED THE IDEA THAT COMMON PERIODICITY MUST APPEAR IN ALL OF THE CIRCUITS EMPLOYED.

NOR DID MR. JUSTICE PARKER'S OPINION FIND THE INVENTIVE FEATURE IN THE USE OF TWO CIRCUITS INSTEAD OF ONE, AT ANY RATE IN THE RECEIVER. FOR HE EXPRESSLY NOTES THIS IN LODGE. BUT HE POINTS OUT THAT LODGE ADDED THE SEPARATE CIRCUIT "TO LEAVE HIS RECEIVING AERIAL FREER TO VIBRATE ELECTRICALLY WITHOUT DISTURBANCE FROM ATTACHED WIRES." AND HE GOES ON TO NOTE THAT THIS SECONDARY (OR CLOSED) CIRCUIT NOT ONLY WAS NOT, BUT COULD NOT BE, "TUNED" TO THE AERIAL CIRCUIT. AND THIS HE FINDS "EXCEEDINGLY STRONG EVIDENCE" THAT "MARCONI'S 1900 INVENTION WAS NOT SO OBVIOUS AS TO DEPRIVE IT OF SUBJECT MATTER." LODGE HAD "TUNED" THE ANTENNA CIRCUIT, BY PLACING IN IT A VARIABLE INDUCTANCE. BUT HE DID NOT DO THIS OR ACCOMPLISH THE SAME THING BY ANY OTHER DEVICE, SUCH AS A CONDENSER, IN THE CLOSED CIRCUIT. AND THE FACT THAT SO EMINENT A SCIENTIST, THE ONE WHO IN FACT POSED THE PROBLEM AND ITS DIFFICULTIES, DID NOT SEE THE NEED FOR EXTENDING THIS "INDEPENDENT TUNING" (TO USE MARCONI'S PHRASE) TO THE CLOSED CIRCUIT, SO AS TO BRING IT THUS IN TUNE WITH THE OPEN ONE, WAS ENOUGH TO CONVINCE MR. JUSTICE PARKER, AND I THINK RIGHTLY, THAT WHAT MARCONI DID OVER LODGE WAS NOT SO OBVIOUS AS TO BE WITHOUT SUBSTANCE.

IN SHORT, MR. JUSTICE PARKER FOUND THE GIST OF MARCONI'S INVENTION, NOT IN MERE APPLICATION OF THE GENERAL PRINCIPLE OR PRINCIPLES OF RESONANCE TO A FOUR-CIRCUIT SYSTEM, OR IN THE USE OF FOUR CIRCUITS OR THE SUBSTITUTION OF TWO FOR ONE IN EACH OR EITHER STATION; BUT, AS PETITIONER NOW CONTENDS, IN RECOGNITION OF THE PRINCIPLE THAT, WHETHER IN THE TRANSMITTER OR THE RECEIVER, ATTAINMENT OF THE MAXIMUM RESONANCE REQUIRED THAT MEANS FOR TUNING THE CLOSED TO THE OPEN CIRCUIT BE INSERTED IN BOTH. THAT RECOGNIZED, THE METHOD OF ACCOMPLISHING THE ADJUSTMENT WAS OBVIOUS, AND DIFFERENT METHODS, AS BY USING VARIABLE INDUCTANCE OR A CONDENSER, WERE AVAILABLE. AS PETITIONER'S REPLY BRIEF STATES THE MATTER, "THE

MARCONI INVENTION WAS NOT THE USE OF A VARIABLE INDUCTANCE, NOR INDEED ANY OTHER SPECIFIC WAY OF TUNING AN ANTENNA - BEFORE MARCONI IT WAS KNOWN THAT ELECTRICAL CIRCUITS COULD BE TUNED OR NOT TUNED, BY INDUCTANCE COILS OR CONDENSERS. HIS BROAD INVENTION WAS THE COMBINATION OF A TUNED ANTENNA CIRCUIT AND A TUNED CLOSED CIRCUIT." AND IT IS ONLY IN THIS VIEW THAT THE ACTION OF THE PATENT OFFICE IN FINALLY AWARDING THE PATENT TO MARCONI CAN BE EXPLAINED OR SUSTAINED, FOR IT ALLOWED CLAIMS BOTH LIMITED TO AND NOT SPECIFYING VARIABLE INDUCTANCE. THAT FEATURE WAS ESSENTIAL FOR BOTH CIRCUITS IN PRINCIPLE, BUT NOT IN THE PARTICULAR METHOD BY WHICH MARCONI ACCOMPLISHED IT. AND IT WAS RECOGNITION OF THIS WHICH EVENTUALLY INDUCED ALLOWANCE OF THE CLAIMS, NOTWITHSTANDING THE PREVIOUS REJECTIONS ON LODGE, STONE AND OTHER REFERENCES, INCLUDING ALL IN ISSUE HERE.

IN THE PERSPECTIVE OF THIS DECADE, MARCONI'S ADVANCE, IN REQUIRING "INDEPENDENT TUNING," THAT IS, POSITIVE MEANS OF TUNING LOCATED IN BOTH CLOSED AND OPEN CIRCUITS, SEEMS SIMPLE AND OBVIOUS. IT WAS SIMPLE. BUT, AS IS OFTEN TRUE WITH GREAT INVENTIONS, THE SIMPLEST AND THEREFORE GENERALLY THE BEST SOLUTION IS NOT OBVIOUS AT THE TIME, THOUGH IT BECOMES SO IMMEDIATELY IT IS SEEN AND STATED. LOOKING BACK NOW AT EDISON'S LIGHT BULB ONE MIGHT THINK IT ABSURD THAT THAT HIGHLY USEFUL AND BENEFICIAL IDEA HAD NOT BEEN WORKED OUT LONG BEFORE, BY ANYONE WHO KNEW THE ELEMENTARY LAWS OF RESISTANCE IN THE FIELD OF ELECTRIC CONDUCTION. BUT IT WOULD BE SHOCKING, NOTWITHSTANDING THE PRESENTLY OBVIOUS CHARACTER OF WHAT EDISON DID, FOR ANY COURT NOW TO RULE HE MADE NO INVENTION.

THE SAME THING APPLIES TO MARCONI. THOUGH WHAT HE DID WAS SIMPLE, IT WAS BRILLIANT, AND IT BROUGHT BIG RESULTS. ADMITTEDLY THE MARGIN OF DIFFERENCE BETWEEN HIS CONCEPTION AND THOSE OF THE REFERENCES, ESPECIALLY LODGE AND STONE, WAS SMALL. IT CAME DOWN TO THIS, THAT LODGE SAW THE NEED FOR AND USED MEANS FOR PERFORMING THE FUNCTION WHICH VARIABLE INDUCTANCE ACHIEVES IN THE ANTENNA OR OPEN CIRCUIT, STONE DID THE SAME THING IN THE CLOSED CIRCUIT, BUT MARCONI FIRST DID IT IN BOTH. SLIGHT AS EACH OF THESE STEPS MAY SEEM NOW, IN DEPARTURE FROM THE OTHERS, IT IS AS TRUE AS IT WAS IN 1911, WHEN MR. JUSTICE PARKER WROTE, THAT THE VERY FACT MEN OF THE EMINENCE OF LODGE AND STONE SAW THE NECESSITY OF TAKING THE STEP FOR ONE CIRCUIT BUT NOT FOR THE OTHER IS STRONG, IF NOT CONCLUSIVE, EVIDENCE THAT TAKING IT FOR BOTH CIRCUITS WAS NOT OBVIOUS. IF THIS WAS SO CLEARLY INDICATED THAT ANYONE SKILLED IN THE ART SHOULD HAVE SEEN IT, THE UNANSWERED AND I THINK UNANSWERABLE QUESTION REMAINS, WHY DID NOT LODGE AND STONE, BOTH ASSIDUOUSLY SEARCHING FOR THE SECRET AND BOTH PREEMINENT IN THE FIELD, RECOGNIZE THE FACT AND MAKE THE APPLICATION? THE BEST EVIDENCE OF THE NOVELTY OF MARCONI'S ADVANCE LIES NOT IN ANY JUDGMENT, SCIENTIFIC OR LAY, WHICH COULD NOW BE FORMED ABOUT IT. IT IS RATHER IN THE CAREFUL, CONSIDERED AND SUBSTANTIALLY CONTEMPORANEOUS JUDGMENTS, FORMED AND RENDERED BY BOTH THE PATENT TRIBUNALS AND THE COURTS WHEN YEARS HAD NOT DISTORTED EITHER THE SCIENTIFIC OR THE LEGAL PERSPECTIVE OF THE DAY WHEN THE INVENTION WAS MADE. ALL OF THE REFERENCES NOW USED TO INVALIDATE MARCONI WERE IN ISSUE, AT ONE TIME OR ANOTHER, BEFORE THESE TRIBUNALS, THOUGH NOT ALL OF THEM WERE PRESENTED TO EACH. THEIR UNANIMOUS CONCLUSION, BACKED BY THE FACTS WHICH HAVE BEEN STATED, IS MORE PERSUASIVE THAN THE MOST COMPETENT CONTRARY OPINION FORMED NOW ABOUT THE MATTER COULD BE.

IT REMAINS TO GIVE FURTHER ATTENTION CONCERNING STONE. ADMITTEDLY HIS ORIGINAL APPLICATION DID NOT REQUIRE TUNING, IN MARCONI'S SENSE, OF THE ANTENNA CIRCUIT, THOUGH IT SPECIFIED THIS FOR THE CLOSED ONE. HE INCLUDED VARIABLE INDUCTANCE IN THE LATTER, BUT NOT IN THE FORMER. HIS DEVICE THEREFORE WAS, IN THIS RESPECT, EXACTLY THE CONVERSE OF LODGE. BUT IT IS SAID HIS OMISSION TO SPECIFY THE FUNCTION (AS DISTINGUISHED FROM THE APPARATUS WHICH PERFORMED IT) FOR THE

ANTENNA CIRCUIT WAS NOT IMPORTANT, BECAUSE THE FUNCTION WAS IMPLICIT IN THE SPECIFICATION AND THEREFORE SUPPORTED HIS LATER AMENDMENT, FILED MORE THAN A YEAR FOLLOWING MARCONI'S DATE, EXPRESSLY SPECIFYING THIS FEATURE FOR THE OPEN CIRCUIT.

SUBSTANTIALLY THE SAME ANSWER MAY BE MADE TO THIS AS MR. JUSTICE PARKER MADE TO THE CLAIM BASED ON LODGE. TUNING BOTH CIRCUITS, THAT IS, INCLUDING IN EACH INDEPENDENT MEANS FOR VARIABLE ADJUSTMENT, WAS THE VERY GIST OF MARCONI'S INVENTION. AND IT WAS WHAT MADE POSSIBLE THE HIGHLY SUCCESSFUL RESULT. IT SEEMS STRANGE THAT ONE WHO SAW NOT ONLY THE PROBLEM, BUT THE COMPLETE SOLUTION, SHOULD SPECIFY ONLY HALF WHAT WAS NECESSARY TO ACHIEVE IT, NEGLECTING TO MENTION THE OTHER AND EQUALLY IMPORTANT HALF AS WELL, PARTICULARLY WHEN, AS IS CLAIMED, THE TWO WERE SO NEARLY IDENTICAL EXCEPT FOR LOCATION. THE VERY OMISSION OF EXPLICIT STATEMENT OF SO IMPORTANT AND, IT IS CLAIMED, SO OBVIOUS A FEATURE IS EVIDENCE IT WAS NEITHER OBVIOUS NOR CONCEIVED. AND THE FORCE OF THE OMISSION IS MAGNIFIED BY THE FACT THAT ITS AUTHOR, WHEN HE FULLY RECOGNIZED ITS EFFECT, FOUND IT NECESSARY TO MAKE AMENDMENT TO INCLUDE IT, AFTER THE FEATURE WAS EXPRESSLY AND FULLY DISCLOSED BY ANOTHER. AMENDMENT UNDER SUCH CIRCUMSTANCES, PARTICULARLY WITH RESPECT TO A MATTER WHICH GOES TO THE ROOT RATHER THAN AN INCIDENT OR A DETAIL OF THE INVENTION, IS ALWAYS TO BE REGARDED CRITICALLY AND, WHEN THE FOUNDATION CLAIMED FOR IT IS IMPLICIT EXISTENCE IN THE ORIGINAL APPLICATION, AS IT MUST BE, THE CLEAREST AND MOST CONVINCING EVIDENCE SHOULD BE REQUIRED WHEN THE EFFECT IS TO GIVE PRIORITY, BY BACKWARD RELATION, OVER ANOTHER APPLICATION INTERMEDIATELY FILED.

APART FROM THE SIGNIFICANCE OF OMITTING TO EXPRESS A FEATURE SO IMPORTANT, I AM UNABLE TO FIND CONVINCING EVIDENCE THE IDEA WAS IMPLICIT IN STONE AS HE ORIGINALLY FILED. HIS DISTINCTION BETWEEN "NATURAL" AND "FORCED" OSCILLATIONS SEEMS TO ME TO PROVE, IN THE LIGHT OF HIS ORIGINAL DISCLOSURE, NOT THAT "TUNING" OF THE ANTENNA CIRCUIT AS MARCONI REQUIRED THIS WAS IMPLICIT, BUT RATHER THAT IT WAS NOT PRESENT IN THAT APPLICATION AT ALL. IT IS TRUE HE SOUGHT, AS MARCONI DID, TO MAKE THE ANTENNA CIRCUIT AT THE TRANSMITTER THE SOURCE OF WAVES OF BUT A SINGLE PERIODICITY AND THE SAME CIRCUIT AT THE RECEIVER AN ABSORBER ONLY OF THE WAVES SO TRANSMITTED. BUT THE METHODS THEY USED WERE NOT THE SAME. STONE'S METHOD WAS TO PROVIDE "WHAT ARE SUBSTANTIALLY FORCED VIBRATIONS" IN THE TRANSMITTER'S ANTENNA CIRCUIT AND, AT THE RECEIVER, TO IMPOSE "BETWEEN THE VERTICAL CONDUCTOR (THE ANTENNA) ... AND THE TRANSLATING DEVICES (IN THE CLOSED CIRCUIT)(OTHER) RESONANT CIRCUITS ATTUNED TO THE PARTICULAR FREQUENCY OF THE ELECTRO-MAGNETIC WAVES WHICH IT IS DESIRED TO HAVE OPERATE THE TRANSLATING DEVICES." IN SHORT, HE PROVIDED FOR "TUNING," AS MARCONI DID, THE TRANSMITTER'S CLOSED CIRCUIT, THE RECEIVER'S CLOSED CIRCUIT AND THE INTERMEDIATE CIRCUITS WHICH HE INTERPOSED IN THE RECEIVER BETWEEN THE OPEN OR ANTENNA ONE AND THE CLOSED ONE. BUT NOWHERE DID HE PROVIDE FOR OR SUGGEST "TUNING," AS MARCONI DID AND IN HIS MEANING, THE ANTENNA CIRCUIT OF THE TRANSMITTER OR THE ANTENNA CIRCUIT OF THE RECEIVER. FOR RESONANCE IN THE FORMER HE DEPENDED UPON THE INTRODUCTION, FROM THE CLOSED CIRCUIT, OF "SUBSTANTIALLY FORCED ELECTRIC VIBRATIONS" AND FOR SELECTIVITY IN THE LATTER HE USED THE INTERMEDIATE TUNED CIRCUITS. STONE AND MARCONI USED THE SAME MEANS FOR CREATING PERSISTENT OSCILLATION, NAMELY, THE USE OF THE SEPARATE CLOSED CIRCUIT; AND IN THIS BOTH ALSO DEVELOPED SINGLE PERIODICITY TO THE EXTENT THE VARIABLE INDUCTANCE INCLUDED THERE AND THERE ONLY COULD DO SO. BUT WHILE BOTH CREATED PERSISTENT OSCILLATION IN THE SAME WAY, MARCONI WENT FARTHER THAN STONE WITH SINGLE PERIODICITY AND SECURED ENHANCEMENT OF THIS BY PLACING MEANS FOR TUNING IN THE ANTENNA CIRCUIT, WHICH ADMITTEDLY STONE NOWHERE EXPRESSLY

REQUIRED IN HIS ORIGINAL APPLICATION. AND, SINCE THIS IS THE GIST OF THE INVENTION IN ISSUE AND OF THE DIFFERENCE BETWEEN THE TWO, IT WILL NOT DO TO DISMISS THIS OMISSION MERELY WITH THE STATEMENT THAT THERE IS NOTHING TO SUGGEST THAT STONE "DID NOT DESIRE TO HAVE THOSE CIRCUITS TUNED." NOR IN MY OPINION DO THE PASSAGES IN THE SPECIFICATIONS RELIED UPON AS "SUGGESTING" THE "INDEPENDENT" TUNING OF THE ANTENNA CIRCUITS BEAR OUT THIS INFERENCE.

WHEN STONE STATES THAT "THE VERTICAL CONDUCTOR AT THE TRANSMITTER STATION IS MADE THE SOURCE OF ... WAVES OF BUT A SINGLE PERIODICITY," I FIND NOTHING TO SUGGEST THAT THIS IS ACCOMPLISHED BY SPECIALLY TUNING THAT CIRCUIT, OR, IN FACT, ANYTHING MORE THAN THAT THIS CIRCUIT IS A GOOD CONDUCTOR SENDING OUT THE SINGLE PERIOD WAVES FORCED INTO IT FROM THE CLOSED CIRCUIT. THE SAME IS TRUE OF THE FURTHER STATEMENT THAT "THE TRANSLATING APPARATUS AT THE RECEIVING STATION IS CAUSED TO BE SELECTIVELY RESPONSIVE TO WAVES OF BUT A SINGLE PERIODICITY" (WHICH TUNING THE INTERMEDIATE AND/OR CLOSED CIRCUITS THERE ACCOMPLISHES), SO THAT "THE TRANSMITTING APPARATUS CORRESPONDS TO A TUNING FORK SENDING BUT A SINGLE MUSICAL TONE, AND THE RECEIVING APPARATUS CORRESPONDS TO AN ACOUSTIC RESONATOR CAPABLE OF ABSORBING THE ENERGY OF THAT SINGLE SIMPLE MUSICAL TONE ONLY." THIS MEANS NOTHING MORE THAN THAT THE TRANSMITTER, WHICH INCLUDES THE ANTENNA, AND THE RECEIVER, WHICH ALSO INCLUDES THE ANTENNA, SEND OUT AND RECEIVE RESPECTIVELY A SINGLE PERIOD WAVE. IT DOES NOT MEAN THAT THE ANTENNA, IN EITHER STATION, WAS TUNED, IN MARCONI'S SENSE, NOR DOES IT SUGGEST THIS.

THE SAME IS TRUE OF THE OTHER PASSAGES RELIED UPON BY THE COURT FOR SUGGESTION. NO WORD OR HINT CAN BE FOUND IN THEM THAT STONE INTENDED OR CONTEMPLATED INDEPENDENTLY TUNING THE ANTENNA. THEY MERELY SUGGESTED, ON THE ONE HAND, THAT WHEN "THE APPARATUS" AT THE RECEIVING STATION IS PROPERLY TUNED TO A PARTICULAR TRANSMITTER, IT WILL RECEIVE SELECTIVELY MESSAGES FROM THE LATTER AND, FURTHER, THAT THE OPERATOR MAY AT WILL ADJUST "THE APPARATUS AT HIS COMMAND" SO AS TO COMMUNICATE WITH ANY ONE OF SEVERAL SENDING STATIONS; ON THE OTHER HAND, THAT "ANY SUITABLE DEVICE" MAY BE USED AT THE TRANSMITTER "TO DEVELOP THE SIMPLE HARMONIC FORCE IMPRESSED UPON" THE ANTENNA. "THE APPARATUS," AS USED IN THE STATEMENTS CONCERNING THE ADJUSTMENTS AT THE RECEIVING STATION, CLEARLY MEANS "THE APPARATUS AT HIS COMMAND," THAT IS, THE WHOLE OF THAT STATION'S EQUIPMENT, WHICH CONTAINED IN THE INTERMEDIATE AND CLOSED CIRCUITS, BUT NOT IN THE OPEN ONE, THE MEANS FOR MAKING THE ADJUSTMENTS DESCRIBED. THERE IS NOTHING WHATEVER TO SUGGEST INCLUDING A TUNING DEVICE ALSO IN THE OPEN CIRCUIT. THE STATEMENT CONCERNING THE USE OF "ANY SUITABLE DEVICE" TO "DEVELOP THE SIMPLE HARMONIC FORCE IMPRESSED UPON THE VERTICAL WIRE" MIGHT BE TAKEN, IN OTHER CONTEXT, POSSIBLY TO SUGGEST MAGNIFYING THE IMPRESSED FORCE BY INSERTING A DEVICE FOR THAT PURPOSE IN THE OPEN CIRCUIT AND THEREFORE TO COME MORE CLOSELY THAN THE OTHER PASSAGES TO SUGGESTING MARCONI'S IDEA. BUT SUCH A CONSTRUCTION WOULD BE WHOLLY STRAINED IN THE ABSENCE OF ANY OTHER REFERENCE OR SUGGESTION IN THE LONG APPLICATION TO SUCH A PURPOSE. STANDING WHOLLY ALONE AS IT DOES, IT WOULD BE GOING FAR TO BASE ANTICIPATION OF MARCONI'S IDEA UPON THIS LANGUAGE ONLY. THE MORE REASONABLE AND, IN VIEW OF THE TOTAL ABSENCE OF SUGGESTION ELSEWHERE, THE ONLY TENABLE VIEW IS THAT THE LANGUAGE WAS INTENDED TO SAY, NOT THAT STONE CONTEMPLATED INCLUDING ANY DEVICE FOR TUNING IN THE OPEN CIRCUIT, BUT THAT HE LEFT TO THE MECHANIC OR BUILDER THE CHOICE OF THE VARIOUS DEVICES WHICH MIGHT BE USED, ACCORDING TO PREFERENCE, TO CREATE OR "DEVELOP," IN THE CLOSED CIRCUIT, THE FORCE TO BE IMPRESSED UPON THE ANTENNA.

FINALLY, STONE WAS NO NOVICE. HE TOO WAS "A VERY EXPERT PERSON AND ONE OF THE BEST MEN IN THE ART." NATIONAL ELECTRIC SIGNALLING CO. V. TELEFUNKEN WIRELESS TEL.

CO., 209 F. 856, 864(D.C.). HE KNEW THE DIFFERENCE BETWEEN TUNED AND UNTUNED CIRCUITS, HOW TO DESCRIBE THEM, AND HOW TO APPLY THEM WHEN HE WANTED TO DO SO. HE USED THIS KNOWLEDGE WHEN HE SPECIFIED INCLUDING MEANS FOR TUNING IN HIS CLOSED CIRCUIT. HE DID NOT USE IT TO SPECIFY SIMILARLY TUNING THE OPEN ONE. THE OMISSION, IN SUCH CIRCUMSTANCES, COULD HARDLY HAVE BEEN INTENTIONAL. IN MY OPINION HE DELIBERATELY SELECTED AN APERIODIC AERIAL, ONE TO WHICH THE MANY RECEIVING CIRCUITS HIS APPLICATION CONTEMPLATED COULD BE ADJUSTED AND ONE WHICH WOULD CARRY TO THEM, FROM HIS TRANSMITTER'S TUNED PERIODICITY AND BY ITS FORCE ALONE, WHAT IT SENT FORWARD. IN SHORT, STONE DELIBERATELY SELECTED AN UNTUNED ANTENNA, A TUNED CLOSED CIRCUIT, AND CONTROLLED THE PERIODICITY OF BOTH, NOT BY INDEPENDENT MEANS IN EACH MAKING THEM MUTUALLY AND RECIPROCALLY ADJUSTABLE, BUT BY IMPRESSING UPON THE UNTUNED ANTENNA THE FORCED PERIODICITY OF THE CLOSED CIRCUIT. IT MAY BE THAT BY HIS METHOD HE ATTAINED RESULTS COMPARABLE, OR NEARLY SO, TO THOSE MARCONI ACHIEVED. THE RECORD DOES NOT SHOW THAT HE DID SO PRIOR TO HIS AMENDMENT. IF HE DID, THAT ONLY GOES TO SHOW HE ACCOMPLISHED IN CONSEQUENCE WHAT MARCONI DID, BUT BY A DIFFERENT METHOD. THAT BOTH HAD THE SAME "BROAD PURPOSE" OF PROVIDING A HIGH DEGREE OF TUNING AT BOTH STATIONS, AND THAT BOTH MAY HAVE ACCOMPLISHED THIS OBJECT SUBSTANTIALLY, DOES NOT SHOW THAT THEY DID SO IN THE SAME WAY OR THAT STONE, BY HIS DIFFERENT METHOD, ANTICIPATED MARCONI.

IN MY OPINION THEREFORE STONE'S AMENDMENT WAS NOT SUPPORTED BY ANYTHING IN HIS ORIGINAL APPLICATION AND SHOULD NOT HAVE BEEN ALLOWED. AS PETITIONER SAYS, IT ADDED THE NEW FEATURE OF TUNING THE ANTENNA AND IN THAT RESPECT RESEMBLED THE AMENDMENT OF A FESSENDEN APPLICATION "TO INCLUDE THE TUNING OF THE CLOSED CIRCUIT." NATIONAL ELECTRIC SIGNALLING CO. V. TELEFUNKEN WIRELESS TEL. CO., SUPRA. THE AMENDMENT HERE SHOULD RECEIVE THE SAME FATE AS BEFELL THE ONE THERE INVOLVED.

STONE'S LETTERS TO BAKER, QUOTED IN THE COURT'S OPINION, SHOW NO MORE THAN HIS ORIGINAL APPLICATION DISCLOSED. THERE IS NO HINT OR SUGGESTION IN THEM OF TUNING THE ANTENNA CIRCUITS "INDEPENDENTLY" AS MARCONI DID. AND THE CORRESPONDENCE GIVES FURTHER PROOF HE CONTEMPLATED INTRODUCING THE INDUCTANCE COIL (OR A DEVICE EQUIVALENT IN FUNCTION) INTO THE CLOSED CIRCUIT, BUT EXPRESSED NO IDEA OF DOING THE SAME THING IN THE OPEN ONE.

IN MY OPINION THEREFORE THE JUDGMENT SHOULD BE REVERSED, IN SO FAR AS IT HOLDS MARCONI'S BROAD CLAIMS INVALID.

FN1 MARCONI V. BRITISH RADIO TEL. & TEL. CO., 27 T.L.R. 274; MARCONI V. HELSBY WIRELESS TEL. CO., 30 T.L.R. 688; SOCIETE MARCONI V. SOCIETE GENERALE, ETC., CIVIL TRIBUNAL OF THE SEINE, 3D CHAMBER, DEC. 24, 1912; MARCONI WIRELESS TELEGRAPH CO. V. NATIONAL ELECTRIC SIGNALLING CO., 213 F. 815(D.C.); MARCONI WIRELESS TELEGRAPH CO. V. KILBOURNE & CLARK MFG. CO., 265 F. 644(C.C.A.), AFF'G 239 F. 328(D.C.).

FN2 CF., E.G., 14 ENCYC. BRITANNICA (14TH ED.) 869.

FN3 HIS EARLIEST AMERICAN PATENT, U.S. PATENT NO. 586,193, GRANTED ON JULY 13, 1897, LATER BECOMING REISSUE PATENT NO. 11,913, IS NOT IN SUIT HERE. THAT PATENT DID NOT EMBRACE MANY OF THE CRUCIAL CLAIMS HERE INVOLVED AND ITS PRODUCT CANNOT COMPARE IN COMMERCIAL USEFULNESS WITH THAT OF THE PATENT IN SUIT.

FN4 COURTS CLOSER TO IT CHRONOLOGICALLY THAN WE ARE HAVE CHARACTERIZED IT AS A "CONSPICUOUS ADVANCE IN WIRELESS TELEGRAPHY"; "A REAL ACCOMPLISHMENT" AND THE IDEAS INVOLVED IN THE PATENT WERE SAID TO "HAVE PROVEN OF GREAT VALUE TO THE WORLD," TO HAVE BROUGHT ABOUT "AN ENTIRELY NEW AND USEFUL RESULT," "A NEW AND VERY IMPORTANT INDUSTRIAL RESULT" AND "A WONDERFUL CONQUEST." "THE MARCONI PATENT STANDS OUT AS AN UNASSAILABLE MONUMENT UNTIL NEW DISCOVERIES ARE MADE." CF. THE AUTHORITIES CITED IN NOTE 1, SUPRA.

FN5 HE WAS ONLY TWENTY-SIX YEARS OLD AT THE TIME HE APPLIED FOR THE PATENT IN SUIT, BUT HE HAD ALREADY MADE SUBSTANTIAL CONTRIBUTIONS TO THE FIELD.

FN6 U.S. PATENT NO. 763,772; BRITISH PATENT NO. 7777 OF 1900; FRENCH PATENT NO. 305,060 OF NOV. 3, 1900.

FN7 BRITISH PATENT TO LODGE NO. 29,505.

FN8 CF. NOTE 3 SUPRA.

FN9 U.S. PATENT TO TESLA NO. 649,621, MAY 15, 1900, DIVISION OF 645,576, MARCH 20, 1900(FILED SEPT. 2, 1897).

FN10 CF. TEXT INFRA.

FN11 CF. NOTE 1 SUPRA.

FN12 IBID.

FN13 TESLA IN FACT DID NOT USE HERTZIAN WAVES. HIS IDEA WAS TO MAKE THE ETHER A CONDUCTOR FOR LONG DISTANCES BY USING EXTREMELY HIGH VOLTAGE, 20,000,000 TO 30,000,000 VOLTS, AND EXTREMELY HIGH ALTITUDES, 30,000 TO 40,000 FEET OR MORE, TO SECURE TRANSMISSION FROM AERIAL TO AERIAL. BALLOONS, WITH WIRES ATTACHED REACHING TO THE GROUND, WERE HIS SUGGESTED AERIALS. HIS SYSTEM WAS REALLY ONE FOR TRANSMITTING POWER FOR MOTORS, LIGHTING, ETC., TO "ANY TERRESTRIAL DISTANCE," THOUGH HE INCIDENTALLY MENTIONS "INTELLIGIBLE MESSAGES." AS HE DID NOT USE HERTZIAN WAVES, HE HAD NO SUCH PROBLEM OF SELECTIVITY AS MARCONI, LODGE, STONE AND OTHERS WERE WORKING ON LATER. END.

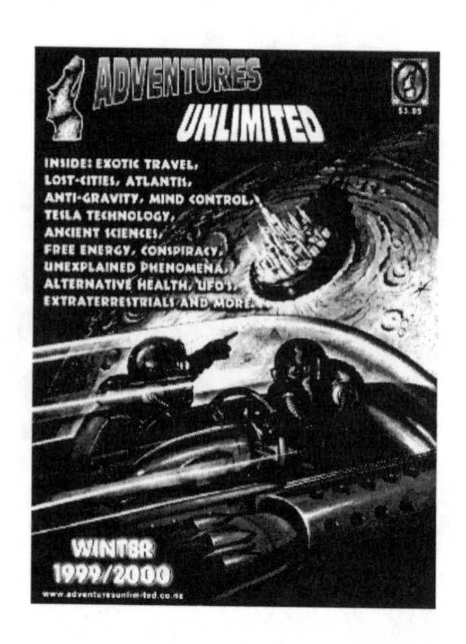

The Adventures Unlimited Catalog

Visit us online at:
www.wexclub.com/aup

HARNESSING THE WHEELWORK OF NATURE
Tesla's Science of Energy
by Thomas Valone, Ph.D., P.E.

A compilation of essays, papers and technical briefings on the emerging Tesla Technology and Zero Point Energy engineering that will soon change the entire way we live. Chapters include: Tesla: Scientific Superman who Launched the Westinghouse Industrial Firm by John Shatlan; Nikola Tesla—Electricity's Hidden Genius, excerpt from The Search for Free Energy; Tesla's History at Niagara Falls; Non-Hertzian Waves: True Meaning of the Wireless Transmission of Power by Toby Grotz; On the Transmission of Electricity Without Wires by Nikola Tesla; Tesla's Magnifying Transmitter by Andrija Puharich; Tesla's Self-Sustaining Electrical Generator and the Ether by Oliver Nichelson; Self-Sustaining Non-Hertzian Longitudinal Waves by Dr. Robert Bass; Modification of Maxwell's Equations in Free Space; Scalar Electromagnetic Waves; Disclosures Concerning Tesla's Operation of an ELF Oscillator; A Study of Tesla's Advanced Concepts & Glossary of Tesla Technology Terms; Electric Weather Forces: Tesla's Vision by Charles Yost; The New Art of Projecting Concentrated Non-Dispersive Energy Through Natural Media; The Homopolar Generator: Tesla's Contribution by Thomas Valone; Tesla's Ionizer and Ozonator: Implications for Indoor Air Pollution by Thomas Valone; How Cosmic Forces Shape Our Destiny by Nikola Tesla; Tesla's Death Ray plus Selected Tesla Patents; more.

288 PAGES. 6X9 PAPERBACK. ILLUSTRATED. $16.95. CODE: HWWN

HITLER'S FLYING SAUCERS
A Guide to German Flying Discs of the Second World War
by Henry Stevens

Learn why the Schriever-Habermohl project was actually two projects and read the written statement of a German test pilot who actually flew one of these saucers; about the Leduc engine, the key to Dr. Miethe's saucer designs; how U.S. government officials kept the truth about foo fighters hidden for almost sixty years and how they were finally forced to "come clean" about the foo fighter's German origin. Learn of the Peenemuende saucer project and how it was slated to "go atomic." Read the testimony of a German eyewitness who saw "magnetic discs." Read the U.S. government's own reports on German field propulsion saucers. Read how the post-war German KM-2 field propulsion "rocket" worked. Learn details of the work of Karl Schappeller and Viktor Schauberger. Learn how their ideas figure in the quest to build field propulsion flying discs. Find out what happened to this technology after the war. Find out how the Canadians got saucer technology directly from the SS. Find out about the surviving "Third Power" of former Nazis. Learn of the U.S. government's methods of UFO deception and how they used the German "Sonderbueroll" as the model for Project Blue Book.

388 PAGES. 6X9 PAPERBACK. ILLUSTRATED. INDEX. $18.95. CODE: HFS

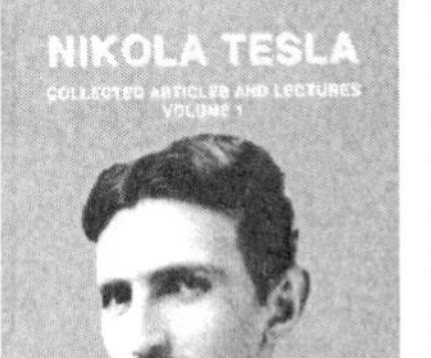

LEY LINE & EARTH ENERGIES
An Extraordinary Journey into the Earth's Natural Energy System
by David Cowan & Chris Arnold

The mysterious standing stones, burial grounds and stone circles that lace Europe, the British Isles and other areas have intrigued scientists, writers, artists and travellers through the centuries. They pose so many questions: Why do some places feel special? How do ley lines work? How did our ancestors use Earth energy to map their sacred sites and burial grounds? How do ghosts and poltergeists interact with Earth energy? How can Earth spirals and black spots affect our health? This exploration shows how natural forces affect our behavior, how they can be used to enhance our health and well being, and ultimately, how they bring us closer to penetrating one of the deepest mysteries being explored. A fascinating and visual book about subtle Earth energies and how they affect us and the world around them.

368 PAGES. 6X9 PAPERBACK. ILLUSTRATED. BIBLIOGRAPHY. INDEX. $18.95. CODE: LLEE

COLLECTED ARTICLES & LECTURES VOL. 1
by Nikola Tesla

This deluxe hardback imported from Germany features the collected articles and lectures of the amazing genuis known as Nikola Tesla. This volume starts with Tesla's very first lecture on May 16, 1888: A New System of Alternating Current Motors and Transformers. Also in this volume: On Light and Other High Frequency Phenomena; High Frequency Oscillators for Electro-Therapeutic and Other Purposes; The Tesla Alternate Current Motor; Experiments with Alternating Currents of High Frequency; Electric Discharge in Vacuum Tubes; Tesla Describes His Efforts in Various Fields of Work; Tesla's Wireless Light; Tesla on Electrostatic Generators; Roentgen Ray Investigation; Tesla Writes About His Experiments in Electrical Healing; Sleep from Electricy; tons more.

464 PAGES. 6X9 HARDBACK. ILLUSTRATED. $24.95. CODE: CAV1

COLLECTED ARTICLES & LECTURES VOL. 2
by Nikola Tesla

Volume 2 features more of the collected articles and lectures of Tesla. This volume starts with Tesla's Reply to Edison, July 14, 1905 and ends with Developments in the Practice and Art of Telephotography, a bizarre and detailed article on "telepathic photography" alone worth the price of the book. Also in this volume: Tesla on Wireless; Tesla and Marconi; Tesla Sees a Wireless Vision; Tesla Tells of New Radio Theories; Talking with Planets; How to Signal Mars; Interplanetary Communication; The Transmission of Electrical Energy Without Wires; Wireless Power Can Bridge Gap to Mars; The Wonder World to be Created by Electricity; A Lighting Machine of Novel Principles; Our Future Motive Power; Mr. Tesla's Vision; Famous Scientific Illusions; Tesla on Einstein's Theories; Tesla's Tidal Wave to Make War Impossible; My Submarine Destroyer; Electric Autos; tons more.

499 PAGES. 6X9 HARDBACK. ILLUSTRATED. $24.95. CODE: CAV2

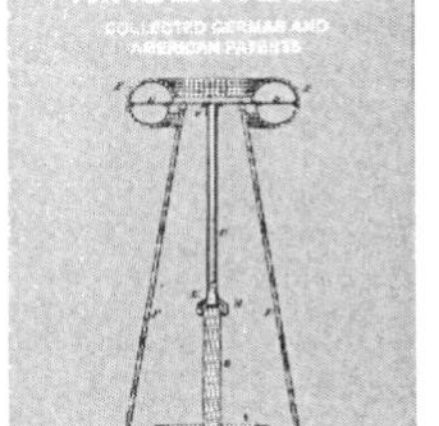

COLLECTED GERMAN AND AMERICAN PATENTS
by Nikola Tesla

This thick volume starts with Tesla's first German patent in 1888, and has descriptions and diagrams of virtually all of Tesla's patents right up to his last patent on Oct. 4, 1927 for an "Apparatus for Aerial Transportation." Patents include: Electric-Arc Lamp (1885); Pyromagneto-Electric Generator (1887); System of Electric Distribution (1888); Method of Operating Arc-Lamps(1890); Electrical Condenser (1891); Apparatus for Producing Currents of High Frequency (1896); Art of Transmitting Electrical Energy Through the Natural Mediums (1900); Method of Utilizing Radiant Energy (1901); Fluid Propulsion (1909); Method of Aerial Transportation (1921); tons more.

801 PAGES. 6X9 HARDBACK. ILLUSTRATED. $34.95. CODE: CGAP

NEW BOOKS

THE LAND OF OSIRIS

An Introduction to Khemitology

by Stephen S. Mehler

Was there an advanced prehistoric civilization in ancient Egypt? Were they the people who built the great pyramids and carved the Great Sphinx? Did the pyramids serve as energy devices and not as tombs for kings? Independent Egyptologist Stephen S. Mehler has spent over 30 years researching the answers to these questions and believes the answers are yes! Mehler has uncovered an indigenous oral tradition that still exists in Egypt, and has been fortunate to have studied with a living master of this tradition, Abd'El Hakim Awyan. Mehler has also been given permission to present these teachings to the Western world, teachings that unfold a whole new understanding of ancient Egypt and have only been presented heretofore in fragments by other researchers. Chapters include: Egyptology and Its Paradigms; Khemitology—New Paradigms; Asgat Nefer—The Harmony of Water; Khemit and the Myth of Atlantis; The Extraterrestrial Question; 17 chapters in all.

272 PAGES. 6X9 PAPERBACK. ILLUSTRATED. COLOR SECTION. BIBLIOGRAPHY. $18.95. CODE: LOOS

QUEST FOR ZERO-POINT ENERGY

Engineering Principles for "Free Energy"

by Moray B. King

King expands, with diagrams, on how free energy and anti-gravity are possible. The theories of zero point energy maintain there are tremendous fluctuations of electrical field energy embedded within the fabric of space. King explains the following topics: Tapping the Zero-Point Energy as an Energy Source; Fundamentals of a Zero-Point Energy Technology; Vacuum Energy Vortices; The Super Tube; Charge Clusters: The Basis of Zero-Point Energy Inventions; Vortex Filaments, Torsion Fields and the Zero-Point Energy; Transforming the Planet with a Zero-Point Energy Experiment; Dual Vortex Forms: The Key to a Large Zero-Point Energy Coherence. Packed with diagrams, patents and photos. With power shortages now a daily reality in many parts of the world, this book offers a fresh approach very rarely mentioned in the mainstream media.

224 PAGES. 6X9 PAPERBACK. ILLUSTRATED. $14.95. CODE: QZPE NOVEMBER PUBLICATION

ATLANTIS & THE POWER SYSTEM OF THE GODS

Mercury Vortex Generators & the Power System of Atlantis

by David Hatcher Childress and Bill Clendenon

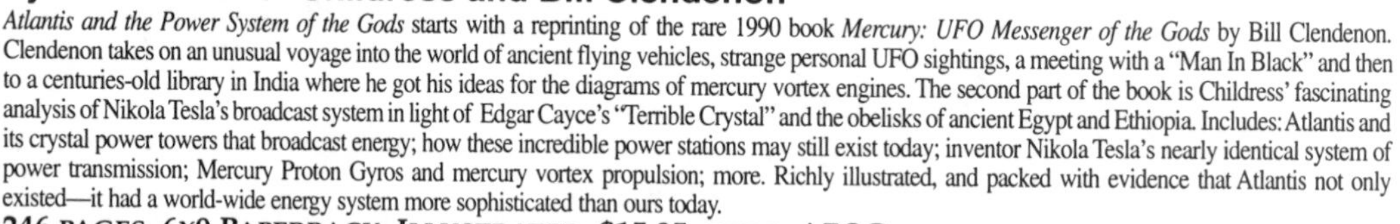

Atlantis and the Power System of the Gods starts with a reprinting of the rare 1990 book *Mercury: UFO Messenger of the Gods* by Bill Clendenon. Clendenon takes on an unusual voyage into the world of ancient flying vehicles, strange personal UFO sightings, a meeting with a "Man In Black" and then to a centuries-old library in India where he got his ideas for the diagrams of mercury vortex engines. The second part of the book is Childress' fascinating analysis of Nikola Tesla's broadcast system in light of Edgar Cayce's "Terrible Crystal" and the obelisks of ancient Egypt and Ethiopia. Includes: Atlantis and its crystal power towers that broadcast energy; how these incredible power stations may still exist today; inventor Nikola Tesla's nearly identical system of power transmission; Mercury Proton Gyros and mercury vortex propulsion; more. Richly illustrated, and packed with evidence that Atlantis not only existed—it had a world-wide energy system more sophisticated than ours today.

246 PAGES. 6X9 PAPERBACK. ILLUSTRATED. $15.95. CODE: APSG

THE GIZA DEATH STAR

The Paleophysics of the Great Pyramid & the Military Complex at Giza

by Joseph P. Farrell

Physicist Joseph Farrell's amazing book on the secrets of Great Pyramid of Giza. *The Giza Death Star* starts where British engineer Christopher Dunn leaves off in his 1998 book, *The Giza Power Plant*. Was the Giza complex part of a military installation over 10,000 years ago? Chapters include: An Archaeology of Mass Destruction, Thoth and Theories; The Machine Hypothesis; Pythagoras, Plato, Planck, and the Pyramid; The Weapon Hypothesis; Encoded Harmonics of the Planck Units in the Great Pyramid; High Fregquency Direct Current "Impulse" Technology; The Grand Gallery and its Crystals: Gravito-acoustic Resonators; The Other Two Large Pyramids; the "Causeways," and the "Temples"; A Phase Conjugate Howitzer; Evidence of the Use of Weapons of Mass Destruction in Ancient Times; more.

290 PAGES. 6X9 PAPERBACK. ILLUSTRATED. $16.95. CODE: GDS NOVEMBER PUBLICATION

THE ORION PROPHECY

Egyptian & Mayan Prophecies on the Cataclysm of 2012

by Patrick Geryl and Gino Ratinckx

In the year 2012 the Earth awaits a super catastrophe: its magnetic field reverse in one go. Phenomenal earthquakes and tidal waves will completely destroy our civilization. Europe and North America will shift thousands of kilometers northwards into polar climes. Nearly everyone will perish in the apocalyptic happenings. These dire predictions stem from the Mayans and Egyptians—descendants of the legendary Atlantis. The Atlanteans had highly evolved astronomical knowledge and were able to exactly predict the previous world-wide flood in 9792 BC. They built tens of thousands of boats and escaped to South America and Egypt. In the year 2012 Venus, Orion and several others stars will take the same 'code-positions' as in 9792 BC! For thousands of years historical sources have told of a forgotten time capsule of ancient wisdom located in a mythical labyrinth of secret chambers filled with artifacts and documents from the previous flood. We desperately need this information now—and this book gives one possible location.

324 PAGES. 6X9 PAPERBACK. ILLUSTRATED. BIBLIOGRAPHY. $16.95. CODE: ORP

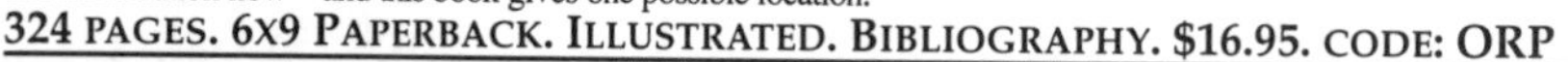

ALTAI-HIMALAYA

A Travel Diary

by Nicholas Roerich

Nicholas Roerich's classic 1929 mystic travel book is back in print in this deluxe paperback edition. The famous Russian-American explorer's expedition through Sinkiang, Altai-Mongolia and Tibet from 1924 to 1928 is chronicled in 12 chapters and reproductions of Roerich's inspiring paintings. Roerich's "Travel Diary" style incorporates various mysteries and mystical arts of Central Asia including such arcane topics as the hidden city of Shambala, Agartha, more. Roerich is recognized as one of the great artists of this century and the book is richly illustrated with his original drawings.

407 PAGES. 6X9 PAPERBACK. ILLUSTRATED. $18.95. CODE: AHIM NOVEMBER PUBLICATION

24 hour credit card orders—call: 815-253-6390 fax: 815-253-6300
email: auphq@frontiernet.net www.adventuresunlimitedpress.com www.wexclub.com

ANTI-GRAVITY

THE FREE-ENERGY DEVICE HANDBOOK

A Compilation of Patents and Reports

by David Hatcher Childress

A large-format compilation of various patents, papers, descriptions and diagrams concerning free-energy devices and systems. *The Free-Energy Device Handbook* is a visual tool for experimenters and researchers into magnetic motors and other "over-unity" devices. With chapters on the Adams Motor, the Hans Coler Generator, cold fusion, superconductors, "N" machines, space-energy generators, Nikola Tesla, T. Townsend Brown, and the latest in free-energy devices. Packed with photos, technical diagrams, patents and fascinating information, this book belongs on every science shelf. With energy and profit being a major political reason for fighting various wars, free-energy devices, if ever allowed to be mass distributed to consumers, could change the world! Get your copy now before the Department of Energy bans this book!

292 PAGES. 8X10 PAPERBACK. ILLUSTRATED. BIBLIOGRAPHY. $16.95. CODE: FEH

THE ANTI-GRAVITY HANDBOOK

edited by David Hatcher Childress, with Nikola Tesla, T.B. Paulicki, Bruce Cathie, Albert Einstein and others

The new expanded compilation of material on Anti-Gravity, Free Energy, Flying Saucer Propulsion, UFOs, Suppressed Technology, NASA Cover-ups and more. Highly illustrated with patents, technical illustrations and photos. This revised and expanded edition has more material, including photos of Area 51, Nevada, the government's secret testing facility. This classic on weird science is back in a 90s format!

- **How to build a flying saucer.**
- **Arthur C. Clarke on Anti-Gravity.**
- **Crystals and their role in levitation.**
- **Secret government research and development.**
- **Nikola Tesla on how anti-gravity airships could draw power from the atmosphere.**
- **Bruce Cathie's Anti-Gravity Equation.**
- **NASA, the Moon and Anti-Gravity.**

230 PAGES. 7X10 PAPERBACK. BIBLIOGRAPHY/INDEX/APPENDIX. HIGHLY ILLUSTRATED. $14.95. CODE: AGH

ANTI–GRAVITY & THE WORLD GRID

Is the earth surrounded by an intricate electromagnetic grid network offering free energy? This compilation of material on ley lines and world power points contains chapters on the geography, mathematics, and light harmonics of the earth grid. Learn the purpose of ley lines and ancient megalithic structures located on the grid. Discover how the grid made the Philadelphia Experiment possible. Explore the Coral Castle and many other mysteries, including acoustic levitation, Tesla Shields and scalar wave weaponry. Browse through the section on anti-gravity patents, and research resources.

274 PAGES. 7X10 PAPERBACK. ILLUSTRATED. $14.95. CODE: AGW

ANTI–GRAVITY & THE UNIFIED FIELD

edited by David Hatcher Childress

Is Einstein's Unified Field Theory the answer to all of our energy problems? Explored in this compilation of material is how gravity, electricity and magnetism manifest from a unified field around us. Why artificial gravity is possible; secrets of UFO propulsion; free energy; Nikola Tesla and anti-gravity airships of the 20s and 30s; flying saucers as superconducting whirls of plasma; anti-mass generators; vortex propulsion; suppressed technology; government cover-ups; gravitational pulse drive; spacecraft & more.

240 PAGES. 7X10 PAPERBACK. ILLUSTRATED. $14.95. CODE: AGU

ETHER TECHNOLOGY

A Rational Approach to Gravity Control

by Rho Sigma

This classic book on anti-gravity and free energy is back in print and back in stock. Written by a well-known American scientist under the pseudonym of "Rho Sigma," this book delves into international efforts at gravity control and discoid craft propulsion. Before the Quantum Field, there was "Ether." This small, but informative book has chapters on John Searle and "Searle discs;" T. Townsend Brown and his work on anti-gravity and ether-vortex turbines. Includes a forward by former NASA astronaut Edgar Mitchell.

108 PAGES. 6X9 PAPERBACK. ILLUSTRATED. $12.95. CODE: ETT

TAPPING THE ZERO POINT ENERGY

Free Energy & Anti-Gravity in Today's Physics

by Moray B. King

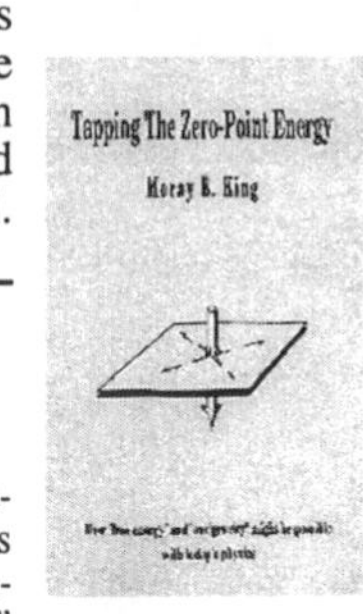

King explains how free energy and anti-gravity are possible. The theories of the zero point energy maintain there are tremendous fluctuations of electrical field energy imbedded within the fabric of space. This book tells how, in the 1930s, inventor T. Henry Moray could produce a fifty kilowatt "free energy" machine; how an electrified plasma vortex creates anti-gravity; how the Pons/Fleischmann "cold fusion" experiment could produce tremendous heat without fusion; and how certain experiments might produce a gravitational anomaly.

170 PAGES. 5X8 PAPERBACK. ILLUSTRATED. $9.95. CODE: TAP

24 hour credit card orders—call: 815-253-6390 fax: 815-253-6300
email: auphq@frontiernet.net www.adventuresunlimitedpress.com www.wexclub.com

ANTI-GRAVITY

COSMIC MATRIX

Piece for a Jig-Saw, Part Two

by Leonard G. Cramp

Leonard G. Cramp, a British aerospace engineer, wrote his first book *Space Gravity and the Flying Saucer* in 1954. Cosmic Matrix is the long-awaited sequel to his 1966 book *UFOs & Anti-Gravity: Piece for a Jig-Saw.* Cramp has had a long history of examining UFO phenomena and has concluded that UFOs use the highest possible aeronautic science to move in the way they do. Cramp examines anti-gravity effects and theorizes that this super-science used by the craft—described in detail in the book—can lift mankind into a new level of technology, transportation and understanding of the universe. The book takes a close look at gravity control, time travel, and the interlocking web of energy between all planets in our solar system with Leonard's unique technical diagrams. A fantastic voyage into the present and future!

364 PAGES. 6X9 PAPERBACK. ILLUSTRATED. BIBLIOGRAPHY. $16.00. CODE: CMX

UFOS AND ANTI-GRAVITY

Piece For A Jig-Saw

by Leonard G. Cramp

Leonard G. Cramp's 1966 classic book on flying saucer propulsion and suppressed technology is a highly technical look at the UFO phenomena by a trained scientist. Cramp first introduces the idea of 'anti-gravity' and introduces us to the various theories of gravitation. He then examines the technology necessary to build a flying saucer and examines in great detail the technical aspects of such a craft. Cramp's book is a wealth of material and diagrams on flying saucers, anti-gravity, suppressed technology, G-fields and UFOs. Chapters include Crossroads of Aerodymanics, Aerodynamic Saucers, Limitations of Rocketry, Gravitation and the Ether, Gravitational Spaceships, G-Field Lift Effects, The Bi-Field Theory, VTOL and Hovercraft, Analysis of UFO photos, more.

388 PAGES. 6X9 PAPERBACK. ILLUSTRATED. $16.95. CODE: UAG

THE HARMONIC CONQUEST OF SPACE

by Captain Bruce Cathie

Chapters include: Mathematics of the World Grid; the Harmonics of Hiroshima and Nagasaki; Harmonic Transmission and Receiving; the Link Between Human Brain Waves; the Cavity Resonance between the Earth; the Ionosphere and Gravity; Edgar Cayce—the Harmonics of the Subconscious; Stonehenge; the Harmonics of the Moon; the Pyramids of Mars; Nikola Tesla's Electric Car; the Robert Adams Pulsed Electric Motor Generator; Harmonic Clues to the Unified Field; and more. Also included are tables showing the harmonic relations between the earth's magnetic field, the speed of light, and anti-gravity/gravity acceleration at different points on the earth's surface. New chapters in this edition on the giant stone spheres of Costa Rica, Atomic Tests and Volcanic Activity, and a chapter on Ayers Rock analysed with Stone Mountain, Georgia.

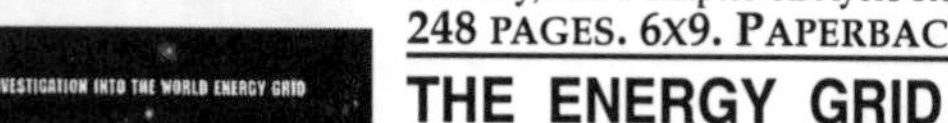

248 PAGES. 6X9. PAPERBACK. ILLUSTRATED. BIBLIOGRAPHY. $16.95. CODE: HCS

THE ENERGY GRID

Harmonic 695, The Pulse of the Universe

by Captain Bruce Cathie.

This is the breakthrough book that explores the incredible potential of the Energy Grid and the Earth's Unified Field all around us. Cathie's first book, *Harmonic 33*, was published in 1968 when he was a commercial pilot in New Zealand. Since then, Captain Bruce Cathie has been the premier investigator into the amazing potential of the infinite energy that surrounds our planet every microsecond. Cathie investigates the Harmonics of Light and how the Energy Grid is created. In this amazing book are chapters on UFO Propulsion, Nikola Tesla, Unified Equations, the Mysterious Aerials, Pythagoras & the Grid, Nuclear Detonation and the Grid, Maps of the Ancients, an Australian Stonehenge examined, more.

255 PAGES. 6X9 TRADEPAPER. ILLUSTRATED. $15.95. CODE: TEG

THE BRIDGE TO INFINITY

Harmonic 371244

by Captain Bruce Cathie

Cathie has popularized the concept that the earth is crisscrossed by an electromagnetic grid system that can be used for anti-gravity, free energy, levitation and more. The book includes a new analysis of the harmonic nature of reality, acoustic levitation, pyramid power, harmonic receiver towers and UFO propulsion. It concludes that today's scientists have at their command a fantastic store of knowledge with which to advance the welfare of the human race.

204 PAGES. 6X9 TRADEPAPER. ILLUSTRATED. $14.95. CODE: BTF

MAN-MADE UFOS 1944—1994

Fifty Years of Suppression

by Renato Vesco & David Hatcher Childress

A comprehensive look at the early "flying saucer" technology of Nazi Germany and the genesis of man-made UFOs. This book takes us from the work of captured German scientists to escaped battalions of Germans, secret communities in South America and Antarctica to todays state-of-the-art "Dreamland" flying machines. Heavily illustrated, this astonishing book blows the lid off the "government UFO conspiracy" and explains with technical diagrams the technology involved. Examined in detail are secret underground airfields and factories; German secret weapons; "suction" aircraft; the origin of NASA; gyroscopic stabilizers and engines; the secret Marconi aircraft factory in South America; and more. Introduction by W.A. Harbinson, author of the Dell novels *GENESIS* and *REVELATION.*

318 PAGES. 6X9 PAPERBACK. ILLUSTRATED. INDEX & FOOTNOTES. $18.95. CODE: MMU

FREE ENERGY SYSTEMS

LOST SCIENCE

by Gerry Vassilatos

Rediscover the legendary names of suppressed scientific revolution—remarkable lives, astounding discoveries, and incredible inventions which would have produced a world of wonder. How did the aura research of Baron Karl von Reichenbach prove the vitalistic theory and frighten the greatest minds of Germany? How did the physiophone and wireless of Antonio Meucci predate both Bell and Marconi by decades? How does the earth battery technology of Nathan Stubblefield portend an unsuspected energy revolution? How did the geoaetheric engines of Nikola Tesla threaten the establishment of a fuel-dependent America? The microscopes and virus-destroying ray machines of Dr. Royal Rife provided the solution for every world-threatening disease. Why did the FDA and AMA together condemn this great man to Federal Prison? The static crashes on telephone lines enabled Dr. T. Henry Moray to discover the reality of radiant space energy. Was the mysterious "Swedish stone," the powerful mineral which Dr. Moray discovered, the very first historical instance in which stellar power was recognized and secured on earth? Why did the Air Force initially fund the gravitational warp research and warp-cloaking devices of T. Townsend Brown and then reject it? When the controlled fusion devices of Philo Farnsworth achieved the "break-even" point in 1967 the FUSOR project was abruptly cancelled by ITT.

304 PAGES. 6X9 PAPERBACK. ILLUSTRATED. BIBLIOGRAPHY. $16.95. CODE: LOS

SECRETS OF COLD WAR TECHNOLOGY

Project HAARP and Beyond

by Gerry Vassilatos

Vassilatos reveals that "Death Ray" technology has been secretly researched and developed since the turn of the century. Included are chapters on such inventors and their devices as H.C. Vion, the developer of auroral energy receivers; Dr. Selim Lemstrom's pre-Tesla experiments; the early beam weapons of Grindell-Mathews, Ulivi, Turpain and others; John Hettenger and his early beam power systems. Learn about Project Argus, Project Teak and Project Orange; EMP experiments in the 60s; why the Air Force directed the construction of a huge Ionospheric "backscatter" telemetry system across the Pacific just after WWII; why Raytheon has collected every patent relevant to HAARP over the past few years; more.

250 PAGES. 6X9 PAPERBACK. ILLUSTRATED. $15.95. CODE: SCWT

THE A.T. FACTOR

A Scientists Encounter with UFOs: Piece For A Jigsaw Part 3

by Leonard Cramp

British aerospace engineer Cramp began much of the scientific anti-gravity and UFO propulsion analysis back in 1955 with his landmark book *Space, Gravity & the Flying Saucer* (out-of-print and rare). His next books (available from Adventures Unlimited) *UFOs & Anti-Gravity: Piece for a Jig-Saw* and *The Cosmic Matrix: Piece for a Jig-Saw Part 2* began Cramp's in depth look into gravity control, free-energy, and the interlocking web of energy that pervades the universe. In this final book, Cramp brings to a close his detailed and controversial study of UFOs and Anti-Gravity.

324 PAGES. 6X9 PAPERBACK. ILLUSTRATED. BIBLIOGRAPHY. INDEX. $16.95. CODE: ATF

THE TIME TRAVEL HANDBOOK

A Manual of Practical Teleportation & Time Travel

edited by David Hatcher Childress

In the tradition of *The Anti-Gravity Handbook* and *The Free-Energy Device Handbook*, science and UFO author David Hatcher Childress takes us into the weird world of time travel and teleportation. Not just a whacked-out look at science fiction, this book is an authoritative chronicling of real-life time travel experiments, teleportation devices and more. *The Time Travel Handbook* takes the reader beyond the government experiments and deep into the uncharted territory of early time travellers such as Nikola Tesla and Guglielmo Marconi and their alleged time travel experiments, as well as the Wilson Brothers of EMI and their connection to the Philadelphia Experiment—the U.S. Navy's forays into invisibility, time travel, and teleportation. Childress looks into the claims of time travelling individuals, and investigates the unusual claim that the pyramids on Mars were built in the future and sent back in time. A highly visual, large format book, with patents, photos and schematics. Be the first on your block to build your own time travel device!

316 PAGES. 7X10 PAPERBACK. ILLUSTRATED. $16.95. CODE: TTH

THE TESLA PAPERS

Nikola Tesla on Free Energy & Wireless Transmission of Power

by Nikola Tesla, edited by David Hatcher Childress

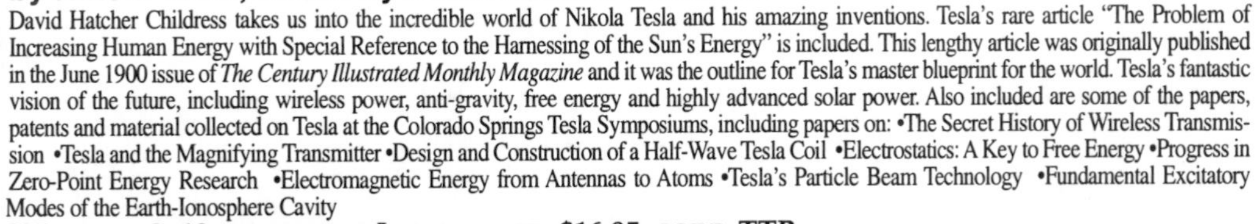

David Hatcher Childress takes us into the incredible world of Nikola Tesla and his amazing inventions. Tesla's rare article "The Problem of Increasing Human Energy with Special Reference to the Harnessing of the Sun's Energy" is included. This lengthy article was originally published in the June 1900 issue of *The Century Illustrated Monthly Magazine* and it was the outline for Tesla's master blueprint for the world. Tesla's fantastic vision of the future, including wireless power, anti-gravity, free energy and highly advanced solar power. Also included are some of the papers, patents and material collected on Tesla at the Colorado Springs Tesla Symposiums, including papers on: •The Secret History of Wireless Transmission •Tesla and the Magnifying Transmitter •Design and Construction of a Half-Wave Tesla Coil •Electrostatics: A Key to Free Energy •Progress in Zero-Point Energy Research •Electromagnetic Energy from Antennas to Atoms •Tesla's Particle Beam Technology •Fundamental Excitatory Modes of the Earth-Ionosphere Cavity

325 PAGES. 8X10 PAPERBACK. ILLUSTRATED. $16.95. CODE: TTP

THE FANTASTIC INVENTIONS OF NIKOLA TESLA

by Nikola Tesla with additional material by David Hatcher Childress

This book is a readable compendium of patents, diagrams, photos and explanations of the many incredible inventions of the originator of the modern era of electrification. In Tesla's own words are such topics as wireless transmission of power, death rays, and radio-controlled airships. In addition, rare material on German bases in Antarctica and South America, and a secret city built at a remote jungle site in South America by one of Tesla's students, Guglielmo Marconi. Marconi's secret group claims to have built flying saucers in the 1940s and to have gone to Mars in the early 1950s! Incredible photos of these Tesla craft are included. The Ancient Atlantean system of broadcasting energy through a grid system of obelisks and pyramids is discussed, and a fascinating concept comes out of one chapter: that Egyptian engineers had to wear protective metal head-shields while in these power plants, hence the Egyptian Pharoah's head covering as well as the Face on Mars! •His plan to transmit free electricity into the atmosphere. •How electrical devices would work using only small antennas. •Why unlimited power could be utilized anywhere on earth. •How radio and radar technology can be used as death-ray weapons in Star Wars.

342 PAGES. 6X9 PAPERBACK. ILLUSTRATED. $16.95. CODE: FINT

TESLA TECHNOLOGY

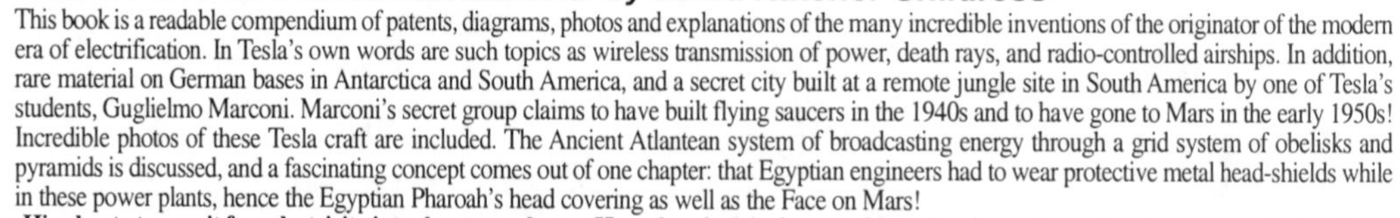

THE FANTASTIC INVENTIONS OF NIKOLA TESLA

Nikola Tesla with additional material by David Hatcher Childress

This book is a readable compendium of patents, diagrams, photos and explanations of the many incredible inventions of the originator of the modern era of electrification. In Tesla's own words are such topics as wireless transmission of power, death rays, and radio-controlled airships. In addition, rare material on German bases in Antarctica and South America, and a secret city built at a remote jungle site in South America by one of Tesla's students, Guglielmo Marconi. Marconi's secret group claims to have built flying saucers in the 1940s and to have gone to Mars in the early 1950s! Incredible photos of these Tesla craft are included. The Ancient Atlantean system of broadcasting energy through a grid system of obelisks and pyramids is discussed, and a fascinating concept comes out of one chapter: that Egyptian engineers had to wear protective metal head-shields while in these power plants, hence the Egyptian Pharoah's head covering as well as the Face on Mars!

•His plan to transmit free electricity into the atmosphere. •How electrical devices would work using only small antennas mounted on them. •Why unlimited power could be utilized anywhere on earth. •How radio and radar technology can be used as death-ray weapons in Star Wars. •Includes an appendix of Supreme Court documents on dismantling his free energy towers. •Tesla's Death Rays, Ozone generators, and more…

342 PAGES. 6X9 PAPERBACK. ILLUSTRATED. BIBLIOGRAPHY AND APPENDIX. $16.95. CODE: FINT

THE TESLA PAPERS

Nikola Tesla on Free Energy & Wireless Transmission of Power

by Nikola Tesla, edited by David Hatcher Childress

In the tradition of *The Fantastic Inventions of Nikola Tesla, The Anti-Gravity Handbook* and *The Free-Energy Device Handbook,* science and UFO author David Hatcher Childress takes us into the incredible world of Nikola Tesla and his amazing inventions. Tesla's rare article "The Problem of Increasing Human Energy with Special Reference to the Harnessing of the Sun's Energy" is included. This lengthy article was originally published in the June 1900 issue of *The Century Illustrated Monthly Magazine* and it was the outline for Tesla's master blueprint for the world. Tesla's fantastic vision of the future, including wireless power, anti-gravity, free energy and highly advanced solar power.

Also included are some of the papers, patents and material collected on Tesla at the Colorado Springs Tesla Symposiums, including papers on:
•The Secret History of Wireless Transmission •Tesla and the Magnifying Transmitter
•Design and Construction of a half-wave Tesla Coil •Electrostatics: A Key to Free Energy
•Progress in Zero-Point Energy Research •Electromagnetic Energy from Antennas to Atoms
•Tesla's Particle Beam Technology •Fundamental Excitatory Modes of the Earth-Ionosphere Cavity

325 PAGES. 8X10 PAPERBACK. ILLUSTRATED. $16.95. CODE: TTP

LOST SCIENCE

by Gerry Vassilatos

Secrets of Cold War Technology author Vassilatos on the remarkable lives, astounding discoveries, and incredible inventions of such famous people as Nikola Tesla, Dr. Royal Rife, T.T. Brown, and T. Henry Moray. Read about the aura research of Baron Karl von Reichenbach, the wireless technology of Antonio Meucci, the controlled fusion devices of Philo Farnsworth, the earth battery of Nathan Stubblefield, and more. What were the twisted intrigues which surrounded the often deliberate attempts to stop this technology? Vassilatos claims that we are living hundreds of years behind our intended level of technology and we must recapture this "lost science."

304 PAGES. 6X9 PAPERBACK. ILLUSTRATED. BIBLIOGRAPHY. $16.95. CODE: LOS

SECRETS OF COLD WAR TECHNOLOGY

Project HAARP and Beyond

by Gerry Vassilatos

Vassilatos reveals that "Death Ray" technology has been secretly researched and developed since the turn of the century. Included are chapters on such inventors and their devices as H.C. Vion, the developer of auroral energy receivers; Dr. Selim Lemstrom's pre-Tesla experiments; the early beam weapons of Grindell-Mathews, Ulivi, Turpain and others; John Hettenger and his early beam power systems. Learn about Project Argus, Project Teak and Project Orange; EMP experiments in the 60s; why the Air Force directed the construction of a huge Ionospheric "backscatter" telemetry system across the Pacific just after WWII; why Raytheon has collected every patent relevant to HAARP over the past few years; more.

250 PAGES. 6X9 PAPERBACK. ILLUSTRATED. $15.95. CODE: SCWT

HAARP

The Ultimate Weapon of the Conspiracy

by Jerry Smith

The HAARP project in Alaska is one of the most controversial projects ever undertaken by the U.S. Government. Jerry Smith gives us the history of the HAARP project and explains how works, in technically correct yet easy to understand language. At best, HAARP is science out-of-control; at worst, HAARP could be the most dangerous device ever created, a futuristic technology that is everything from super-beam weapon to world-wide mind control device. Topics include Over-the-Horizon Radar and HAARP, Mind Control, ELF and HAARP, The Telsa Connection, The Russian Woodpecker, GWEN & HAARP, Earth Penetrating Tomography, Weather Modification, Secret Science of the Conspiracy, more. Includes the complete 1987 Eastlund patent for his pulsed super-weapon that he claims was stolen by the HAARP Project.

256 PAGES. 6X9 PAPERBACK. ILLUSTRATED. $14.95. CODE: HARP

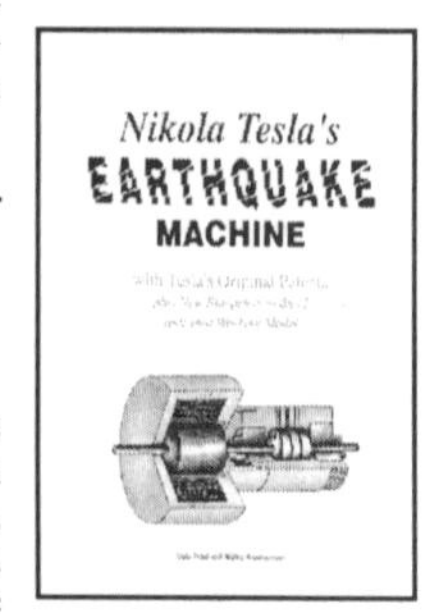

NIKOLA TESLA'S EARTHQUAKE MACHINE

with Tesla's Original Patents

by Dale Pond and Walter Baumgartner

Now, for the first time, the secrets of Nikola Tesla's Earthquake Machine are available. Although this book discusses in detail Nikola Tesla's 1894 "Earthquake Oscillator," it is also about the new technology of sonic vibrations which produce a resonance effect that can be used to cause earthquakes. Discussed are Tesla Oscillators, Vibration Physics, Amplitude Modulated Additive Synthesis, Tele-Geo-dynamics, Solar Heat Pump Apparatus, Vortex Tube Coolers, the Serogodsky Motor, more. Plenty of technical diagrams. Be the first on your block to have a Tesla Earthquake Machine!

175 PAGES. 9X11 PAPERBACK. ILLUSTRATED. BIBLIOGRAPHY & INDEX. $16.95. CODE: TEM

LOST CITIES

TECHNOLOGY OF THE GODS

The Incredible Sciences of the Ancients

by David Hatcher Childress

Popular *Lost Cities* author David Hatcher Childress takes us into the amazing world of ancient technology, from computers in antiquity to the "flying machines of the gods." Childress looks at the technology that was allegedly used in Atlantis and the theory that the Great Pyramid of Egypt was originally a gigantic power station. He examines tales of ancient flight and the technology that it involved; how the ancients used electricity; megalithic building techniques; the use of crystal lenses and the fire from the gods; evidence of various high tech weapons in the past, including atomic weapons; ancient metallurgy and heavy machinery; the role of modern inventors such as Nikola Tesla in bringing ancient technology back into modern use; impossible artifacts; and more.

356 PAGES. 6X9 PAPERBACK. ILLUSTRATED. BIBLIOGRAPHY. $16.95. CODE: TGOD

VIMANA AIRCRAFT OF ANCIENT INDIA & ATLANTIS

by David Hatcher Childress, introduction by Ivan T. Sanderson

Did the ancients have the technology of flight? In this incredible volume on ancient India, authentic Indian texts such as the *Ramayana* and the *Mahabharata* are used to prove that ancient aircraft were in use more than four thousand years ago. Included in this book is the entire Fourth Century BC manuscript *Vimaanika Shastra* by the ancient author Maharishi Bharadwaaja, translated into English by the Mysore Sanskrit professor G.R. Josyer. Also included are chapters on Atlantean technology, the incredible Rama Empire of India and the devastating wars that destroyed it. Also an entire chapter on mercury vortex propulsion and mercury gyros, the power source described in the ancient Indian texts. Not to be missed by those interested in ancient civilizations or the UFO enigma.

334 PAGES. 6X9 PAPERBACK. RARE PHOTOGRAPHS, MAPS AND DRAWINGS. $15.95. CODE: VAA

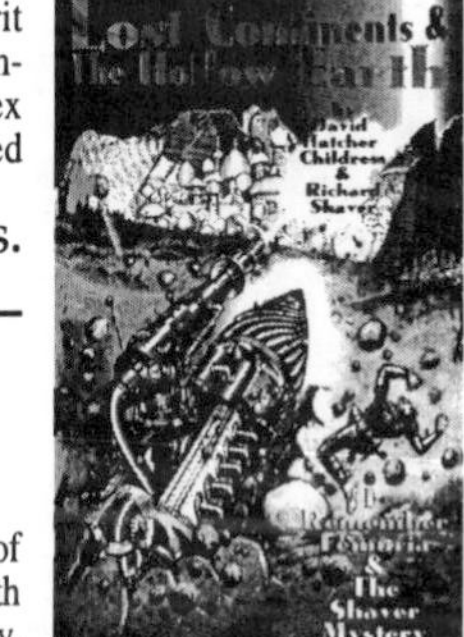

LOST CONTINENTS & THE HOLLOW EARTH

I Remember Lemuria and the Shaver Mystery

by David Hatcher Childress & Richard Shaver

Lost Continents & the Hollow Earth is Childress' thorough examination of the early hollow earth stories of Richard Shaver and the fascination that fringe fantasy subjects such as lost continents and the hollow earth have had for the American public. Shaver's rare 1948 book *I Remember Lemuria* is reprinted in its entirety, and the book is packed with illustrations from Ray Palmer's *Amazing Stories* magazine of the 1940s. Palmer and Shaver told of tunnels running through the earth—tunnels inhabited by the Deros and Teros, humanoids from an ancient spacefaring race that had inhabited the earth, eventually going underground, hundreds of thousands of years ago. Childress discusses the famous hollow earth books and delves deep into whatever reality may be behind the stories of tunnels in the earth. Operation High Jump to Antarctica in 1947 and Admiral Byrd's bizarre statements, tunnel systems in South America and Tibet, the underground world of Agartha, the belief of UFOs coming from the South Pole, more.

344 PAGES. 6X9 PAPERBACK. ILLUSTRATED. $16.95. CODE: LCHE

LOST CITIES OF NORTH & CENTRAL AMERICA

by David Hatcher Childress

Down the back roads from coast to coast, maverick archaeologist and adventurer David Hatcher Childress goes deep into unknown America. With this incredible book, you will search for lost Mayan cities and books of gold, discover an ancient canal system in Arizona, climb gigantic pyramids in the Midwest, explore megalithic monuments in New England, and join the astonishing quest for lost cities throughout North America. From the war-torn jungles of Guatemala, Nicaragua and Honduras to the deserts, mountains and fields of Mexico, Canada, and the U.S.A., Childress takes the reader in search of sunken ruins, Viking forts, strange tunnel systems, living dinosaurs, early Chinese explorers, and fantastic lost treasure. Packed with both early and current maps, photos and illustrations.

590 PAGES. 6X9 PAPERBACK. ILLUSTRATED. FOOTNOTES & BIBLIOGRAPHY. $14.95. CODE: NCA

LOST CITIES & ANCIENT MYSTERIES OF SOUTH AMERICA

by David Hatcher Childress

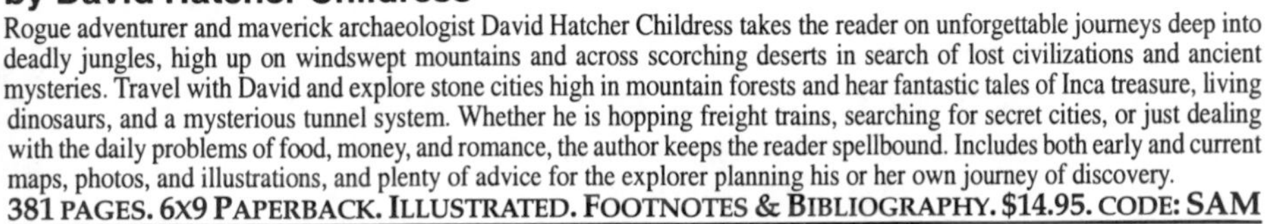

Rogue adventurer and maverick archaeologist David Hatcher Childress takes the reader on unforgettable journeys deep into deadly jungles, high up on windswept mountains and across scorching deserts in search of lost civilizations and ancient mysteries. Travel with David and explore stone cities high in mountain forests and hear fantastic tales of Inca treasure, living dinosaurs, and a mysterious tunnel system. Whether he is hopping freight trains, searching for secret cities, or just dealing with the daily problems of food, money, and romance, the author keeps the reader spellbound. Includes both early and current maps, photos, and illustrations, and plenty of advice for the explorer planning his or her own journey of discovery.

381 PAGES. 6X9 PAPERBACK. ILLUSTRATED. FOOTNOTES & BIBLIOGRAPHY. $14.95. CODE: SAM

LOST CITIES & ANCIENT MYSTERIES OF AFRICA & ARABIA

by David Hatcher Childress

Across ancient deserts, dusty plains and steaming jungles, maverick archaeologist David Childress continues his world-wide quest for lost cities and ancient mysteries. Join him as he discovers forbidden cities in the Empty Quarter of Arabia; "Atlantean" ruins in Egypt and the Kalahari desert; a mysterious, ancient empire in the Sahara; and more. This is the tale of an extraordinary life on the road: across war-torn countries, Childress searches for King Solomon's Mines, living dinosaurs, the Ark of the Covenant and the solutions to some of the fantastic mysteries of the past.

423 PAGES. 6X9 PAPERBACK. ILLUSTRATED. FOOTNOTES & BIBLIOGRAPHY. $14.95. CODE: AFA

LOST CITIES

LOST CITIES OF ATLANTIS, ANCIENT EUROPE & THE MEDITERRANEAN

by David Hatcher Childress

Atlantis! The legendary lost continent comes under the close scrutiny of maverick archaeologist David Hatcher Childress in this sixth book in the internationally popular *Lost Cities* series. Childress takes the reader in search of sunken cities in the Mediterranean; across the Atlas Mountains in search of Atlantean ruins; to remote islands in search of megalithic ruins; to meet living legends and secret societies. From Ireland to Turkey, Morocco to Eastern Europe, and around the remote islands of the Mediterranean and Atlantic, Childress takes the reader on an astonishing quest for mankind's past. Ancient technology, cataclysms, megalithic construction, lost civilizations and devastating wars of the past are all explored in this book. Childress challenges the skeptics and proves that great civilizations not only existed in the past, but the modern world and its problems are reflections of the ancient world of Atlantis.

524 PAGES. 6X9 PAPERBACK. ILLUSTRATED WITH 100S OF MAPS, PHOTOS AND DIAGRAMS. BIBLIOGRAPHY & INDEX. $16.95. CODE: MED

LOST CITIES OF CHINA, CENTRAL INDIA & ASIA

by David Hatcher Childress

Like a real life "Indiana Jones," maverick archaeologist David Childress takes the reader on an incredible adventure across some of the world's oldest and most remote countries in search of lost cities and ancient mysteries. Discover ancient cities in the Gobi Desert; hear fantastic tales of lost continents, vanished civilizations and secret societies bent on ruling the world; visit forgotten monasteries in forbidding snow-capped mountains with strange tunnels to mysterious subterranean cities! A unique combination of far-out exploration and practical travel advice, it will astound and delight the experienced traveler or the armchair voyager.

429 PAGES. 6X9 PAPERBACK. ILLUSTRATED. FOOTNOTES & BIBLIOGRAPHY. $14.95. CODE: CHI

LOST CITIES OF ANCIENT LEMURIA & THE PACIFIC

by David Hatcher Childress

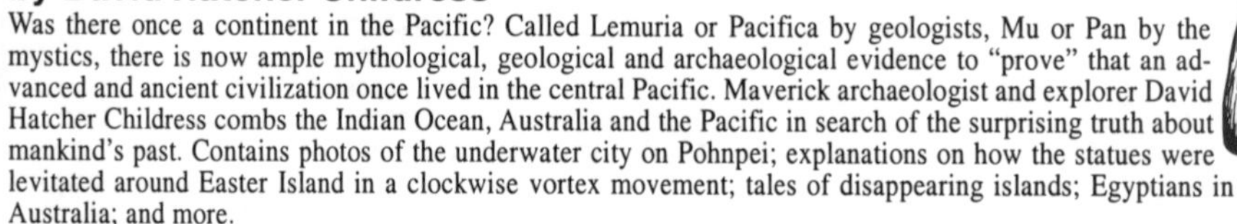

Was there once a continent in the Pacific? Called Lemuria or Pacifica by geologists, Mu or Pan by the mystics, there is now ample mythological, geological and archaeological evidence to "prove" that an advanced and ancient civilization once lived in the central Pacific. Maverick archaeologist and explorer David Hatcher Childress combs the Indian Ocean, Australia and the Pacific in search of the surprising truth about mankind's past. Contains photos of the underwater city on Pohnpei; explanations on how the statues were levitated around Easter Island in a clockwise vortex movement; tales of disappearing islands; Egyptians in Australia; and more.

379 PAGES. 6X9 PAPERBACK. ILLUSTRATED. FOOTNOTES & BIBLIOGRAPHY. $14.95. CODE: LEM

ANCIENT TONGA

& the Lost City of Mu'a

by David Hatcher Childress

Lost Cities series author Childress takes us to the south sea islands of Tonga, Rarotonga, Samoa and Fiji to investigate the megalithic ruins on these beautiful islands. The great empire of the Polynesians, centered on Tonga and the ancient city of Mu'a, is revealed with old photos, drawings and maps. Chapters in this book are on the Lost City of Mu'a and its many megalithic pyramids, the Ha'amonga Trilithon and ancient Polynesian astronomy, Samoa and the search for the lost land of Havai'iki, Fiji and its wars with Tonga, Rarotonga's megalithic road, and Polynesian cosmology. Material on Egyptians in the Pacific, earth changes, the fortified moat around Mu'a, lost roads, more.

218 PAGES. 6X9 PAPERBACK. ILLUSTRATED. COLOR PHOTOS. BIBLIOGRAPHY. $15.95. CODE: TONG

ANCIENT MICRONESIA

& the Lost City of Nan Madol

by David Hatcher Childress

Micronesia, a vast archipelago of islands west of Hawaii and south of Japan, contains some of the most amazing megalithic ruins in the world. Part of our *Lost Cities* series, this volume explores the incredible conformations on various Micronesian islands, especially the fantastic and little-known ruins of Nan Madol on Pohnpei Island. The huge canal city of Nan Madol contains over 250 million tons of basalt columns over an 11 square-mile area of artificial islands. Much of the huge city is submerged, and underwater structures can be found to an estimated 80 feet. Islanders' legends claim that the basalt rocks, weighing up to 50 tons, were magically levitated into place by the powerful forefathers. Other ruins in Micronesia that are profiled include the Latte Stones of the Marianas, the menhirs of Palau, the megalithic canal city on Kosrae Island, megaliths on Guam, and more.

256 PAGES. 6X9 PAPERBACK. ILLUSTRATED. INCLUDES A COLOR PHOTO SECTION. BIBLIOGRAPHY. $16.95. CODE: AMIC

ATLANTIS REPRINT SERIES

ATLANTIS: MOTHER OF EMPIRES

Atlantis Reprint Series

by Robert Stacy-Judd

Robert Stacy-Judd's classic 1939 book on Atlantis is back in print in this large-format paperback edition. Stacy-Judd was a California architect and an expert on the Mayas and their relationship to Atlantis. He was an excellent artist and his work is lavishly illustrated. The eighteen comprehensive chapters in the book are: The Mayas and the Lost Atlantis; Conjectures and Opinions; The Atlantean Theory; Cro-Magnon Man; East is West; And West is East; The Mormons and the Mayas; Astrology in Two Hemispheres; The Language of Architecture; The American Indian; Pre-Panamanians and Pre-Incas; Columns and City Planning; Comparisons and Mayan Art; The Iberian Link; The Maya Tongue; Quetzalcoatl; Summing Up the Evidence; The Mayas in Yucatan.

340 PAGES. 8X11 PAPERBACK. ILLUSTRATED. INDEX. $19.95. CODE: AMOE

MYSTERIES OF ANCIENT SOUTH AMERICA

Atlantis Reprint Series

by Harold T. Wilkins

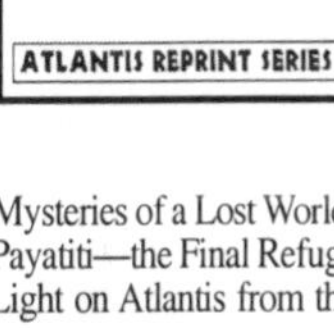

The reprint of Wilkins' classic book on the megaliths and mysteries of South America. This book predates Wilkin's book *Secret Cities of Old South America* published in 1952. *Mysteries of Ancient South America* was first published in 1947 and is considered a classic book of its kind. With diagrams, photos and maps, Wilkins digs into old manuscripts and books to bring us some truly amazing stories of South America: a bizarre subterranean tunnel system; lost cities in the remote border jungles of Brazil; legends of Atlantis in South America; cataclysmic changes that shaped South America; and other strange stories from one of the world's great researchers. Chapters include: Our Earth's Greatest Disaster, Dead Cities of Ancient Brazil, The Jungle Light that Shines by Itself, The Missionary Men in Black: Forerunners of the Great Catastrophe, The Sign of the Sun: The World's Oldest Alphabet, Sign-Posts to the Shadow of Atlantis, The Atlanean "Subterraneans" of the Incas, Tiahuanacu and the Giants, more.

236 PAGES. 6X9 PAPERBACK. ILLUSTRATED. INDEX. $14.95. CODE: MASA

SECRET CITIES OF OLD SOUTH AMERICA

Atlantis Reprint Series

by Harold T. Wilkins

The reprint of Wilkins' classic book, first published in 1952, claiming that South America was Atlantis. Chapters include Mysteries of a Lost World; Atlantis Unveiled; Red Riddles on the Rocks; South America's Amazons Existed!; The Mystery of El Dorado and Gran Payatiti—the Final Refuge of the Incas; Monstrous Beasts of the Unexplored Swamps & Wilds; Weird Denizens of Antediluvian Forests; New Light on Atlantis from the World's Oldest Book; The Mystery of Old Man Noah and the Arks; and more.

438 PAGES. 6X9 PAPERBACK. ILLUSTRATED. BIBLIOGRAPHY & INDEX. $16.95. CODE: SCOS

THE SHADOW OF ATLANTIS

The Echoes of Atlantean Civilization Tracked through Space & Time

by Colonel Alexander Braghine

First published in 1940, *The Shadow of Atlantis* is one of the great classics of Atlantis research. The book amasses a great deal of archaeological, anthropological, historical and scientific evidence in support of a lost continent in the Atlantic Ocean. Braghine covers such diverse topics as Egyptians in Central America, the myth of Quetzalcoatl, the Basque language and its connection with Atlantis, the connections with the ancient pyramids of Mexico, Egypt and Atlantis, the sudden demise of mammoths, legends of giants and much more. Braghine was a linguist and spends part of the book tracing ancient languages to Atlantis and studying little-known inscriptions in Brazil, deluge myths and the connections between ancient languages. Braghine takes us on a fascinating journey through space and time in search of the lost continent.

288 PAGES. 6X9 PAPERBACK. ILLUSTRATED. $16.95. CODE: SOA

RIDDLE OF THE PACIFIC

by John Macmillan Brown

Oxford scholar Brown's classic work on lost civilizations of the Pacific is now back in print! John Macmillan Brown was an historian and New Zealand's premier scientist when he wrote about the origins of the Maoris. After many years of travel thoughout the Pacific studying the people and customs of the south seas islands, he wrote *Riddle of the Pacific* in 1924. The book is packed with rare turn-of-the-century illustrations. Don't miss Brown's classic study of Easter Island, ancient scripts, megalithic roads and cities, more. Brown was an early believer in a lost continent in the Pacific.

460 PAGES. 6X9 PAPERBACK. ILLUSTRATED. $16.95. CODE: ROP

THE RIDDLE OF THE PACIFIC

JOHN MACMILLAN BROWN

This rare 1924 book is back in print!

THE HISTORY OF ATLANTIS

by Lewis Spence

Lewis Spence's classic book on Atlantis is now back in print! Spence was a Scottish historian (1874-1955) who is best known for his volumes on world mythology and his five Atlantis books. *The History of Atlantis* (1926) is considered his finest. Spence does his scholarly best in chapters on the Sources of Atlantean History, the Geography of Atlantis, the Races of Atlantis, the Kings of Atlantis, the Religion of Atlantis, the Colonies of Atlantis, more. Sixteen chapters in all.

240 PAGES. 6X9 PAPERBACK. ILLUSTRATED WITH MAPS, PHOTOS & DIAGRAMS. $16.95. CODE: HOA

ATLANTIS IN SPAIN

A Study of the Ancient Sun Kingdoms of Spain

by E.M. Whishaw

First published by Rider & Co. of London in 1928, this classic book is a study of the megaliths of Spain, ancient writing, cyclopean walls, sun worshipping empires, hydraulic engineering, and sunken cities. An extremely rare book, it was out of print for 60 years. Learn about the Biblical Tartessus; an Atlantean city at Niebla; the Temple of Hercules and the Sun Temple of Seville; Libyans and the Copper Age; more. Profusely illustrated with photos, maps and drawings.

284 PAGES. 6X9 PAPERBACK. ILLUSTRATED. TABLES OF ANCIENT SCRIPTS. $15.95. CODE: AIS

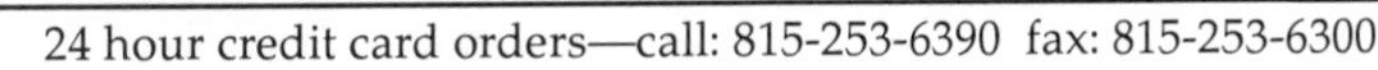

24 hour credit card orders—call: 815-253-6390 fax: 815-253-6300

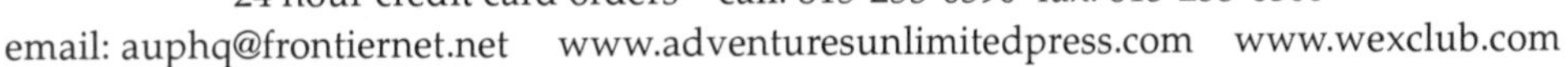

email: auphq@frontiernet.net www.adventuresunlimitedpress.com www.wexclub.com

ATLANTIS STUDIES

MAPS OF THE ANCIENT SEA KINGS

Evidence of Advanced Civilization in the Ice Age

by Charles H. Hapgood

Charles Hapgood's classic 1966 book on ancient maps produces concrete evidence of an advanced world-wide civilization existing many thousands of years before ancient Egypt. He has found the evidence in the Piri Reis Map that shows Antarctica, the Hadji Ahmed map, the Oronteus Finaeus and other amazing maps. Hapgood concluded that these maps were made from more ancient maps from the various ancient archives around the world, now lost. Not only were these unknown people more advanced in mapmaking than any people prior to the 18th century, it appears they mapped all the continents. The Americas were mapped thousands of years before Columbus. Antarctica was mapped when its coasts were free of ice.

316 PAGES. 7X10 PAPERBACK. ILLUSTRATED. BIBLIOGRAPHY & INDEX. $19.95. CODE: MASK

PATH OF THE POLE

Cataclysmic Pole Shift Geology

by Charles Hapgood

Maps of the Ancient Sea Kings author Hapgood's classic book *Path of the Pole* is back in print! Hapgood researched Antarctica, ancient maps and the geological record to conclude that the Earth's crust has slipped in the inner core many times in the past, changing the position of the pole. *Path of the Pole* discusses the various "pole shifts" in Earth's past, giving evidence for each one, and moves on to possible future pole shifts. Packed with illustrations, this is the sourcebook for many other books on cataclysms and pole shifts.

356 PAGES. 6X9 PAPERBACK. ILLUSTRATED. $16.95. CODE: POP.

ATLANTIS: THE ANDES SOLUTION

The Theory and the Evidence

by J.M. Allen, forward by John Blashford-Snell

Imported from Britain, this deluxe hardback is J.M. Allen's fascinating research into the lost world that exists on the Bolivian Plateau and his theory that it is Atlantis. Allen looks into Lake Titicaca, the ruins of Tiahuanaco and the mysterious Lake PooPoo. The high plateau of Bolivia must contain the remains of Atlantis, claims Allen. Lots of fascinating stuff here with Allen discovering the remains of huge ancient canals that once crisscrossed the vast plain southwest of Tiahuanaco. A must-read for all researchers into South America, Atlantis and mysteries of the past. With lots of illustrations, some in color.

196 PAGES. 6X9 HARDBACK. ILLUSTRATED. BIBLIOGRAPHY. $25.99. CODE: ATAS

ATLANTIS IN AMERICA

Navigators of the Ancient World

by Ivar Zapp and George Erikson

This book is an intensive examination of the archeological sites of the Americas, an examination that reveals civilization has existed here for tens of thousands of years. Zapp is an expert on the enigmatic giant stone spheres of Costa Rica, and maintains that they were sighting stones similar to those found throughout the Pacific as well as in Egypt and the Middle East. They were used to teach star-paths and sea navigation to the world-wide navigators of the ancient world. While the Mediterranean and European regions "forgot" world-wide navigation and fought wars, the Mesoamericans of diverse races were building vast interconnected cities without walls. This Golden Age of ancient America was merely a myth of suppressed history—until now. Profusely illustrated, chapters are on Navigators of the Ancient World; Pyramids & Megaliths: Older Than You Think; Ancient Ports and Colonies; Cataclysms of the Past; Atlantis: From Myth to Reality; The Serpent and the Cross: The Loss of the City States; Calendars and Star Temples; and more.

360 PAGES. 6X9 PAPERBACK. ILLUSTRATED. BIBLIOGRAPHY & INDEX. $17.95. CODE: AIA

FAR-OUT ADVENTURES *REVISED EDITION*

The Best of World Explorer Magazine

This is a compilation of the first nine issues of *World Explorer* in a large-format paperback. Authors include: David Hatcher Childress, Joseph Jochmans, John Major Jenkins, Deanna Emerson, Katherine Routledge, Alexander Horvat, Greg Deyermenjian, Dr. Marc Miller, and others. Articles in this book include Smithsonian Gate, Dinosaur Hunting in the Congo, Secret Writings of the Incas, On the Trail of the Yeti, Secrets of the Sphinx, Living Pterodactyls, Quest for Atlantis, What Happened to the Great Library of Alexandria?, In Search of Seamonsters, Egyptians in the Pacific, Lost Megaliths of Guatemala, the Mystery of Easter Island, Comacalco: Mayan City of Mystery, Professor Wexler and plenty more.

580 PAGES. 8X11 PAPERBACK. ILLUSTRATED. REVISED EDITION. $25.00. CODE: FOA

RETURN OF THE SERPENTS OF WISDOM

by Mark Amaru Pinkham

According to ancient records, the patriarchs and founders of the early civilizations in Egypt, India, China, Peru, Mesopotamia, Britain, and the Americas were the Serpents of Wisdom—spiritual masters associated with the serpent—who arrived in these lands after abandoning their beloved homelands and crossing great seas. While bearing names denoting snake or dragon (such as Naga, Lung, Djedhi, Amaru, Quetzalcoatl, Adder, etc.), these Serpents of Wisdom oversaw the construction of magnificent civilizations within which they and their descendants served as the priest kings and as the enlightened heads of mystery school traditions. *The Return of the Serpents of Wisdom* recounts the history of these "Serpents"—where they came from, why they came, the secret wisdom they disseminated, and why they are returning now.

400 PAGES. 6X9 PAPERBACK. ILLUSTRATED. REFERENCES. $16.95. CODE: RSW

MYSTIC TRAVELLER SERIES

THE MYSTERY OF EASTER ISLAND

by Katherine Routledge

The reprint of Katherine Routledge's classic archaeology book which was first published in London in 1919. The book details her journey by yacht from England to South America, around Patagonia to Chile and on to Easter Island. Routledge explored the amazing island and produced one of the first-ever accounts of the life, history and legends of this strange and remote place. Routledge discusses the statues, pyramid-platforms, Rongo Rongo script, the Bird Cult, the war between the Short Ears and the Long Ears, the secret caves, ancient roads on the island, and more. This rare book serves as a sourcebook on the early discoveries and theories on Easter Island.

432 PAGES. 6X9 PAPERBACK. ILLUSTRATED. $16.95. CODE: MEI

MYSTERY CITIES OF THE MAYA

Exploration and Adventure in Lubaantun & Belize

by Thomas Gann

First published in 1925, *Mystery Cities of the Maya* is a classic in Central American archaeology-adventure. Gann was close friends with Mike Mitchell-Hedges, the British adventurer who discovered the famous crystal skull with his adopted daughter Sammy and Lady Richmond Brown, their benefactress. Gann battles pirates along Belize's coast and goes upriver with Mitchell-Hedges to the site of Lubaantun where they excavate a strange lost city where the crystal skull was discovered. Lubaantun is a unique city in the Mayan world as it is built out of precisely carved blocks of stone without the usual plaster-cement facing. Lubaantun contained several large pyramids partially destroyed by earthquakes and a large amount of artifacts. Gann shared Mitchell-Hedges belief in Atlantis and lost civilizations (pre-Mayan) in Central America and the Caribbean. Lots of good photos, maps and diagrams.

252 PAGES. 6X9 PAPERBACK. ILLUSTRATED. $16.95. CODE: MCOM

IN SECRET TIBET

by Theodore Illion

Reprint of a rare 30s adventure travel book. Illion was a German wayfarer who not only spoke fluent Tibetan, but travelled in disguise as a native through forbidden Tibet when it was off-limits to all outsiders. His incredible adventures make this one of the most exciting travel books ever published. Includes illustrations of Tibetan monks levitating stones by acoustics.

210 PAGES. 6X9 PAPERBACK. ILLUSTRATED. $15.95. CODE: IST

DARKNESS OVER TIBET

by Theodore Illion

In this second reprint of Illion's rare books, the German traveller continues his journey through Tibet and is given directions to a strange underground city. As the original publisher's remarks said, "this is a rare account of an underground city in Tibet by the only Westerner ever to enter it and escape alive! "

210 PAGES. 6X9 PAPERBACK. ILLUSTRATED. $15.95. CODE: DOT

DANGER MY ALLY

The Amazing Life Story of the Discoverer of the Crystal Skull

by "Mike" Mitchell-Hedges

The incredible life story of "Mike" Mitchell-Hedges, the British adventurer who discovered the Crystal Skull in the lost Mayan city of Lubaantun in Belize. Mitchell-Hedges has lived an exciting life: gambling everything on a trip to the Americas as a young man, riding with Pancho Villa, questing for Atlantis, fighting bandits in the Caribbean and discovering the famous Crystal Skull.

374 PAGES. 6X9 PAPERBACK. ILLUSTRATED. BIBLIOGRAPHY & INDEX. $16.95. CODE: DMA

IN SECRET MONGOLIA

by Henning Haslund

First published by Kegan Paul of London in 1934, Haslund takes us into the barely known world of Mongolia of 1921, a land of god-kings, bandits, vast mountain wilderness and a Russian army running amok. Starting in Peking, Haslund journeys to Mongolia as part of the Krebs Expedition—a mission to establish a Danish butter farm in a remote corner of northern Mongolia. Along the way, he smuggles guns and nitroglycerin, is thrown into a prison by the new Communist regime, battles the Robber Princess and more. With Haslund we meet the "Mad Baron" Ungern-Sternberg and his renegade Russian army, the many characters of Urga's fledgling foreign community, and the last god-king of Mongolia, Seng Chen Gegen, the fifth reincarnation of the Tiger god and the "ruler of all Torguts." Aside from the esoteric and mystical material, there is plenty of just plain adventure: Haslund encounters a Mongolian werewolf; is ambushed along the trail; escapes from prison and fights terrifying blizzards; more.

374 PAGES. 6X9 PAPERBACK. ILLUSTRATED. BIBLIOGRAPHY & INDEX. $16.95. CODE: ISM

MEN & GODS IN MONGOLIA

by Henning Haslund

First published in 1935 by Kegan Paul of London, Haslund takes us to the lost city of Karakota in the Gobi desert. We meet the Bodgo Gegen, a god-king in Mongolia similar to the Dalai Lama of Tibet. We meet Dambin Jansang, the dreaded warlord of the "Black Gobi." There is even material in this incredible book on the Hi-mori, an "airhorse" that flies through the sky (similar to a Vimana) and carries with it the sacred stone of Chintamani. Aside from the esoteric and mystical material, there is plenty of just plain adventure: Haslund and companions journey across the Gobi desert by camel caravan; are kidnapped and held for ransom; withness initiation into Shamanic societies; meet reincarnated warlords; and experience the violent birth of "modern" Mongolia.

358 PAGES. 6X9 PAPERBACK. ILLUSTRATED. INDEX. $15.95. CODE: MGM

CONSPIRACY & HISTORY

LIQUID CONSPIRACY
JFK, LSD, the CIA, Area 51 & UFOs
by George Piccard

Underground author George Piccard on the politics of LSD, mind control, and Kennedy's involvement with Area 51 and UFOs. Reveals JFK's LSD experiences with Mary Pinchot-Meyer. The plot thickens with an ever expanding web of CIA involvement, from underground bases with UFOs seen by JFK and Marilyn Monroe (among others) to a vaster conspiracy that affects every government agency from NASA to the Justice Department. This may have been the reason that Marilyn Monroe and actress-columnist Dorothy Kilgallen were both murdered. Focusing on the bizarre side of history, *Liquid Conspiracy* takes the reader on a psychedelic tour de force. This is your government on drugs!

264 PAGES. 6X9 PAPERBACK. ILLUSTRATED. $14.95. CODE: LIQC

INSIDE THE GEMSTONE FILE
Howard Hughes, Onassis & JFK
by Kenn Thomas & David Hatcher Childress

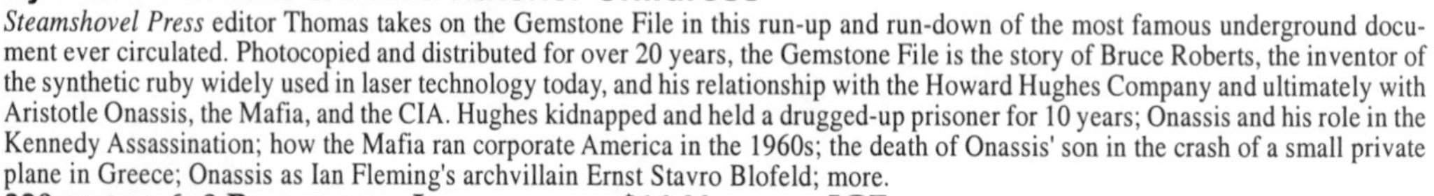

Steamshovel Press editor Thomas takes on the Gemstone File in this run-up and run-down of the most famous underground document ever circulated. Photocopied and distributed for over 20 years, the Gemstone File is the story of Bruce Roberts, the inventor of the synthetic ruby widely used in laser technology today, and his relationship with the Howard Hughes Company and ultimately with Aristotle Onassis, the Mafia, and the CIA. Hughes kidnapped and held a drugged-up prisoner for 10 years; Onassis and his role in the Kennedy Assassination; how the Mafia ran corporate America in the 1960s; the death of Onassis' son in the crash of a small private plane in Greece; Onassis as Ian Fleming's archvillain Ernst Stavro Blofeld; more.

320 PAGES. 6X9 PAPERBACK. ILLUSTRATED. $16.00. CODE: IGF

THE ARCH CONSPIRATOR
Essays and Actions
by Len Bracken

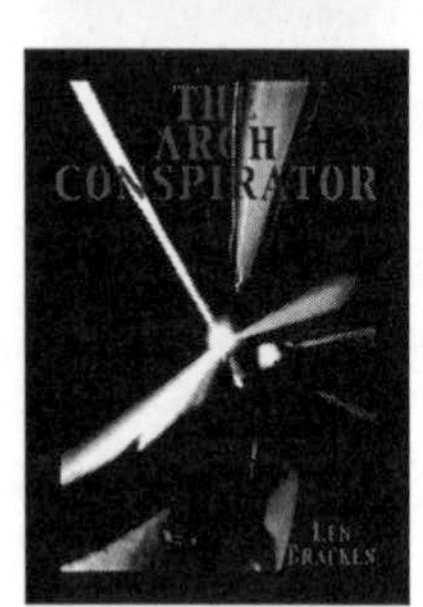

Veteran conspiracy author Len Bracken's witty essays and articles lead us down the dark corridors of conspiracy, politics, murder and mayhem. In 12 chapters Bracken takes us through a maze of interwoven tales from the Russian Conspiracy (and a few "extra notes" on conspiracies) to his interview with Costa Rican novelist Joaquin Gutierrez and his Psychogeographic Map into the Third Millennium. Other chapters in the book are A General Theory of Civil War; A False Report Exposes the Dirty Truth About South African Intelligence Services; The New-Catiline Conspiracy for the Cancellation of Debt; Anti-Labor Day; 1997 with selected Aphorisms Against Work; Solar Economics; and more. Bracken's work has appeared in such pop-conspiracy publications as *Paranoia, Steamshovel Press* and the *Village Voice.* Len Bracken lives in Arlington, Virginia and haunts the back alleys of Washington D.C., keeping an eye on the predators who run our country. With a gun to his head, he cranks out his rants for fringe publications and is the editor of *Extraphile*, described by *New Yorker Magazine* as "fusion conspiracy theory."

256 PAGES. 6X9 PAPERBACK. ILLUSTRATED. BIBLIOGRAPHY. $14.95. CODE: ACON.

MIND CONTROL, WORLD CONTROL
by Jim Keith

Veteran author and investigator Jim Keith uncovers a surprising amount of information on the technology, experimentation and implementation of mind control. Various chapters in this shocking book are on early CIA experiments such as Project Artichoke and Project R.H.I.C.-EDOM, the methodology and technology of implants, mind control assassins and couriers, various famous Mind Control victims such as Sirhan Sirhan and Candy Jones. Also featured in this book are chapters on how mind control technology may be linked to some UFO activity and "UFO abductions."

256 PAGES. 6X9 PAPERBACK. ILLUSTRATED. FOOTNOTES. $14.95. CODE: MCWC

NASA, NAZIS & JFK:
The Torbitt Document & the JFK Assassination
introduction by Kenn Thomas

This book emphasizes the links between "Operation Paper Clip" Nazi scientists working for NASA, the assassination of JFK, and the secret Nevada air base Area 51. The Torbitt Document also talks about the roles played in the assassination by Division Five of the FBI, the Defense Industrial Security Command (DISC), the Las Vegas mob, and the shadow corporate entities Permindex and Centro-Mondiale Commerciale. The Torbitt Document claims that the same players planned the 1962 assassination attempt on Charles de Gaul, who ultimately pulled out of NATO because he traced the "Assassination Cabal" to Permindex in Switzerland and to NATO headquarters in Brussels. The Torbitt Document paints a dark picture of NASA, the military industrial complex, and the connections to Mercury, Nevada which headquarters the "secret space program."

258 PAGES. 5X8. PAPERBACK. ILLUSTRATED. $16.00. CODE: NNJ

MIND CONTROL, OSWALD & JFK:
Were We Controlled?
introduction by Kenn Thomas

Steamshovel Press editor Kenn Thomas examines the little-known book *Were We Controlled?*, first published in 1968. The book's author, the mysterious Lincoln Lawrence, maintained that Lee Harvey Oswald was a special agent who was a mind control subject, having received an implant in 1960 at a Russian hospital. Thomas examines the evidence for implant technology and the role it could have played in the Kennedy Assassination. Thomas also looks at the mind control aspects of the RFK assassination and details the history of implant technology. A growing number of people are interested in CIA experiments and its "Silent Weapons for Quiet Wars." Looks at the case that the reporter Damon Runyon, Jr. was murdered because of this book.

256 PAGES. 6X9 PAPERBACK. ILLUSTRATED. NOTES. $16.00. CODE: MCOJ

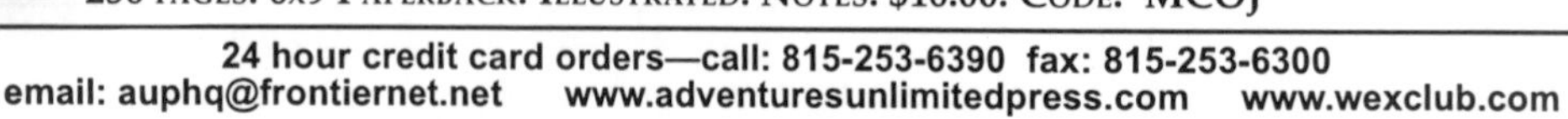

CONSPIRACY & HISTORY

THE HISTORY OF THE KNIGHTS TEMPLARS

The Temple Church and the Temple

by Charles G. Addison, introduction by David Hatcher Childress

Chapters on the origin of the Templars, their popularity in Europe and their rivalry with the Knights of St. John, later to be known as the Knights of Malta. Detailed information on the activities of the Templars in the Holy Land, and the 1312 AD suppression of the Templars in France and other countries, which culminated in the execution of Jacques de Molay and the continuation of the Knights Templars in England and Scotland; the formation of the society of Knights Templars in London; and the rebuilding of the Temple in 1816. Plus a lengthy intro about the lost Templar fleet and its connections to the ancient North American sea routes.

395 PAGES. 6X9 PAPERBACK. ILLUSTRATED. $16.95. CODE: HKT

ECCENTRIC LIVES AND PECULIAR NOTIONS

by John Michell

The first paperback edition of Michell's fascinating study of the lives and beliefs of over 20 eccentric people. Published in hardback by Thames & Hudson in London, *Eccentric Lives and Peculiar Notions* takes us into the bizarre and often humorous lives of such people as Lady Blount, who was sure that the earth is flat; Cyrus Teed, who believed that the earth is a hollow shell with us on the inside; Edward Hine, who believed that the British are the lost Tribes of Israel; and Baron de Guldenstubbe, who was sure that statues wrote him letters. British writer and housewife Nesta Webster devoted her life to exposing international conspiracies, and Father O'Callaghan devoted his to opposing interest on loans. The extraordinary characters in this book were—and in some cases still are—wholehearted enthusiasts for the various causes and outrageous notions they adopted, and John Michell describes their adventures with spirit and compassion. Some of them prospered and lived happily with their obsessions, while others failed dismally. We read, for example, of the hapless inventor of a giant battleship made of ice who died alone and neglected, and of the London couple who achieved peace and prosperity by drilling holes in their heads. Other chapters on the Last of the Welsh Druids; Congressman Ignacius Donnelly, the Great Heretic and Atlantis; Shakespearean Decoders and the Baconian Treasure Hunt; Early Ufologists; Jerusalem in Scotland; Bibliomaniacs; more.

248 PAGES. 6X9 PAPERBACK. ILLUSTRATED. $14.95. CODE: ELPN

ARKTOS

The Myth of the Pole in Science, Symbolism, and Nazi Survival

by Joscelyn Godwin

A scholarly treatment of catastrophes, ancient myths and the Nazi Occult beliefs. Explored are the many tales of an ancient race said to have lived in the Arctic regions, such as Thule and Hyperborea. Progressing onward, the book looks at modern polar legends including the survival of Hitler, German bases in Antarctica, UFOs, the hollow earth, Agartha and Shambala, more.

220 PAGES. 6X9 PAPERBACK. ILLUSTRATED. $16.95. CODE: ARK

THE CHRIST CONSPIRACY

The Greatest Story Ever Sold

by Acharya S.

In this highly controversial and explosive book, archaeologist, historian, mythologist and linguist Acharya S. marshals an enormous amount of startling evidence to demonstrate that Christianity and the story of Jesus Christ were created by members of various secret societies, mystery schools and religions in order to unify the Roman Empire under one state religion. In developing such a fabrication, this multinational cabal drew upon a multitude of myths and rituals that existed long before the Christian era, and reworked them for centuries into the religion passed down to us today. Contrary to popular belief, there was no single man who was at the genesis of Christianity; Jesus was many characters rolled into one. These characters personified the ubiquitous solar myth, and their exploits were well known, as reflected by such popular deities as Mithras, Heracles/Hercules, Dionysos and many others throughout the Roman Empire and beyond. The story of Jesus as portrayed in the Gospels is revealed to be nearly identical in detail to that of the earlier savior-gods Krishna and Horus, who for millennia preceding Christianity held great favor with the people. *The Christ Conspiracy* shows the Jesus character as neither unique nor original, not "divine revelation." Christianity re-interprets the same extremely ancient body of knowledge that revolved around the celestial bodies and natural forces.

256 PAGES. 6X9 PAPERBACK. ILLUSTRATED. $14.95. CODE: CHRC

CONVERSATIONS WITH THE GODDESS

by Mark Amaru Pinkham

Return of the Serpents of Wisdom author Pinkham tells us that "The Goddess is returning!" Pinkham gives us an alternative history of Lucifer, the ancient King of the World, and the Matriarchal Tradition he founded thousands of years ago. The name Lucifer means "Light Bringer" and he is the same as the Greek god Prometheus, and is different from Satan, who was based on the Egyptian god Set. Find out how the branches of the Matriarchy—the Secret Societies and Mystery Schools—were formed, and how they have been receiving assistance from the Brotherhoods on Sirius and Venus to evolve the world and overthrow the Patriarchy. Learn about the revival of the Goddess Tradition in the New Age and why the Goddess wants us all to reunite with Her now! An unusual book from an unusual writer!

296 PAGES. 7X10 PAPERBACK. ILLUSTRATED. BIBLIOGRAPHY. $14.95. CODE: CWTG.

THE AQUARIAN GOSPEL OF JESUS THE CHRIST

Transcribed from the Akashic Records

by Levi

First published in 1908, this is the amazing story of Jesus, the man from Galilee, and how he attained the Christ consciousness open to all men. It includes a complete record of the "lost" 18 years of his life, a time on which the New Testament is strangely silent. During this period Jesus travelled widely in India, Tibet, Persia, Egypt and Greece, learning from the Masters, seers and wisemen of the East and the West in their temples and schools. Included is information on the Council of the Seven Sages of the World, Jesus with the Chinese Master Mencius (Meng Tzu) in Tibet, the ministry, trial, execution and resurrection of Jesus.

270 PAGES. 6X9 PAPERBACK. INDEX. $14.95. CODE: AGJC

One Adventure Place
P.O. Box 74
Kempton, Illinois 60946
United States of America
•Tel.: 1-800-718-4514 or 815-253-6390
•Fax: 815-253-6300
Email: auphq@frontiernet.net
http://www.adventuresunlimitedpress.com
or www.adventuresunlimited.nl

10% Discount when you order 3 or more items!

ORDERING INSTRUCTIONS

- ✓ Remit by US$ Check, Money Order or Credit Card
- ✓ Visa, Master Card, Discover & AmEx Accepted
- ✓ Prices May Change Without Notice
- ✓ 10% Discount for 3 or more Items

SHIPPING CHARGES

United States

- ✓ Postal Book Rate { $3.00 First Item / 50¢ Each Additional Item
- ✓ Priority Mail { $4.50 First Item / $2.00 Each Additional Item
- ✓ UPS { $5.00 First Item / $1.50 Each Additional Item

NOTE: UPS Delivery Available to Mainland USA Only

Canada

- ✓ Postal Book Rate { $6.00 First Item / $2.00 Each Additional Item
- ✓ Postal Air Mail { $8.00 First Item / $2.50 Each Additional Item
- ✓ Personal Checks or Bank Drafts MUST BE US$ and Drawn on a US Bank
- ✓ Canadian Postal Money Orders in US$ OK
- ✓ Payment MUST BE US$

All Other Countries

- ✓ Surface Delivery { $10.00 First Item / $4.00 Each Additional Item
- ✓ Postal Air Mail { $14.00 First Item / $5.00 Each Additional Item
- ✓ Checks and Money Orders MUST BE US$ and Drawn on a US Bank or branch.

✓ Payment by credit card preferred!

SPECIAL NOTES

- ✓ RETAILERS: Standard Discounts Available
- ✓ BACKORDERS: We Backorder all Out-of-Stock Items Unless Otherwise Requested
- ✓ PRO FORMA INVOICES: Available on Request
- ✓ VIDEOS: NTSC Mode Only. Replacement only.
- ✓ For PAL mode videos contact our other offices:

European Office:
Adventures Unlimited, Pannewal 22,
Enkhuizen, 1602 KS, The Netherlands
http: www.adventuresunlimited.nl
Check Us Out Online at:
www.adventuresunlimitedpress.com

Please check: ☑

☐ This is my first order ☐ I have ordered before ☐ This is a new address

Name
Address
City
State/Province | Postal Code
Country
Phone day | Evening
Fax | Email

Item Code	Item Description	Price	Qty	Total

Please check: ☑

☐ Postal-Surface
☐ Postal-Air Mail (Priority in USA)
☐ UPS (Mainland USA only)

Subtotal ➠	
Less Discount-10% for 3 or more items ➠	
Balance ➠	
Illinois Residents 6.25% Sales Tax ➠	
Previous Credit ➠	
Shipping ➠	
Total (check/MO in USD$ only) ➠	

☐ Visa/MasterCard/Discover/Amex

Card Number

Expiration Date

10% Discount When You Order 3 or More Items!

Comments & Suggestions

Share Our Catalog with a Friend